DISCARDED

THE
UNIVERSITY OF WINNIPEG
PORTAGE & BALMORAL
WINNIPEG, MAN.   R3B 2E9
CANADA

Land degradation
Problems and policies

HD
1036
L36
1987

Land degradation
Problems and policies

edited by
Anthony Chisholm
Department of Economics, Australian National University

and
Robert Dumsday
Department of Agricultural Economics, La Trobe University

Published in association with
Centre for Resource and Environmental Studies,
Australian National University

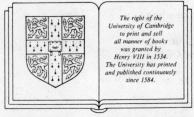

The right of the
University of Cambridge
to print and sell
all manner of books
was granted by
Henry VIII in 1534.
The University has printed
and published continuously
since 1584.

CAMBRIDGE UNIVERSITY PRESS

Cambridge
London   New York   New Rochelle
Melbourne   Sydney

Published by the Press Syndicate of the University of
  Cambridge
The Pitt Building, Trumpington Street, Cambridge
  CB2 1RP, United Kingdom
32 East 57th Street, New York, NY 10022, USA
10 Stamford Road, Oakleigh, Melbourne 3166, Australia

© Centre for Resource and Environmental Studies, 1987

This book is copyright. Apart from any fair dealing for the
purpose of study, research, criticism, or review, as
permitted under the Copyright Act, no part may be
reproduced by any process without written permission.
Inquiries should be made to the publisher.

Land degradation: problems and policies

  Bibliography.
  Includes index.
  ISBN 0 521 34079 9.
  ISSN 0725-3400.

  1. Soil degradation — Australia. 2. Land use —
    Australia. I. Chisholm, Anthony H. (Anthony Hewlings),
  1939-   . II. Dumsday, R.G. (Robert Graeme),
  1944-   . III. Australia National University. Centre
  for Resource and Environmental Studies. (Series: CRES
  monograph; no. 18).

333.73'13'0994

Designed by Stephen Cole/ANU Graphic Design
Typeset by Press Etching (Qld.) Pty Ltd, Brisbane

Cover design based on a photograph by Dr Diana Day

Foreword

Land degradation in all its forms is one of the most serious problems facing the Australian agricultural, horticultural and pastoral industries. Not only is it a worry for the present generation, but there must also be a concern for future generations.

The Workshop from which this volume derives brought together people from the rural industry, natural and social scientists from many disciplines, lawyers and representatives from Commonwealth and state governments and from various government or semi-government agencies. The response was so great that a number of people who wished to attend unfortunately could not be invited. This book gives an opportunity for access to the papers and some of the comments made on them.

In keeping with the multi-disciplinary nature of the Workshop the contributions were valuable in giving information across the disciplinary borderlines. But the objectives were more ambitious than this: there was an attempt to define the biological and physical boundaries of the problem, to identify some of the important social as well as physical issues involved, to consider some of the central scientific, political, legal, economic, social and bureaucratic difficulties in developing an Australian land management strategy or series of strategies and to offer some alternatives.

One thing that came through, and it is reflected in the chapters and commentaries, was that there was less common ground about the extent of the problem than perhaps could have been expected. For example, the economists appeared to argue that the problems are less than the physical scientists assert. There certainly is room for improvement of the information base. However, I would argue that, although data collection is important, investigation must not be a substitute for action.

The earlier chapters identify some of the causes and extent of the problem of land degradation. In particular, they note that neither the rural sector nor government has adopted an integrated approach to the question of what approaches are appropriate to this whole area. A number of options is then canvassed, but no firm solution is offered.

However, what is clear is that whatever approach is finally adopted, it must be firmly based on principles of cooperation. This cooperation must not only be between the Commonwealth and state governments, but between all the parties involved in this complex issue.

What the Workshop did show to this privileged participant was that all the disciplines involved are not only able to talk with one another but are also very willing to do so. It is hoped that the present volume is a useful contribution to tackling the most pressing national problem of land degradation.

Douglas Whalan
Chairman CRES
Advisory Committee

Contents

The Contributors

John Ballard is a Senior Lecturer and Head of the Department of Political Science in the Faculties at the Australian National University

Michael Barker is a Lecturer in the Law Faculty at the Australian National University

Michael Blyth is a Principal Research Officer in the Rural Resources Economics Section of the Bureau of Agricultural Economics

John Bradsen is a Lecturer in the Faculty of Law at the University of Adelaide

Gordon Burch is a Senior Research Scientist in the CSIRO Division of Water and Land Resources

Colin Chartres is a Senior Research Scientist in the CSIRO Division of Soils

Anthony Chisholm is a Visiting Fellow at CRES on leave from his position as Reader in the Faculty of Economics at the Australian National University

Bruce Davis is Head of the Department of Political Science and Dean of the Faculty of Arts at The University of Tasmania

Robert Dumsday is a Visiting Fellow at CRES on leave from his position as a Senior Lecturer in Agricultural Economics at La Trobe University

Geoff Edwards is a Senior Lecturer in Agricultural Economics at La Trobe University

Robert Fowler is a Senior Lecturer in the Faculty of Law at the University of Adelaide

Dean Graetz is a Senior Research Scientist in the CSIRO Division of Wildlife and Rangelands Research

Peter Greig is a Corporate Analyst in the Corporate Development Unit of the Melbourne and Metropolitan Board of Works

Robert Junor is the Assistant Commissioner of the Soil Conservation Service of NSW

John Kerin is the Federal Minister for Primary Industry

Michael Kirby is Acting Assistant Director Economic and Policy Analysis Branch of the Bureau of Agricultural Economics

Andrew McCallum is a Research Officer in the Rural Resources Economics Section of the Bureau of Agricultural Economics

Geoff Mosley is a Director of the Australian Conservation Foundation

Ian Noble is a Fellow in Environmental Biology in the Research School of Biological Sciences at the Australian National University

John Quiggin is a Research Fellow in CRES at the Australian National University on leave from his position as a Senior Economist in the Bureau of Agricultural Economics

John Paterson is Director-General of the Department of Water Resources in Victoria

Roy Rickson is a Senior Lecturer and Deputy Dean in the School of Australian Environmental Studies at Griffith University

Andrew Robb is Executive Director of the National Farmers' Federation

Graeme Robertson is Commissioner of Soil Conservation in the Western Australia Department of Agriculture

Garrett Upstill is an economist in the Department of Arts, Heritage and Environment

Robert Wasson is a Senior Research Scientist in the CSIRO Division of Water and Land Resources

Warwick Watkins is Chief of Regions in the Soil Conservation Service of New South Wales

Adrian Webb is Assistant Director of the Soil Conservation Research Branch of the Queensland Department of Primary Industries

Lance Woods is Executive Officer, Research and Information, in the Department of Resources and Energy

Timothy Yapp is an economist in the Department of Arts, Heritage and Environment

Michael Young is a Senior Reseach Scientist in the CSIRO Division of Wildlife and Rangelands Research

Figures

Tables

Glossary

AAC	—	Australian Agricultural Council
ABS	—	Australian Bureau of Statistics
ACF	—	Australian Conservation Foundation
ACT	—	Australian Capital Territory
ACTU	—	Australian Council of Trade Unions
AGPS	—	Australian Government Publishing Service
AIPS	—	Australian Institute of Political Science
ANU	—	Australian National University
ANZAAS	—	Australia and New Zealand Association for the Advancement of Science
AWRC	—	Australian Water Resources Council
BAE	—	Bureau of Agricultural Economics
BMR	—	Bureau of Mineral Resources
BOD	—	Biochemical Oxygen Demand C-Carbon
CAP	—	Common Agricultural Policy (of the EEC)
CREAMS	—	Chemicals, Runoff and Erosion from Agricultural Management Systems
Cs	—	Caesium
CSIRO	—	Commonwealth Scientific and Industrial Research Organization
deflation	—	Removal of soil particles by wind
DEHCD	—	Department of Environment, Housing and Community Development
EEC	—	European Economic Community
EPIC	—	Erosion-Productivity Impact Calculator
FAO	—	Food and Agriculture Organisation of the United Nations
IAC	—	Industries Assistance Commission
IDC	—	Interdepartmental Committee
LIBRIS	—	Land Image-Based Resource Information System
LPS	—	Land Protection Service (Victoria)
MMBW	—	Melbourne and Metropolitan Board of Works
MUSLE	—	Modified Universal Soil Loss Equation

NFF	—	National Farmer's Federation
NLURMC	—	National Land Use and Resource Management Council
NRI	—	National Resources Inventory (USA)
NSCP	—	National Soil Conservation Program
NSW	—	New South Wales (Australia)
NSWSCS	—	New South Wales Soil Conservation Service
NTP	—	National Tree Program
Pb	—	Lead
pers comm	—	personal communication
Pu	—	Plutonium
QLD	—	Queensland (Australia)
QDPI	—	Queensland Department of Primary Industries
QRRE	—	Quarterly Review of the Rural Economy
RAINEI	—	Rainfall Erosivity Indices
RMC	—	Royal Military College (Duntroon, ACT)
SA	—	South Australia
SCA	—	Standing Committee on Agriculture
SCAV	—	Soil Conservation Authority (Victoria)
SCC	—	Soil Conservation Council
SCSC	—	Standing Committee on Soil Conservation
SOILEC	—	Soil Economic Conservation Model
UNE	—	University of New England
UNEP	—	United Nations Environmental Program
UNESCO	—	United Nations Educational, Scientific and Cultural Organisation
US or USA	—	United States of America
USDA	—	United States Department of Agriculture
USLE	—	Universal Soil Loss Equation
WA	—	Western Australia

Preface

There has been a resurgence of interest in land degradation and conservation issues during the 1970s and 1980s, both in Australia and internationally. Rural, community, conservation, scientific and other professional groups have joined in the widening debate which has consequently attracted significant political attention. The impetus for this interest and concern derives from many ideas. They include the perceptions that land in its natural state is becoming increasingly scarce and socially valuable and that more land should therefore be set aside for state and national parks, recreation areas, and flora and fauna reserves; that the quality of land and water has steadily deteriorated in many regions, particularly as a consequence of the soil erosion and salinity associated with agricultural activities; that new technology and increased inputs have masked the effects of degradation which reduced the innate productivity of land in agricultural use; and that the aesthetic appeal of the rural landscape has declined because of unsightly erosion and loss of trees and other native vegetation.

The purpose of the workshop, in which this volume has its origins, was to provide a conceptual and institutional perspective and to analyse in some depth the issues of degrading lands and soil conservation in a way bearing directly on policy. Land degradation is defined broadly to include all adverse onsite and offsite effects that land uses may have on the services provided by land and water. Three objectives were subsequently identified by the workshop organising committee: to determine the socially important elements of the problem of land degradation; to define the biological and physical setting of the problem; and to set out ways of pursuing land management in the national interest taking into account legal, economic, social, political and organisational issues.

To attain these objectives, it was necessary to bring together a group of people comprising researchers from a number of disciplinary areas, representatives of farm and conservation organisations, and policy advisers and decision makers, and to promote effective communication and interaction among them. Contributions to the Land Degradation and Public Policy Workshop, hosted by the Centre for Resource and Environmental Studies (CRES) at the Australian National University, 3-4 September 1985, were invited on six main themes which form the main sections of this volume.

The primary focus of the book is on social and policy aspects of land degradation. The contributions by scientists in Section 1 are designed to provide the biological and physical setting and not the fine detail of scientific and survey research which can be found in other works. The contributions on economic, legal and social aspects in Sections II, III, IV and the appendixes, are intended to provide empirical information and a conceptual and analytical framework to aid informed policy choice and decision making. Section V focuses on institutional, organisational and political concerns. Finally, overviews are given in Section VI of the contributions from the biological, physical and social sciences and of the practicalities of policy proposals.

Not all forms of land degradation are specifically dealt with and it is not the purpose of this volume to present detailed case studies. The sociological case study presented in Chapter 9, which was commissioned immediately after the workshop, is an exception. The principles and policies developed in the book are applicable to such topics as soil erosion and salinity of land and waterways, tree dieback, loss of native flora and fauna habitats, desertification and damage to land through recreational uses. It is hoped that the book will provide a useful foundation and source of reference for future multi-disciplinary case studies on soil and water conservation. The reader should not be left with an impression that there was a consensus among workshop participants on the issues. There was often a vigorous exchange of views reflecting the vitality of multi-disciplinary debate in this area.

The original inspiration for the workshop came from Professor Stuart Harris while he was Director of the Centre for Resource and Environmental Studies (he is now head of Australia's Department of Foreign Affairs). Professor Harris chaired the early meetings of the workshop's organising committee and he clearly saw the interrelationships of the agricultural, environmental, scientific, economic, political and social issues involved as being a critical component of the workshop. Professor Warren Musgrave succeeded Professor Harris as chairman of the committee and chaired the workshop itself. His chairmanship contributed greatly to the smooth conduct of the workshop. We should like to acknowledge the valuable work of the organising committee leading up to the workshop. The other members of the organising committee were: Michael Barker, Michael Blyth, Chris Buller, Gordon Burch, Geoffrey Downes, Gary Goucher, Tony Jakeman, Ian Noble, John Quiggin, Dingle Smith and Warwick Watkins.

Sadly, Geoffrey Downes died before arrangements for the workshop were completed. Geoffrey Downes was an outstanding soil scientist, administrator and leader in the field of soil conservation. His achievements, especially during his long service as Chairman of the Soil Conservation Authority of Victoria, are recognised nationally and internationally.

We also regret that Dr Alec Costin, well known for his scientific and practical contributions to soil conservation and applied ecology and a former research fellow at CRES, was, after some early participation, unable to be a contributor at the workshop.

Many people have been involved in making possible this volume and the multi-disciplinary exercise it represents. Thanks are due to the contributors and the commentators for their efforts and patience with the editing process. We should also like to thank the external referees for their careful reading of the manuscript and their detailed and constructive comments.

We are especially grateful to CRES for organisational support and for our Visiting Fellowships which, by freeing us of teaching and other administrative duties, allowed us to devote our full energies to the workshop and to editing and contributing to the volume. Special thanks are due to Chris Buller and Valda Stipnieks, respectively, the former and present Secretary, and to Tony Jakeman and Dingle Smith who have alternated as Acting Head of CRES over the last two years. The smooth running of the workshop was ensured by the thorough organisational lead-up work by Pamela Soding and the supportive work of Suzanne Ridley, Mark Greenaway, Eric Ward, Margaret Mahoney, Janine Corey and Gerry Vandermey.

We are indebted to Gordon Sheldon, the Centre's Publications Officer, for the time and skills he has devoted to the publication process, to Rosemary Boyden who compiled the extensive bibliography, and to Christina Jankovic for assistance with proof-reading. Our special thanks go to Kathy Parkes who worked so hard and competently in typing the many drafts and editorial corrections to produce the final manuscript.

> Anthony Chisholm
> Robert Dumsday
> May 1986

Land degradation and government

The Hon John Kerin MP,
Minister for Primary Industry

In dealing with a complex multi-disciplinary issue such as land degradation there is often a tendency for scientists and researchers to concentrate on particular problems in isolation from the wider political and social context. This tendency reflects the nature of the work being undertaken and the need for specialisation if worthwhile advances are to be made. However, it does not hurt to step back occasionally and take a wider view.

I believe a minister is reasonably well placed to undertake such a task. The very nature of the job demands that the competing interests of various sectors of society are taken into account in decision making. Thus, it is possible to obtain a good appreciation of all aspects of a particular issue and come to an understanding of how effective or ineffective policy initiatives can be.

There are good reasons why government has become involved in the control of land degradation. There has been a realisation by society in recent years that land degradation is the most significant threat to the continued viability of rural industry in Australia. The justification for such a statement is not difficult. Various studies have indicated that over half the rural lands in Australia are experiencing a significant level of land degradation.

Statements of concern could undoubtedly apply to many of Australia's natural resources. However, to place the problem of land degradation in context with other environmental issues competing for public recognition and resources, the sobering facts are that:
- soil is the fundamental resource on which primary production is dependent; it is the origin of a food chain that terminates with human consumption of rural products.
- there have been no serious suggestions that an alternative source of food supply is likely to render our continued requirement for soil obsolete.

This opening address was presented to the Workshop on Land Degradation and Public Policy by Mr John Brumby, MP, on behalf of Mr Kerin.

- in view of the insignificant rate of soil formation and the scarcity of topsoil in Australia, soil erosion can realistically be considered as an irreversible withdrawal from the nation's resource bank.

The accelerating rate of land degradation since European settlement can be directly attributed to the land use practices imposed on an environment that is insufficiently robust to withstand such stress.

Given the historically short period during which agriculture has been undertaken in Australia, coupled with the alarming decline in the quality and quantity of our soils over that period, governments, both state and commonwealth, have been left with little choice but to intervene in maintaining this vital resource.

In our society it is appropriate that government policies address the issue of current land management with a view to influencing land users to adopt more suitable systems and techniques. The alternative of inaction now would inevitably lead to a far stronger government response in the future.

Several means of achieving more appropriate land use are available to government. First, governments could determine suitable land use strategies to be implemented and legislate for their adoption. Second, they can provide incentives for land users, usually in the form of financial assistance, at a level at which investment in land degradation control is competitive with alternative investment options. And third, governments can support public awareness and education programs to influence the ethics of land users in regard to degradation, with the objective of achieving voluntary adoption of stable productive systems.

A role exists, in varying degrees of application, for all these approaches. Regulation of land use practices has proved necessary in marginal areas, particularly where leasehold title is involved. It is becoming apparent that where farmers persist with a land use that is clearly the cause of irreversible damage, governments have a responsibility to resort to regulatory measures.

Financial incentives are commonly justified in terms of the gap between reasonable community and landholder requirements for resource protection. Incentives may also be justified where the risk of production penalties during the learning process of a developing technology is sufficient to discourage interest from land users.

It would be ideal if all landholders were to adopt the stewardship ethic. The level of responsibility for land conservation accepted by primary producers is undoubtedly on the increase, and this can perhaps be partly attributed to the activity generated by recent government initiatives. There is, however, great potential for influencing landholders further to think and act in an environmentally and socially responsible manner. Against this there will be consistently strong pressure for individual landholders to take short-term business decisions which are inconsistent with society's longer term interests. I can appreciate the dilemma that many of our primary producers are in— they understand very well the need for soil protection but at the same time are under increasing financial pressure in managing their enterprises.

There have been suggestions from time to time that what is required is a massive injection of funds and resources to the cause of land degradation control. This displays an inherent failure by those making such suggestions to appreciate the underlying cause of degradation processes. The proposal is totally inappropriate. A coordinated attack by both federal and state governments on the land use problem is, however, appropriate.

Australia has a federal system of government with legislative authority being divided between the Commonwealth and the states. The Commonwealth has a primary interest in the management of the economy, the efficient allocation of resources and the equitable distribution of national income. The Commonwealth's constitutional powers in relation to the use and management of land are limited. However, in the interest of national prosperity through the maintenance of our agricultural resource base, it is important that the Commonwealth government be involved in soil conservation.

Against this background the Commonwealth government has initiated a National Soil Conservation Program (NSCP). This program was established in formal recognition that the Commonwealth had a responsibility to provide direction and leadership in an effort to combat land degradation. Through it, the Commonwealth is firmly committed to cooperating with the states in the expansion of their soil conservation activities. The program is designed to act as a catalyst for encouraging all sectors of the community with an interest or involvement in land management to practise soil conservation. It is consistent with the National Conservation Strategy for Australia.

Further, the Commonwealth is also committed to encouraging community groups, private organisations and non-state institutions, through the national component of the program, to research and promote priority action to combat land degradation.

The Commonwealth recognises the inherent long-term nature of land degradation and the many difficulties involved in anticipating future land use requirements. Accordingly, the program is concerned with ensuring that options remain available to future generations by maintaining the existing land resource in a stable, non-degraded and productive state.

NSCP is directed at all sectors of the community with an interest or involvement in land management. Landholders, with whom main responsibility for erosion control rests, are the major target but community groups, researchers, local government and various agencies in the state and federal governments also have important roles to play. The emphasis is on cooperation and coordination as the fragmentation of responsibility among many government agencies has, in the past been a real limitation.

Financial assistance, although just one element of the overall program, is an essential ingredient for the support of a range of other policy measures to be employed such as education, training, demonstration, research, publicity, provision of technical assistance and construction of works. Funds have been provided for projects in these broad areas of activity.

NSCP seeks to identify and, where possible, eliminate or minimise the constraints on achieving effective control of land degradation. Particular priority is given to increasing land user perception of the implications for continued degradation of rural land to a level where they are motivated to adopt sustainable productive systems.

I commend to you the five goals of NSCP. Briefly, these goals address the issues of:

1. sustainable land uses appropriate to land capability;
2. total catchment planning;
3. recognition and acceptance by governments and land users of their responsibilities for land conservation;
4. cooperation and coordination between all involved with land management; and
5. adoption of a land conservation ethic.

I believe NSCP goals provide an excellent blueprint from which sound policies and strategies for the control of land degradation can be developed.

I

Physical and biological aspects of land degradation

1 Australia's land resources at risk

Colin Chartres

Introduction

In its broadest sense 'land degradation can result from any causative factor or combination of factors, which damage the physical, chemical or biological status of the land and which may also restrict the land's productive capacity. In the agricultural sense, for example, soil erosion reduces the land's ability to sustain yields or stock numbers, and thus may reduce economic returns. A large number of factors may contribute to land degradation. The Food and Agriculture Organisation (FAO), for example, lists erosion, salts and alkali, organic wastes, pesticides, radioactivity, heavy metals, fertilisers and detergents as all being potentially causative agents of degradation (Rauschkolb 1971). In an Australian context, the causes and responses to land degradation have been comprehensively reviewed by Woods (1983). He lists a number of environmental problems including loss of sustainable production because of degradation, declining water quality, loss of genetic diversity, land use conflict etc, as all being major issues worthy of further study.

Table 1: Forms of degradation in non-arid areas of Australia, 1975

Form of degradation	Area	
	(000 km²)	(%)
Area in use	1 804	—
Area not requiring treatment	987	55
Water erosion	577	32
Wind erosion	57	3
Combined wind and water erosion	55	3
Vegetation degradation	92	5
Dryland salinity-sometimes in combination with water erosion	10	< 1
Irrigation area salinity	9	< 1
Other	14	< 1
Total area requiring treatment	815	45

All values are approximate and have been rounded

Source: Adapted from Woods (1983, Table 5.1)

Table 2: Forms of degradation in arid areas of Australia, 1975

Form of degradation	Area	
	(000 km²)	(%)
Area in use	3 356	—
Area not requiring treatment	1 506	45
Area affected by:		
Vegetation degradation and little erosion	950	29
Vegetation degradation and some erosion	467	14
Vegetation degradation and substantial erosion	284	8
Vegetation degradation and severe erosion	148	4
Dryland salinity—sometimes in combination with water erosion	1	<1
Total area needing treatment	1 850	55

All values are approximate and have been rounded

Source: Adapted from Woods (1983, Table 7.3)

Land degradation, however, is by no means limited to specific geographic zones, nor does it occur at a uniform rate. Degradation can be insidious: for example, when perhaps a few millimetres of topsoil are blown off cultivated fields every year; or castastrophic: severe erosion associated with the washing out of the Daintree road in northern Queensland during the 1985 monsoon season. Furthermore, its effects are often not simply limited to the locality where damage is noted. Many effects may be offsite - for example, when transported soil particles are deposited in a water storage.

In an environment without humans, some erosion occurs as a result of natural processes. This erosion is usually termed geologic erosion. When humans and their various technologies appear on the scene, the geologic erosion rate is greatly exceeded in many cases and that erosion is termed 'accelerated erosion'. Consequently, human influence, through agricultural, pastoral and urban activities is critical in determining the extent, rate and risk of degradation. From this it follows that land degradation is essentially a human-induced process. For example, studies of the Sahel droughts of West Africa, which occurred during the last decade, indicate quite clearly that it was not a southerly migration of the Sahara desert which was responsible for desertification, but rather the northern movement of increasing numbers of agriculturalists and pastoralists into an ecologically sensitive environment (UNESCO 1980, Dregne 1983). In a practically uninhabited state, environments similar to the Sahel should be able to withstand and recover from drought without severe erosion losses, but the additional burden of high stock numbers, which remove protective vegetation, cannot be tolerated. What follows includes severe physical degradation and devastating social, economic and life-threatening effects on the human population. In Australia, we have been fortunate that the devastating effects of drought have been less severe. However, serious social and economic consequences of land degradation have been noted (eg see Aust DEHCD 1978a).

Erosion by water and wind, along with salinisation and vegetation degradation account for the majority of degradation in Australia (Tables 1 and 2). The areas in which these degradational processes are currently of most concern are frequently those where land use pressure is placed on lands of only marginal value for agriculture and pastoral uses. These areas are often in the semi-arid and arid climatic zones of the continent. Consequently, for the purposes of this review attention is focused on degradation in the semi-arid and arid zones.

The following sections of this chapter therefore take a historical perspective to look at the magnitude and extent of land degradation primarily in Australia's arid and semi-arid zones. An attempt is then made to focus on specific factors which contribute to land degradation. Some concluding remarks are made on current degradation assessment in Australia and a possible future strategy.

Degradation of pastoral landscapes

Most arid and semi-arid (Fig 1) pastoral lands of Australia have been occupied by European pastoralists for only a little more than 100 years. During this period, however, it is generally conceded that their flocks and herds, abetted by rabbits, have caused substantial degradation of the rangelands (Newman and Condon 1969). Before the advent of Europeans the use of fire by Aborigines would have had some effects on the landscape, but since Aboriginal numbers were small the impact of their fires was probably not great and fires started by lightning have always been a part of the Australian environment under natural conditions. What is clear, however, is that some features such as bare, naturally eroded surfaces were commonplace before European settlement according to accounts of early explorers such as Sturt.

Of the arid zone of Australia, which covers 75 per cent of the continent, half is unsuitable for grazing, the remainder having been developed for sheep and cattle on an extensive leasehold management system using the natural vegetation (Perry 1968). After settlement, stock numbers grew rapidly; for example, there were 10 million sheep west of the River Darling in NSW by 1890 (Fig 2). Similar stock build up took place in central Australia (cattle), Queensland and Western Australia (cattle and sheep) within approximately 30 years of settlement. Generally, the sequence of rising livestock numbers occurred against a background of adequate rainfall and satisfactory markets. In western NSW, however, a combination of drought, low wool prices, rabbits and financial depression in 1895 led to a dramatic collapse in the sheep industry and establishment of a Royal Commission in 1901 to look into the problems of Crown tenants in the Western division. Although events of the late 19th century in NSW affected primarily the livelihood of the graziers themselves, elsewhere overgrazing has been known to induce offsite problems. In February 1961 severe flooding occurred in Carnarvon WA, following heavy rains in the Gascoyne catchment (Williams et al 1980). The flooding and erosion were extensive enough to suggest that runoff from the catchment was excessive (Wilcox and McKinnon 1972). The work of Lightfoot (1961) indicated that the excess runoff was due to degradation of the catchment area attributable to overgrazing.

Studies of the grazing lands of the Northern Territory commissioned in the 1960s (Condon et al 1969 a, b, c, d) also indicated widespread degradation of native pastures and accelerated soil erosion. Similarly, degradational effects of grazing have been noted in South Australia and Queensland (Woods 1983).

The degrading effects of grazing in the arid and semi-arid lands frequently include the removal of natural vegetation, a change in the species composition of vegetative cover, often associated with an increase in the proportion of unpalatable, woody species, and accelerated wind and water erosion. In the early days of settlement, damage was often most severe on river and lake frontage country and adjoining areas, and along stock routes. With fencing of leases and sinking of permanent bores, however, damage became more extensive. In western NSW, for example, the Royal Commission of 1901 noted that many edible shrubs such as saltbush (*Atriplex vesicaria*) and cottonbush (*Maireana aphylla*) were eaten out over extensive areas. In many areas where the shrubs were lost, rabbits also prevented the development of good grass pastures following rain by eating new shoots. Subsequently non-edible species such as pine (*Callitris*), box (*Eucalyptus longiflorens*) and budda (*Eremophila mitchelli*) were observed spreading into former shrub and herb pastures.

The problem of degradation of the vegetative cover has intensified to some extent in more recent times with the spread of woody weeds unpalatable to stock into large areas of coarser textured soils in north west NSW (Booth and Barker 1981). Although these plants provide considerable protection for the soil surface from wind and water erosion, they greatly reduce the grazing capacity of the pasture, in extreme cases by as much as 57 per cent. Not only saltbush country was affected by overgrazing. The semi-arid bimble-box shrub woodlands have suffered a permanent loss of topsoil and nutrients and a dramatic increase in woody inedible shrubs (Harrington et al 1979). Similarly, much of the Acacia shrublands has also experienced widespread

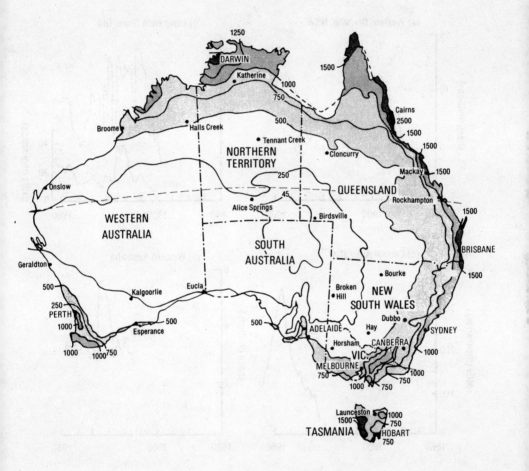

Figure 1 Rainfall in Australia. The boundary of the arid and semi-arid zones (not shaded) coincide approximately with the 500 mm rainfall isohyet.

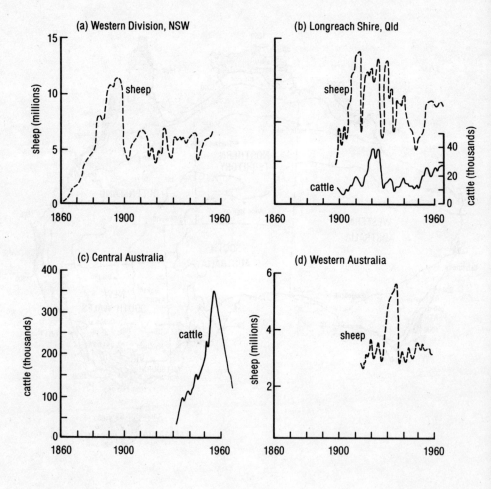

Figure 2 Stock numbers 1860–1960 (after Mabbutt 1978).

deterioration and in places widespread sheet erosion (Williams O B 1979).

Changes in the nature and amount of vegetative cover were accompanied by increasing numbers of dust storms, sand drifting and extension of bare areas. 'Scalding' (shallow stripping by wind and water) and erosion of the coarser textured surface layers of texture contrast soils were widespread effects of overgrazing. Elsewhere gullying increased and previously stable dunes were reactivated. In almost all the rangelands of the continent stock numbers built up, apparently crossed a threshold and then rapidly declined owing to a combination of degradation and economic factors. Damage to the original vegetative cover was probably so great that never again will the high stocking rates of former years be possible.

Degradation in semi-arid cultivated lands

Although it may be true to say that much of pastoral semi-arid and arid Australia is not being eroded as rapidly as in the past, the situation in the cultivated areas of the semi-arid climatic zones is certainly more perilous. Currently, previously pastoral land continues to be cleared in Queensland, NSW and Western Australia, in areas with rainfall as low as 250-300 mm. During the drought years of the early 1980s in southern Australia some of these cultivated lands experienced considerable wind erosion and on one occasion in February 1983 a choking plume of dust from northern Victoria and NSW engulfed Melbourne (Lourensz and Abe 1983). In spite of previous phases of land clearing for cultivation in semi-arid Victoria and South Australia, which resulted in severe land degradation because of sand drifting and wind erosion of soils, there has been a resurgence in land clearance in the last ten years following the greater viability of wheat cropping.

Land degradation in cultivated semi-arid South Australia and Victoria

Expansion of wheat cultivation into the semi-arid northern districts of South Australia began in the 1870s, but much of the newly cultivated area was abandoned 20 years later because of poor winter rainfall and declining prices (Mabbutt 1978, Heathcote 1965). In spite of the problems experienced in the northern districts, the next area to be cleared for cultivation was the Murray Mallee, with northward settlement of the Victorian Mallee also taking place from the early years of the 20th century. The mallee country consisted of linear sandhills with light soils stabilised by Mallee Eucalypts. Agricultural practices adopted involved frequent tillage and exposure of the soil to desiccating westerly winds. Under this system, soil exhaustion was rapid and erosion became a severe problem. Drought and economic problems frequently exacerbated the situation in the 1930s and many bankruptcies followed in the wheat farming community. The forms of erosion caused by inadequate farming procedures were predominantly deflation and sand drifting (Mabbutt 1978). Tree roots were exposed and fences undermined by these processes, with much of the nutrient-rich soil A1 horizons also being lost. Eroded and subsequently transported material blocked roads, railways and water supply channels, and dust storms became increasingly common.

Clearing for cultivation in semi-arid NSW and Queensland

Memories of the events in the mallee lands have evidently been shortlived, for clearing, some of it illegal, has begun in the last 10-20 years in the mallee landscapes of western NSW. Similarly, considerable areas of semi-arid central western Queensland and northern NSW on heavy textured soils are also being cleared for cultivation purposes. In NSW, Fitzpatrick (1982) using remote sensing (see Chapter 2) and other techniques, estimated that the total amount of clearing for wheat-pasture rotations amounted to only three-four per cent of total mallee landscapes in the 15-year period 1964-79. Much of the clearing, however, was concentrated near Balranald north of the Murray-Murrumbidgee confluence and in the area west of Condobolin in the mid west of the state on lands with high wind erosion risk (Mabbutt 1982). Although some legislation does exist to forbid clearing of particularly erosion-susceptible land, enforcement of the legislation has not been easy. Consequently, increased areas of cultivated land in central and western NSW are at risk from wind erosion, while in northern NSW and Queensland a greater threat exists from water erosion.

Salinisation associated with dryland cultivation

Salinisation of dryland areas is usually associated with clearing of native vegetation including deep-rooted eucalypts and their replacement by cereal cropping or pastures. Essentially these shallower rooting systems use up less water and in time there may be increased water accumulation in near-surface horizons and the development of shallow water-tables. Frequently, the local groundwaters are saline because of concentration of salts in the underlying weathered bedrocks. Lateral movement of surface and groundwater downslope often results in seepages forming in lower topographic positions and salts accumulate following evaporation (Conacher and Murray 1973, Peck 1978). In the wheat belt of Western Australia losses of productive land from this form of salinisation have been severe with approximately 260 000 ha being affected by 1978. Furthermore, this type of salinisation can also have offsite effects because of the ingress of salts into the local streams, with a consequent deterioration in river-water quality. Similar problems to those in Western Australia exist in Victoria with 100 000 ha affected. Of these, 40 000 ha are affected by seepage in the mallee, with significant seepage problems occurring also in the highlands, western plains and Dundas Plateau and adjacent areas of the south west. Dryland salinisation is not restricted to Western Australia and Victoria; similar problems have been noticed in South Australia, NSW and Queensland. The whole problem of dryland salting is comprehensively reviewed by the Working Party on Dryland Salinity in Australia (1982) and also dealt with by Peck et al (1983).

Degradation associated with irrigation

The extent and number of degradational problems associated with irrigation in Australia are too great to outline in detail here. However, some of the major problems and areas affected are outlined in following sections. Salinisation in irrigation areas arises principally as a result of rising water tables, which bring saline groundwater into the root zone, but in areas with deep groundwater, similar problems may also arise because of application of insufficient water to leach soluble salts out of the soil profile. Waterlogging associated with poor drainage and seepage from irrigation canals can also be a major problem. Frequently, as the groundwater levels rise, discharge of saline groundwater into the rivers increases and can cause major problems for downstream water users. In Australia the majority of irrigated land is within the Riverine Plain of the Murray-Murrumbidgee Rivers, and it is in these areas that degradation has been most severe. A recent Victorian Parliamentary Committee report (1984) on salinity in Victoria indicates that in the Goulburn-Murray Irrigation District 1400 km^2 are salt affected and a further 4000 km^2 potentially salt prone. Furthermore, the cost of damage from all forms of salinity in Victoria is about $40 million dollars per year, with productivity losses in the Kerang and Shepparton areas responsible for 75 per cent of the total loss.

In the Kerang area, initially the watertable was at a depth of 6-9 m in the north and up to 12 m in the south (Mabbutt 1978), with groundwater salinity increasing to the north and west. By the 1930s, about 50 years after the introduction of irrigation, shallow water tables were widespread and surface salinisation, expressed in the extent of bare ground and halophytic vegetation, had reached a maximum. Subsequent drainage has somewhat alleviated the local problem, but helps to contribute to higher salinity levels in the Murray.

Elsewhere in the irrigation areas on the NSW side of the Murray, salinisation and waterlogging are occurring in several localities, although groundwater salinity levels are often, but not always, well below those of the Kerang area and elsewhere in northern Victoria. In many places in the Murray-Murrumbidgee system ameliorative measures, including evaporating basins, tube-well pumping to lower water tables, and the prevention of saline drainage water from re-entering the river channel, have been taken to minimise the risk of further salinisation. The general problem is, however, of such a scale that it is well beyond the resources of individuals to combat and requires concerted government action. Cole (1985) suggests some remedies and strategies to combat the problems of the irrigation areas.

Erodibility of Australian soils

Australian soils are often cited as being among the most deficient in phosphate and other nutrients on a worldwide basis. Under natural conditions, however, the vegetative cover is well adapted to low nutrient levels, whereas many introduced crops and other plants are not. Physically and chemically, many Australian soils do not bear comparison with those of much of Europe and North America. Principally, this can be attributed to low organic matter contents, which can also be related to aridity and possibly to distinctive vegetative cover. Coupled with low physical and chemical fertility, and often directly linked with these properties, is the fact that many Australian soils have developed from ancient, deeply weathered materials. Unlike much of northern hemisphere, Australia was not glaciated to any significant extent during the Pleistocene period and thus missed out on what is essentially a rejuvenating episode in terms of soil development and nutrient availability. Finally, fairly large areas of the continent are affected by the occurrence of salt associated with groundwater and also trapped in old marine-deposited rocks. Thus low fertility, old highly weathered soils, topographic factors, climatic factors and salinity all combine to make the present soil resources somewhat fragile and particularly susceptible to degradation following land use changes from naturally occurring to artificial conditions.

Traditionally, arid and semi-arid soils are considered to be shallow, sandy and often stony, lacking in fertility and often saline, or sodic, with very low organic matter contents. Although some of the soils of arid and semi-arid Australia undoubtedly possess these unfavourable properties, this is not always the case. Quite extensive areas of deep, clay-rich soils - cracking clays (Stace et al 1968) - are also found in the Australian arid and semi-arid lands. The occurrence of these clays is generally restricted to alluvium, aeolian clays (parna) and weathered basic igneous rocks.

Major factors which influence soil erodibility in the arid and semi-arid regions include the amount, type and annual distribution of rainfall, the percentage vegetative cover, slope, and soil properties. In general, as total annual rainfall decreases, the amount of vegetative cover similarly declines and the degree of protection from wind and rain also diminishes. In the arid zones rainfall is infrequent and sporadic and consequently, even with limited vegetation cover, the total amount of water erosion is limited, particularly when averaged over many years. In the semi-arid regions, however, rainfall is higher than in the arid zones, and the amount of vegetation can be critical as a protective agent for the soil. Frequently, the clearing of perennial plant cover and its replacement by annuals and/or cultivation can lead to catastrophic erosion by both water and wind.

Soils which are sandy and poorly aggregated, such as the red sands, clayey sands and calcareous earths are particularly prone to wind erosion if their protective cover is removed. These soils have low clay and organic matter contents in their A horizons and are therefore also susceptible to structural degradation when cultivated. In parts of sub-humid to semi-arid NSW, South Australia and Western Australia, these soils are often cultivated for wheat and other cereal crops. In a matter of only a few seasons severe sand-drifting and wind erosion can result from such cultivation following the clearance of unsuitable areas, or poor management techniques.

Soils with higher clay contents are generally better aggregated than sandy soils and consequently tend to be less susceptible to wind erosion. The infiltration of water into clays is, however, much slower than into sands, and under heavy rainfall, particularly where there is no vegetative cover, these soils are prone to erosion by sheet-wash and gullying. These forms of erosion are common on some cracking and non-cracking clays and on earths and podsolic soils towards the more humid continental margins. Details of the major soils in the arid and semi-arid zones are given in Stanley (1983) and Stace et al (1968).

Assessment of degradation status, rate, risk and hazard

Degradation status refers to the state or condition existing on a particular piece of land at the time of observation compared to conditions in the past (FAO 1984). Rate of degradation is an assessment of the overall evolutionary trend of land properties, such as salinity, vegetation cover etc. Risk, however, refers to the vulnerability of the landscape to degradational processes, such as erosion, salinisation etc. Finally, degradation hazard can be determined from the status, rate and inherent risk of degradation by examination of the dominant determinative processes, including human and animal pressures on the environment (FAO 1984). The latter definitions were initially used to relate to the process of desertification, which is 'the impoverishment of arid and semi-arid and some sub-humid ecosystems by the combined impact of man's activities and drought' (Dregne 1977). Desertification is therefore one aspect of land degradation particularly pertinent to arid and semi-arid Australia.

Land becomes at risk from degradation when circumstances combine to facilitate changes in land use such as sheep grazing to wheat cultivation. Technological factors, economic demand, and atypical weather patterns have all enabled cultivation to extend into what may be considered marginal land for dryland cultivation, as an example. The damage is often done, however, when economic or weather factors deteriorate. Because of the complex interrelationship of weather, economics and technology, it makes it extremely difficult to pinpoint areas at risk from degradation. Although dryland cultivation is currently increasing, particularly in some marginal lands, it could be argued that a better understanding of the soils involved and appropriate cultivation and management techniques have considerably reduced the risk of future occurrence of the severe problems experienced in the Victorian Mallee 50 years ago. However, management standards vary widely from farm to farm and whereas careful husbandry, such as the use of minimum tillage, may conserve soil resources on one property, inappropriate measures can cause devastating problems on another.

Management and conservation strategies also markedly affect the degree of soil erosion and so these or their lack must also be considered when risk factors are being assessed. Nevertheless, using existing soil maps and information about climatological parameters (eg maps of rainfall distribution and reliability, probability of erosive winds, potential evapotranspiration etc), assessments could be made about potential soil erodibility. A map showing the distribution of saline and sodic soils prone to degradation is shown in Fig 3. Soil depth, texture, topography, slope gradient are among the many factors that can be easily measured in the field to make at least tentative assessments of risk (see for example Chartres 1982). Meteorological data can also be used to assess the risk of drought on a probability basis in marginal cultivation areas (see for example Fitzpatrick 1982), and Lovett (1973) includes several chapters on the significance of drought in the environment.

Basic land inventory data from which some of the above information

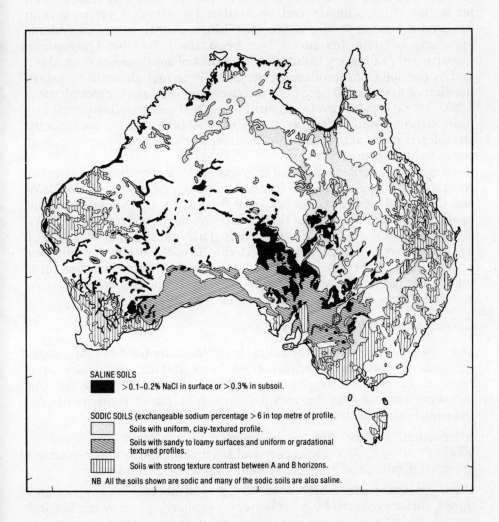

Figure 3 A map of saline and sodic soils in Australia (based on Northcote and Skene 1972).

is obtainable is available over much of the continent, but usually at very small scales. The land system and soil surveys conducted by the Commonwealth Scientific and Industrial Research Organization (CSIRO) between about 1950 and 1975 provide some information on landforms, soils, climate and vegetation for areas of up to several hundred thousand square kilometres. More recently the land systems approach of data inventory has been used by the Queensland Department of Primary Industries in studies of south-west Queensland and by the Soil Conservation Service of NSW to map the entire western division of that state. These land resource inventories at scales of about 1:250 000 or greater can be useful for general planning purposes, but more detailed data, often at scales of 1:50 000 or less, are usually considered more appropriate for producing management strategies at the individual property level.

An example of criteria used for degradation/desertification assessment from a FAO study conducted by the University of NSW and the NSW Soil Conservation Service on behalf of the FAO of the Nymagee (in the central west) and Milparinka (in the extreme north-west) areas of NSW is given in Table 3 (Chartres et al 1982). This investigation was part of a research program funded by FAO/UNESCO with the aim of producing a methodology for assessment and mapping of desertification on a worldwide basis. Attempts were made to determine the current status, rate and risk of desertification processes.

Current status of land desertification

The current status of desertification is intrinsically hard to map unless detailed information is available on land surface conditions in a relatively pristine state. FAO guidelines (FAO 1984) suggest the following criteria may be useful for current status assessment. An example is also given in Table 4.

Salinisation: surface affected by soluble salts, degree of salinity or alkalinity or both, decrease in crop yield owing to salinity and vegetation groups indicating the presence of salts in soils or groundwater.

Wind erosion: surface area covered by sand sheets, hummocks and dunes, surface covered by gravels, decrease of crop yield owing to wind erosion, vegetative groups indicating sand accumulation because of wind.

*Water erosion:*surface status, subsoil presence, surface affected by gullies, characteristic vegetative groups growing on skeletal soils.

Degradation of vegetative cover: range condition, canopy cover, biomass.

Soil crusting and compaction: presence near the surface of a

Table 3: Major criteria selected for the assessment of desertification in the NSW test areas

	Vegetation criteria	Wind erosion criteria	Water erosion criteria
Current status of desertification	Canopy cover degradation % potential productivity Woody shrub invasion Biomass Production of fodder Biomass/rain	Loss of topsoil Type of erosion Aeolian formations	Type of erosion Loss of topsoil Surface affected by gullies Soil deposits Sequence of horizons Thickness of horizons Crop yield (Nymagee only) Organic matter decline (Nymagee only)
Rate of desertification	Range trend line Forest trend line (Nymagee only) Cereal trend line (Nymagee only) Shrub invasion	(as listed in FAO 1984)	Soil loss Removal/deposition
Risk of desertification	Increased arable dryland (Nymagee only) Livestock production Overgrazing Annual unit growth rate Climatic index for biological degradation Potentional for reclamation	Erosivity Wind storm frequency Days wind storm/year	Slope Precipitation Weight of soil loss Rainfall factor Soil erodibility Topographic factor Biotic index Erosivity index

Table 4: Criteria for assessing water erosion

Desertification aspect	Assessment factor	Slight	Moderate	Severe	Very severe
Status	1) Surface status (%)	Gravel and stones < 10	Stones and boulders 10-25	Boulders and rocks 25-50	Boulders, exposures of rocks > 50
	2) Type of erosion	In sheet and rill. (Slight to moderate)	In sheet and rill. (Moderate to severe)	In sheet, rill and gully. (Severe)	In sheet, rill and deep gully. (Very severe)
	3) Subsoil exposed, % of area	< 10	10-25	25-50	> 50
	4) Gully area, % of total area	< 10	10-25	25-50	> 50
	5) Soil thickness (cm)	> 90	90-50	50-10	< 10
	6) Loss of soil depth over root-inhibiting layer, %				
	a. Original soil depth < 1m	< 25	25-50	50-75	> 75
	b. Original soil depth > 1m	< 30	30-60	60-90	> 90
	7) Present productivity % of potential productivity	85-100	65-85	25-65	< 25
Rate	1) Increase in eroded area, % per year	< 1	1-2	2-5	> 5
	2) Soil loss, Mt/ha/year	< 2.0	2.0-3.5	3.5-5.0	> 5.0
	3) Decrease in annual biomass production, % per year	< 1.5	1.5-3.5	3.5-7.5	> 7.5
	4) Sediment deposition in reservoirs, watershed 500km² (m³/km²/yr)	< 60	60-200	200-500	> 500
	Watershed > 500km²	< 40	40-100	100-250	> 250
	5) Annual loss of storage (%)	< 0.2	0.2-0.4	0.4-1.0	> 1.0
Inherent risk	1) Rating of climatic aggressivity*	0.03	0.03-0.06	0.07-0.10	> 0.10
	2) Rating of pedo-topographical conditions	< 1	1-2	2-3	> 3
	3) Rating of potential soil loss in t/ha/ya+	< 5	5-15	15-25	> 25

Source: FAO (1984)
* See FAO methodology
+ Mechanical water erosion with present vegetative cover taking into consideration annual precipitation, coefficient of variation of monthly and annual precipitation, soil texture and slope classes.

cementation resulting from carbonate, gypsum, iron and silicon accumulation and of vegetative groups indicating the presence of cementation in the soil.

Rate of desertification

In the case of the rate of desertification the following criteria were suggested:

Salinisation: rate of increase of salt-affected areas or increase in salinity of the soils.

Wind erosion: expansion of cropland, volume of loose sand, growth rate of affected area, soil loss.

Water erosion: expansion of area with exposed subsoil or surface affected by gullies and sediment in dams.

Degradation of the vegetative cover: decline of biomass production, increase in the ratio of shrub-to-grass cover decrease in the area of woodland, range cover trend line.

Soil crusting and compaction: the rate cannot be readily evaluated.

Risk of desertification

With regard to *risk* of desertification the following factors were considered important:

Salinisation: depth of groundwater table and quality of irrigation water.

Wind erosion: wind erodibility groups, mean annual wind speed, climatic factor for wind erosion.

Water erosion: slope, surface soil erodibility, climatic factor for water erosion.

Degradation of vegetative cover: climatic factor for biological degradation (climatic factors referred to above are given in FAO 1984).

Some of the criteria developed for desertification assessment could also be used with others for assessment of land degradation in other areas.

In the case of the NSW investigation into desertification it was decided to determine whether adequate information about the status, rate, risk and processes of desertification could be obtained using existing land-systems survey reports and maps (Chartres et al 1982). The results demonstrated that some relevant data, particularly for the pastoral areas, could be obtained. This was especially the case if inputs from map makers and local field staff of the agencies entrusted with land management and conservation are available. In the cropped areas the level of detail required is not readily available from land-systems maps. Furthermore, data on rate of desertification are generally very sparse.

When it comes to assessment of land degradation in the humid regions of the continent more data in the form of individual scientific investigations, agricultural land capability maps, soil survey data etc is often available, but this is frequently not in the ideal format for planners.

Future prospects for the assessment of land degradation in Australia

The foregoing sections illustrate the extent and nature of some of the more marked types of land degradation that have been recorded to date. Furthermore, it is apparent that causes and effects of many of these degradational processes are well known and, using appropriate methodology (see Wasson, Chapter 3), can be measured and quantified to some extent. If history is anything to go by, it is also clear that, given appropriate conditions, many of the mistakes of the past will be repeated, particularly if climate, technological factors and economic conditions combine favourably to allow profitable exploitation of a land resource. While some land users will always be prudent and realise the value of conservation measures, there will always be those after the 'fast buck', with no concern for either the ultimate fate of their own properties, or for offsite effects of their own inappropriate management strategies. For this reason it is highly pertinent to ask whether we currently have a suitable system of assessing and mapping land degradation? While individual governmental agencies undoubtedly have developed means of assessment and prevention of certain specific problems (eg the prevention of wash erosion by contour banking) there is, as yet, no all-embracing strategy or methodology for the assessment and mapping of land degradation either at a federal or state level. From the nature of the problems discussed previously, it is clear that many of them are widespread and transcend state boundaries (eg salinity problems in the Murray Valley). Others generate offsite factors, which can affect neighbouring states (eg wind and water erosion). Consequently, the development of strategies to assess and combat land degradation should be tackled primarily at a federal level with considerable collaboration from the states. What is needed is the development of a system of land degradation assessment which identifies and maps the principal causes of land degradation in all environments; determines the magnitude of onsite and offsite effects of the processes in terms of not only physical, but also socio-economic factors; assesses the risk and hazard to the environment of land degradation; and develops and proposes conservation strategies to minimise potential damage along the lines proposed by Anon (1983). Whether such a system could be developed given both financial and constitutional constraints, however, remains to be seen.

This information can then be used as a rational basis for legislative and planning procedures, which can be applied as seen fit. Some data of relevance to the aims, applying to specific problems, are already being collected by various state and federal authorities. However, it is rare that the different problems are being considered together. Consequently, what is required is some form of land degradation assessment which is all embracing. Such a system cannot in reality be divorced from systems of land evaluation (eg see McRae and Burnham 1981).

The type of approach espoused by the Canadian Land Use Inventory, which assesses land suitability for specified uses such as forestry, agriculture, recreation, wildlife etc, is perhaps an example of a fairly successful approach to land evaluation in general. The Canadian approach also uses agroclimatic information and socio-economic data, with the entire system being computerised to assist data retrieval. Elsewhere, other countries have developed reasonably successful systems of land capability assessment (Klingebiel and Montgomery 1961), but these are based frequently on primarily physical data and usually ignore and minimise inputs of socio-economic variables. In Australia several states have agricultural capability assessment schemes, such as the derivative of the United States Department of Agriculture (USDA) system used by the Soil Conservation Service in NSW and in Queensland by the Department of Primary Industries. However, these schemes are usually directed at one specific land use (eg agriculture). Similarly, state governments often commission land evaluation surveys of specific regions, particularly those under pressure from different land uses. At present, however, these surveys are one-off, often piecemeal, and usually still based primarily on physical data inputs, with limited social and economic inputs. What is required is an approach which marries physical and socio-economic data, is based on not only current, but likely future land-use trends, has a sound philosophical background with special reference to the Australian environment, and moreover transcends state boundaries. Furthermore, such an approach would have to have a data storage and retrieval system which is readily accessible and easily manipulated.

The land-systems approach could provide a starting point for the development of an Australian land-evaluation/degradation scheme, but it would need considerable development in terms of mapping units if it were to be used at the large scales necessary, particularly in the more densely populated regions.

Currently, we are at a crossroads. As the following chapters will demonstrate, we know and understand many of the mechanisms of land degradation and their effects. What is more we have the scientific and economic methodologies to measure them. We do not as yet have an adequate database, particularly for measurement of rate and risk and hazard of degradation. Developing a system of measurement and monitoring and a system of data storage and mapping would be expensive, but in terms of potential future risks that face us from land degradation costs, such an expense would be minimal.

2 Biological and physical phenomena in land degradation

Gordon Burch, Dean Graetz and Ian Noble

Introduction

Within Australia we have the paradox of having a detailed understanding of the processes whereby our renewable resources are being destroyed or degraded but lacking a rigorous quantitative assessment of how much has been degraded and how severely, or whether this degradation is accelerating or stable. Nor are the costs known. We understand the detail clearly but have no overview. Unfortunately, it is the latter that is required for instigating changes in management to curb land degradation. Land degradation is taken here as the end result of any factor or combination of factors which damage the land, water or vegetation resources and restrict their use or productive capacity (see Chartres, Chapter 1).

The comprehensive and accurate overview is lacking for two reasons. The first is that we are dealing with a dynamic system of land in much of agricultural Australia; major changes in vegetation, soils and drainage associated with agricultural development are still occurring even though most of the agricultural lands have been occupied for a century or more. For example, extensive clearing of native vegetation for cropping in Western Australia and Queensland has occurred during the previous two decades, and is still in progress. It follows that many of the biological and physical changes due to European settlement have yet to stabilise, or worse still, many consequences of these changes remain unidentified.

Secondly, the scope and duration of scientific investigations of these problems have frequently been inadequate to allow firm conclusions to be drawn or management solutions to be developed for Australian agro-ecosystems. For example, there have been attempts recently to determine the trend in salinity levels in the Murray River. Reliable water quality recording began only in 1963 and because of the extreme variability in flow and salinity concentration scientists have insufficient record to be confident in specifying the current rate of salinity rise. Estimated rates of salinity rise at the same recording location vary between six and 1.5 per cent per annum (Morton and Cunningham 1985,

Close pers comm). Consequently, it is equally difficult, if not impossible, to evaluate the effectiveness of recent expensive engineering structures designed to desalinate return flow to the river from irrigation and drainage schemes.

There are some features of Australian ecosystems which can make them particularly vulnerable to a wide range of degradation processes and these have been introduced in the first chapter, particularly concerning the properties of Australian soils. For more detailed information on soils the reader is referred to texts which include Northcote et al (1975) and a CSIRO, Division of Soils (1983) publication 'Soils: an Australian viewpoint'. The extreme age of Australian landscapes has had a strong influence on soil formation because of the processes of deep weathering, leaching, translocation and truncation due to erosion. The end result has been highly infertile or shallow soils, often low in organic matter and high in residual immobile compounds such as iron, manganese and aluminium oxides. These residual compounds have the added disadvantage of tending to fix any input of labile nutrients such as phosphorus and in high concentrations in soils with a low pH they become toxic to plants (Bromfield et al 1983). By European or North American agricultural standards Australian soils compare very poorly and continued production from them is possible only under sympathetic management systems. It has taken over a century to adapt the agricultural traditions of Europe to the realities of the Australian environment.

The inherent infertility of Australian soils is exemplified by the historic trends in wheat yields (Figure 4) documented by Donald (1982). In addition, the history of superphosphate application in Australia is presented in Figure 4 to indicate that rises in wheat yields since 1900 are associated with the use of fertilisers and the associated introduction of subterranean clover into pasture rotation cropping systems. The early decline in yields due to cropping clearly illustrates the limited natural fertility and structural instability of most Australian soils. Subsequent fertiliser usage over 80 or more years has failed to lift the mean yield above 1.5 tonnes per hectare and this is comparable with yields obtained in the early years of wheat cropping.

The commercial mean yields in Victoria and NSW have remained at about 40 per cent of the peak yields recorded annually between 1946 and 1968 (Gifford et al 1975). Equivalent commercial mean yields recorded in the USA (Crosson and Stout 1983) rose from less than 1.0 tonne per hectare for the period 1930-9 to around 2.2 tonnes per hectare for 1976-80. By comparison Australia is not showing these levels of improvement in grain production despite considerable genetic and agronomic inputs and this possibly reflects underlying degradation of the soil resource. Such production trends can be taken as a paradigm of

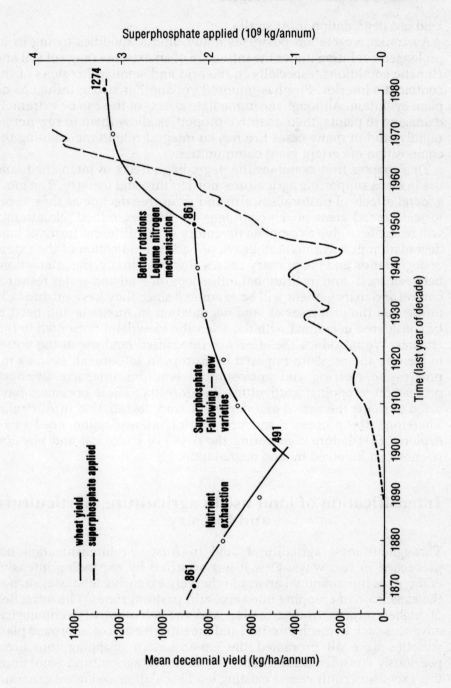

Figure 4 Mean decennial wheat yields 1870–1980 (after Donald 1982) and the history of superphosphate application in Australia (after Gifford et al 1975).

land use degradation in Australia.

Australian vegetation possesses many unique qualities owing to its prolonged evolution under the influence of an extreme range of soil and climatic conditions, especially in the arid and semi-arid regions of the continent's interior. Fire has imposed yet another strong influence on plant evolution. Although the immediate effects of fire can be extremely damaging to plants, their adaptive properties allow them to regenerate rapidly, and in many cases fire has an integral role in maintaining the composition of certain plant communities.

This chapter first examines the degrading effects of intensified land use in areas supporting agriculture, horticulture and forestry. The more general effects of pastoralism, fire and rural tree decline as they apply to widespread areas of the continent are then examined. No attempt will be made to give a complete inventory of the different forms of land degradation in Australia; rather we provide an indication of the extent of degradation and the primary causes at work. Finally, the interactions between social and institutional influences on land and water resource control and management will be examined since they have intrinsically influenced the processes of land degradation in Australia and need to be considered in context with the scientific viewpoint presented in this chapter. We introduce the effects of intensified land use at the outset to identify the pressure imposed by European settlement, such as the progressive clearing and enclosure of land for intensive livestock production, cropping, horticulture and forestry. These pressures have acted to shape the extent and degree of land degradation in Australia. Therefore, the processes and progress of intensification need to be explored first before considering the detail of biological and physical phenomena involved in land degradation.

Intensification of land use in agriculture, horticulture and forestry

Throughout most agricultural land in Australia intensification has proceeded in two ways. First it has occurred by expanding intensive enterprises into marginal areas of relatively extensive land use, such as the expansion of cropping into semi-arid pastoral zones. The attraction of higher returns from cropping and the adoption of technological advances such as mechanisation and the introduction of improved plant varieties have all promoted the expansion of cropping into areas previously considered unsuitable for intensive agriculture. Sometimes this expansion contravenes existing land legislation and more generally there has been little or no research on the long-term consequences of that expansion. Second, intensification has occurred in production methods used in existing agricultural enterprises. That intensification can greatly increase the physical and biological stresses threatening the stability of soil-plant associations in agricultural ecosystems. These stresses can cause serious instabilities which managers frequently combat with equally extreme corrective measures. For example, some high-value crops such as cotton are especially prone to disease and insect attack and require routine applications of highly toxic sprays which destroy the natural balance of soil and plant fauna with potentially deleterious long-term consequences.

Intensification of production, as in the second category mentioned above, is occurring especially in horticulture. Here the high value of crops produced and continuous cost-price pressures promote the intensive use of irrigation, chemicals and fertilisers. The geographic location of many horticultural areas in Australia makes irrigation essential and the increasing demand for water is tending to exhaust the supply of readily available high-quality water, thus leading to the use of more inferior quality water. The long-term results of these practices in many horticultural regions has been to degrade irrigated soils chemically, especially by salinisation. Accessions to groundwaters are increased also, causing watertables to rise to within the root zone of crops thereby inducing waterlogging and, once again, surface salinisation. There are additional onsite and offsite effects of intensive production methods used in the horticultural industry which will be discussed in more detail later.

Both categories of intensification are occurring in forest production in Australia. Management practices adopted in forestry often change the natural forms of native forest communities in ways that need not necessarily be interpreted as land degradation, nevertheless, growth and harvest cycles can be so prolonged that degradation is difficult to detect. Therefore, examples of problems arising in forest production will be mentioned that need not result in serious or permanent degradation, although risks do exist. Previously unused or unproductive areas of land, often vegetated by native species, are now being planted to exotic or preferred tree species and harvested on a regular cycle; for example the establishment of pine plantations at Mt Gambier in South Australia. These areas invariably have infertile or problem soils that can suffer nutrient exhaustion or structural breakdown if the harvesting cycle is too exploitative. At Mt Gambier, second rotation pines are requiring added fertiliser to achieve previously recorded growth rates and additional problems are arising because of long-term compaction of the soil. The second category of intensification is usually associated with attempts to promote productivity in native forests by cyclic harvesting and re-establishment techniques to achieve even-age stands of one or more preferred species of high growth rate. Here some of the management techniques, such as controlled burning to reduce the risk of wildfire, have changed the species composition of the forests, leading to disease problems and unavailability or inadequate cycling of nutrients (Hingston et al 1981). In some cases, such as in the rainforests of north-eastern Australia, harvesting has led to the commercial exhaustion of certain sought-after timber species as well as to serious soil erosion and leaching of nutrients.

Land clearing

Since European settlement first started there has been a progressive and continuing process of wholesale clearing of native vegetation in almost every part of the Australian continent. This process has been especially prevalent in areas with adequate rainfall, fertile soils and tractable topography. So complete has been the removal of trees in some areas that a further decline in the few remaining trees has sparked a vigorous conservationist outcry. Ironically, in other parts of the continent broad acre clearing goes on unabated (Jasper and Richards 1983). The extent and range of effects of land clearing are best considered within three major geographic regions, the south-west, south-east and north-east as defined in Figure 5 (Burch 1986). Approximations of the area occupied by dominant overstorey vegetation types and the areal extent of land cover modification have been obtained by Laut et al (1980) for each of the three geographic regions and are given in Table 5. This Table can be used to provide a gross picture of the magnitude of European impact on native flora.

The south-west region was dominated by eucalypt woodland and shrubland before settlement (Table 5). These vegetation types represent a highly stable perennial cover well adapted to the effects of fire and other extreme elements of the climate in that region, particularly drought and wind. Many of the soils of this region are sandy, poorly structured and infertile which renders them particularly vulnerable to wind erosion after clearing (Pearce and Bradby 1983). Salinisation is the other main degradation process threatening areas that range from those originally covered by wet temperate forests close to the seaboard to those cleared of open woodland at the arid extremities of the inland wheat belt.

Common to all occurrences of secondary salinisation is the removal of perennial vegetation cover which changes the hydrological balance of the landscape (Bettenay et al 1964, Mulcahy 1978). Reduced evapotranspiration due to removal of native vegetation results in increased vertical and/or lateral drainage in hillslopes and the leaching of salt to groundwaters, causing water tables to rise and become more

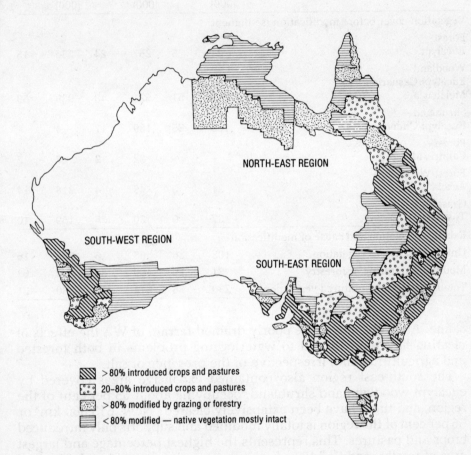

NORTH-EAST REGION

SOUTH-WEST REGION

SOUTH-EAST REGION

▨ >80% introduced crops and pastures

⊙ 20–80% introduced crops and pastures

▥ >80% modified by grazing or forestry

☰ <80% modified — native vegetation mostly intact

Figure 5 Extent and degree of vegetation modification for three geographic regions, the south-west, south-east, and north-east of Australia (after Laut et al 1980).

Table 5: Approximate area occupied by dominant vegetation types before settlement
(after Laut et al 1980)

| | Geographic regions | | | | | |
| | South-west | | South-east | | North-east | |
	km² (000)	%	km² (000)	%	km² (000)	%
Vegetation cover before modification (settlement)						
Forest— Eucalypt	21	5	257	24	236	15
Woodland— Eucalypt/Casuarina/ Melaleuca	214	51	521	48	939	59
Shrubland— Eucalypt/Chenopod	105	25	189	17	4	–
Forest— Callitris/Mixed	–	–	22	2	27	2
Forest/Woodland— Acacia	21	5	49	4	218	14
Grassland— Tussock	37	9	50	5	159	10
Existing vegetation and cause of modification						
Unmodified native vegetation	108	26	65	6	258	16
Modified by grazing or forestry	81	19	409	38	1 009	63
Totally replaced by cultured vegetation	230	55	614	56	330	21

saline. Also, in some of the poorly drained terrain of WA the effects of clearing have contributed to waterlogging problems in both forested and agricultural land, irrespective of the presence of salt.

The south-east region also contained expansive areas covered by eucalypt woodland and shrubland, occupying about 55 per cent of the region, and these have been extensively cleared. Now, 613 860 km² or 56 per cent of the region is totally modified and supports only introduced crops and pastures. This represents the highest percentage and largest area of totally modified flora in the three regions considered. Much of this area comprises a high proportion of the land used for intensive agricultural production and so represents a valuable land resource. This has been emphasised by the statistics presented by Woods (1983) which show that although Australia is a large continent, no more than 10 per cent is arable, about 5.8 per cent is under extensive cropping and only 0.3 per cent is under intensive cropping.

The degrading effects of land clearing and subsequent land use are very diverse in this region, ranging from the well-documented effects of overclearing and cropping on fragile mallee soils in Victoria and South Australia (Williams 1978) to erosion and salinisation in adjacent high rainfall areas. Several degradation processes are at work in the mallee area, including extensive wind erosion (Stoneman 1973), physical deterioration and nutritional exhaustion because of repetitive cropping (Greenland 1971, Hunt 1980), and localised secondary salinisation because of leaching and seepage from old dune formations (Rowan 1971).

Secondary salinisation is occurring on a large scale throughout the central and northern highlands of Victoria as a result of tree clearing (Jenkin 1981). In addition to the immediate effects of tree clearing on groundwater recharge and salinisation, there are associated changes to soil hydraulic properties which materially change the overall hydrology of the landscape (Burch et al 1983). Surface and subsurface soil permeabilities are reduced by clearing and subsequent land management, thereby increasing runoff and promoting erosion, especially where dispersive subsoils are exposed to gully or tunnel erosion. Peak storm discharges from a cleared catchment were found to be as much as a thousand-fold greater than those from a forested catchment. In addition, the soils remained wetter during summer with extensive saturation zones forming around drainage lines, from which percolating water possibly contributes to regional groundwater recharge.

Similar effects of excessive clearing and poor land management are evident throughout parts of the central and southern upland regions of NSW, particularly as extensive gully erosion and localised occurrences of secondary salinisation. Both wind and water erosion are rarely observed as continuous processes in Australia. More often erosion occurs during sporadic extreme climatic events of drought or flood. The threat imposed by such extreme events is very real in this region, as shown by the effects of the 1982 drought. There have been other historic influences contributing to land degradation in this region, especially the early excesses in stock numbers reached during the late 1800s followed by the effects of rabbits through until the 1950s.

The high proportion of natural vegetation cover remaining in the tropical humid to semi-arid north-east region is a distinctive feature of this region (Table 1), although there has been extensive modification of the flora owing to grazing. In addition, quite extensive areas have been cleared in recent years for cultivation and pasture establishment (Graham et al 1981, Hughes 1981). Substantial areas of brigalow forest, vine thickets, eucalypt forests and woodlands were cleared using heavy mechanical equipment causing severe disturbance to the soil which exposed it to accelerated wind and water erosion. The effects of clearing are particularly serious in high rainfall coastal areas covered by rainforest where high intensity storms are prevalent (Gilmour 1977). High intensity rainfall also poses a problem for many sub-humid areas in this region now used for intensive crop production (Sallaway et al 1983).

Effects of tillage

In Australia recognition is growing of the potentially damaging effects of excess mechanical soil disturbance caused by tillage. A 28-year study of different crop rotation treatments at Wagga Wagga, NSW, has shown that the soil eroded from cultivated treatments with an eight per cent slope caused as much as a 30 per cent reduction in wheat yields (Aveyard et al 1983). Surface erosion due to overland flow can be reduced using contour control banks but these have little effect on wind erosion. Unfortunately, very little information is available on the effects of wind erosion on crop productivity despite general recognition of the problem (Marsh 1979). Machinery traffic associated with excessive tillage also creates soil compaction and the formation of plough pans which can restrict root growth and impede drainage, resulting in yield reductions due to waterlogging in seasonally wet environments.

Many Australian soils are especially susceptible to structural deterioration when subject to repetitive mechanical disturbance (Hunt 1980). This is often attributed to inherently low soil organic matter levels which are even further reduced by cropping. The structural stability of many cropping soils can be maintained only by regular pasture rotation. New conservation tillage practices which minimise or eliminate repetitive cultivation offer another means of better maintaining organic matter content and soil structure. However, conservation tillage has not always produced in Australian soils substantial increases in organic matter (Hamblin 1980), such as have been obtained overseas (Cannell and Finney 1973). Nor have these practices always improved soil conditions for crop growth, especially in poorly structured sandy soils (Hamblin et al 1982).

Despite these exceptions, conservation tillage is generally being found to improve soil aggregate stabilities (Packer et al 1984), enhance infiltration and drainage (Burch et al 1986), and maintain or improve crop yields (Mason and Fischer 1986).

Soil acidification

Problems associated with soil acidification in south-eastern Australia include poor establishment and persistence of lucerne and the degeneration of subterranean clover pastures leading to reduced animal productivity and lower crop yields after pasture rotation. Wheat production can become prohibitive in some areas that would otherwise enjoy a high yield potential. Acidification is associated with the long-term use of superphosphate on subterranean clover pastures (Williams 1980). The direct fertiliser contribution to increases in acidity is thought to be small compared to contributions from organic matter accumulation and the process of rhizobial fixation of nitrogen by clover. The decline in soil pH is associated with the release of manganese and aluminium ions into the soil solution at levels that are toxic to susceptible plants.

Recent studies are now identifying quite specific effects of changes in the concentrations of aluminium and manganese in soils that are subject to acidification (Bromfield et al 1983). Increasing soil acidity will lead first to problems with manganese toxicity if sensitive species, such as rape, are grown. In other soils naturally low in reactive manganese, aluminium will assume an increasingly important role as the pH declines, inhibiting legume nodulation and depressing the growth of susceptible plants.

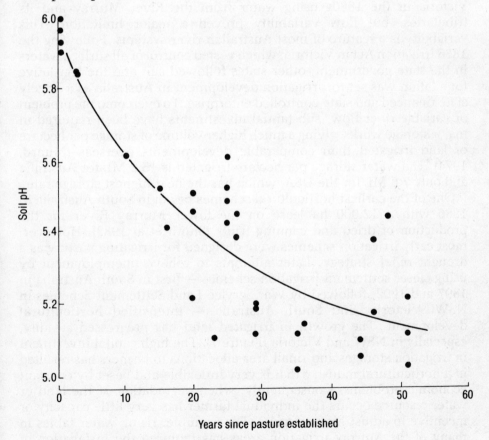

Figure 6 The relation between age of subterranean clover pasture and pH (1:5 water) of the surface 10 cm of soil (after Williams 1980).

Areas studied by Williams (1980) and Bromfield et al (1983) included clover pastures established as early as 1928, with pH observed to decline steadily from about 6.0 for unimproved land to 4.8 or less after 50 or more years (Figure 6). In addition, the pH is reduced throughout the profile to depths of 60 cm or more, with aluminium and manganese concentrations elevated at these depths as well. The implications of these findings are now serious for extensive areas of highly productive agricultural land. Surface liming can be used to combat acidity and provide some relief, but in highly acidified soil deep placement of lime and fertiliser may be required to maintain plant productivity.

Irrigation

Early attempts to establish irrigation in Australia began in northern Victoria in the 1880s using water from the River Murray and its tributaries, but flow variability proved a major limitation. This variability is a feature of most Australian river systems. Following the 1886 Irrigation Act in Victoria, which vested control of all surface waters in the state government, other states followed suit and the legislative foundation was set for irrigation development in Australia as a largely state-financed and state-controlled enterprise. To overcome the problem of variable river flow, substantial investments have been required in major storage works, giving a much higher volume of storage per hectare of land irrigated than comparable developments overseas (Munro, 1974). The water storage per hectare irrigated is 15.2 ML for Australia and only 7.6 ML for the USA, which has the next highest storage rate.

One of the earliest horticultural schemes began in South Australia in 1886 with a 12 000 ha lease on the lower Murray River for the production of dried and canning fruits (Smith et al 1983). However, most early irrigation schemes were designed for irrigating pasture as a drought relief strategy. Later attempts to relieve unemployment by using closer settlement irrigation schemes — first in South Australia in 1897 and 1902, followed by War Service Land Settlement Schemes in NSW, Victoria, and South Australia — intensified horticultural development. The growth in irrigated land has progressed steadily, especially in NSW and Victoria (Figure 7). The high capital investment in irrigation storages and small area allocations to farmers has resulted in a horticultural industry that is very inflexible and beset by constant economic problems. Consequently, when degradation of the land or water resource occurs the individual farmer has very little capacity or incentive to adjust his enterprise. For example, rising water tables in many of the Murray irrigation areas has required the installation of expensive groundwater pumps, a task undertaken by the state

THE
UNIVERSITY OF WINNIPEG
PORTAGE & BALMORAL
WINNIPEG, MAN.   R3B 2E9
CANADA

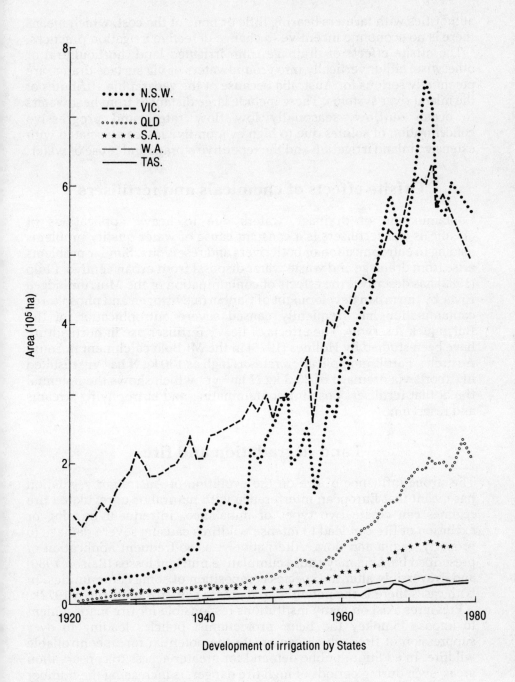

Figure 7   Growth in area irrigated since 1920 for each of the states in Australia (after Smith et al 1983).

authorities with farmers bearing little or none of the cost, which means there is no economic incentive to change defective irrigation practices.

The offsite effects of drainage from irrigated land, horticultural or otherwise, either vertically into groundwaters or via surface drains, are potentially serious for Australia because of the geographic attributes of the inland river systems. These include large distances from headwaters to ocean outflow, seasonally low flow rates, and progressive concentration of solutes due to high evaporative losses associated with extensive inland irrigation and the repetitive storage and reuse of water.

Offsite effects of chemicals and fertilisers

Contamination of drainage waters due to heavy applications of chemicals and fertilisers is a constant cause of water quality problems leading to eutrophication of both rivers and reservoirs. Similar problems arise from drainage and waste water disposal from urban centres. Philp (1982) has described the effects of contamination of the Murrumbidgee River by the urban development of Canberra. Nitrogen and phosphorus contamination has frequently caused severe eutrophication of the Burrinjuck reservoir. The effects of heavy fertiliser use in horticulture have been studied by Holmes (1978) in the Mt Bold catchment in South Australia. Fertiliser application rates as high as 130 kg N ha^{-1} yr^{-1} resulted in exports via drainage of 25.3 kg N ha^{-1} yr^{-1} which shows the potential threat that fertilisers pose for contaminating and eutrophying streams and reservoirs.

Land degradation and fire

The strong influence of fire on the evolution of Australian vegetation has meant that European interference with natural or unregulated fire regimes can create two types of imbalance. Infrequent burning or exclusion of fire can lead to intense wildfires causing severe damage to property, fauna and flora. Alternatively, a too-frequent application of prescribed burning may cause cumulative nutrient losses (Raison 1980) and irreversibly alter the species composition of some communities. In both cases there is the potential to increase soil erosion (Brown 1972).

Pressures exist on many institutions responsible for fire management to impose 'Smokey the Bear' protectionist policies leading to over suppression of fires which can create the potential for uncontrollable wildfire. In addition, public demand for greater access to conservation areas, even during periods of high fire danger, is increasing the number of ignitions in many areas (eg national parks and forest reserves) thus making effective fire management difficult. 'Blow up' days, such as Ash

Wednesday 1983, are often seen as catastrophic events, but are in fact inevitable. The results become catastrophic only if fire regimes before that event have rendered ecosystems vulnerable to irreversible change away from some preferred condition.

Adaptation of the Australian biota to a long history of fire can lead to a false perception that alterations in the fire regime will have little or no effect on plant communities. The processes and response patterns at work in ecosystems adapted to a given fire frequency, intensity and season when burning is most probable, are complex and often poorly understood. Prescribed burning is usually imposed in such ecosystems on the basis of the accumulation of litter. Problems occur when the rate of litter accumulation is rapid compared with the rates of other biological processes, eg germination and establishment of important species, and burning is initiated too frequently. An example of litter accumulation is given in Figure 8 which indicates that litter loads can approach their maximum only two to five years after burning; hence the enticement to prescribe a high-frequency fire regime to maintain low fuel loads.

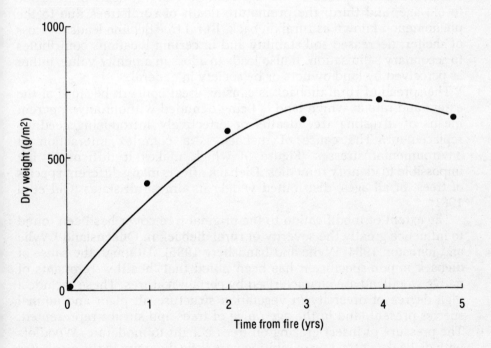

Figure 8   Accumulation of grassy fuel in a tropical grassland (after Walker 1981).

The impact of an inappropriate prescribed burning regime can be indirect as in the case of the jarrah forests in Western Australia (Shea et al 1981). Here, it appears that low-intensity prescribed burning, as used for some time, has favoured the growth of banksia species at the expense of legumes which need high soil temperatures to stimulate germination. The banksia species are hosts for the fungal pathogen *Phytophthora cinnamomi* which is causing dieback in the jarrah forests. Dieback results in the loss of forest products and may affect the quality of Perth's water supply due to increased salinisation resulting from changes in the water table following the loss of jarrah cover. A more appropriate burning regime will require occasional high intensity fires to reduce the density of banksia in the understorey.

Rural tree decline

Rural tree decline is the progressive and fairly rapid dieback and death of native trees on farms. There are three elements of rural tree decline. The first arises with the clearing of existing forests and the prevention of regeneration imposed by management; second, the death of trees due to old age, and third, the premature death of rural trees due to the phenomenon known as rural dieback. Rural tree decline leads to a loss of shelter, decreased soil stability and in certain locations contributes to secondary salinisation. It also leads to a loss in amenity value, either as perceived by land owners or by society in general.

The spread of rural dieback is causing great concern because of the extent and rate at which land is being denuded without any apparent means of arresting tree deaths or effectively introducing seedling replacement. The cause of dieback is a complex interaction of environmental stresses (Figure 9) which makes it difficult if not impossible to identify remedies. Dieback affects many different species of trees, of all ages, distributed widely in almost all states (Old et al 1981).

The extent of modification to the original tree cover has been found to influence greatly the severity of rural dieback in Queensland (Wylie and Johnston 1984, Wylie and Landsberg 1985). Although the cause of dieback is non-specific, it has been noted that 'healthy' remnants of native woodland are characterised by certain features. These include a high degree of diversity in vegetation structure, in plant and animal species present, and in the age range of trees and shrubs represented. The pressure of insect grazing on trees is light to moderate. Woodlots with dieback contrast by exhibiting very little diversity in the attributes listed above and the pressure on trees from grazing by insects is severe and sustained. Saline groundwaters and salt scalds are common in

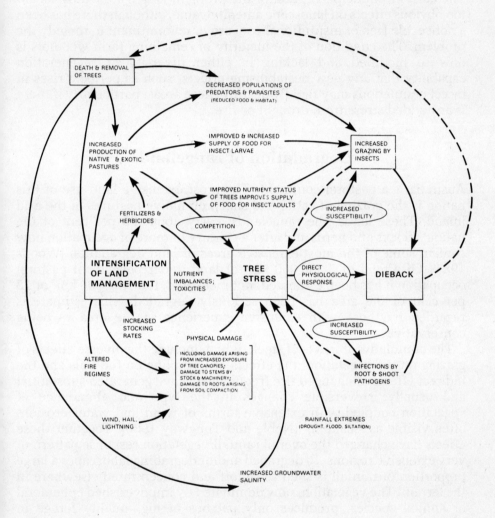

Figure 9 Factors contributing to rural dieback (after Wylie and Landsberg 1985).

vulnerable locations. Once a dieback sequence begins in remnant farm woodlots they tend to regress rapidly through a series of unstable states to treeless grassland (Figure 9).

The present severity of rural tree decline is principally a legacy of the lack of foresight when land was cleared for agricultural or pastoral use. The current public perception of the problem is heightened by its all-too-obvious effects on landscape amenity value, although there has been a noticeable lack of institutional concern or commitment to remedy the problem. The condition of the majority of remaining farm woodlots is now so modified and lacking in either diversity or regeneration capability that any new destabilising effects, such as periodic rises in insect populations may rapidly destroy these areas, particularly if there is any added stress from drought or fire.

Degradation of rangelands

Australia is a pastoral continent. The most extensive land use of this nation is the grazing of cattle and sheep on native pastures of the arid inland. These lands, the rangelands, carry about 25 per cent of the nation's flocks and herds and after a century or more of occupation now contain some of the most extensive areas of land degradation (Woods 1983). Of a total of 3 356 000 km^2 of land under permanent pastoral occupation it has been estimated, not measured, that 432 000 km^2 or 13 per cent of the area has been seriously degraded. This estimate is regarded by those with extensive experience in the area as being conservative.

The cumulative impacts of stock, which include the diverse effects of grazing on the vegetation, the effects of trampling on the soils and the indirect effects of changed fire frequencies, have generated substantial and usually irreversible changes in the type and abundance of vegetation coupled with extensive forms of wind and water erosion, often visible as 'scalding' (Noble and Tongway 1986). Together these effects have changed the overall rainfall/vegetation response pattern of very extensive regions. In degraded and/or degrading landscapes a large proportion of rainfall is shed as runoff and concentrated elsewhere in the terrain. The vegetation, now dominated by impoverished ephemeral or annual species, produces only patches of high-quality forage in contrast to supporting evenly distributed perennial pasture species that characterise an undegraded range (Harrington et al 1984).

Changes to the vegetation resulting from reduced fire frequencies are gradual and have only recently been widely appreciated (Hodgkinson et al 1984) relative to the more rapid and diverse effects of overgrazing which have more serious ramifications, including affecting the survival of native wildlife (Newsome 1975). Australia-wide, woody species (shrubs and trees) have increased at the expense of the herbage layer over extensive areas of the once productive open woodlands. The reduced carrying capacity of this land was often ignored, grazing intensities were not reduced in sympathy and the adverse effects of overgrazing, such as sheet soil erosion, are now widespread (Pressland 1984).

From anecdotal accounts, evidence given to commissions of enquiry and what few quantitative records have been collected for the rangelands it can be deduced that the periods of severe degradation have been episodic, coinciding with severe drought periods. Conservative management of the rangelands is a difficult task (Harrington et al 1984) and was impossible before a reduction in populations of the introduced rabbit was achieved using myxomatosis. Current occurrences of land degradation are not as widespread as pre-myxomatosis, rather they are localised and readily detectable using satellite imagery (Graetz et al 1986). Opportunities now exist to use modern monitoring techniques to ensure that conservative management practices are adopted, especially during periods of peak risk, eg drought, and so reduce the possibility of further rangeland degradation.

Social and institutional influences — a synthesis

A pertinent question to ask at this point is 'why has a wealthy and developed society allowed extreme rates of land degradation to continue?' Is there not sufficient scientific understanding of these ecological processes for degradation to be prevented or at least minimised? The answer is a conditional 'yes'. Earlier in this chapter, evidence has been given of the considerable scientific understanding of degradation processes now at our disposal, though many gaps remain to be filled. Even with incomplete knowledge, enough is known to remedy more effectively many of the degradation scenarios outlined above. Scientific know-how is not necessarily the rate-limiting step in preventing land degradation.

Restrictions exist in the difficult steps needed to employ the knowledge and imperatives arising from research in the natural sciences, particularly if an unreceptive social, economic or political attitude prevails. While the individual landholder has the primary responsibility for soil conservation and land management on freehold land, governments have had the responsibility, both legislative and ethical, to ensure the adequate conservation of land resources. They have frequently failed, for the state of the nation's land resources cannot be regarded as satisfactory. The cumulative depredation of Australian society on the land is high on a comparative world scale. It has been estimated that on a global basis humans have degraded in excess of 430 million hectares of crop land and grazing land over the last 7500 years since the evolution of sedentary agriculture (Kovda 1977). Osborne and Rose (1981) have used a similar approach to estimate that Australia has a degradation ratio of 11 ha per head of population which compares unfavourably with the ratio for North America of 4 ha per head.

Why have our governments failed? The reasons derive from the historical roots of our society and the attitude to land evolved over a short exploitative period since first settlement. Australia as a nation does not profess a conservative land ethic (Roberts 1985a). While this aspect will be discussed in a number of the chapters that follow, it is important to establish some of the links between the natural scientist's viewpoint on land degradation and the social, economic and political forces at work in Australia.

The strongest bond to our land that we seek is the legal one — tenure or ownership. In Europe only the rich and privileged had legal rights to the land. Thus, on reaching Australia the early settlers expected that the ownership of land would surely endow wealth and power. In reality, the first white settlers found Australia a harsh land, unfamiliar and vastly different from that they left behind. At first, there was only disillusionment which soon evaporated, however, following some success at adapting farming methods to Australian environmental conditions. A land hunger was awakened and the rapid spread of settlement began.

The period of history just after the goldrushes had dissipated was a great social crucible in which many of the relationships between society and the land were forged. Governments sold land to raise funds to provide services and so land was allocated inappropriately. Debate on the political control of land use soon arose, but not on how it should be used — stewardship was of no concern — but rather to whom and how much land should be given.

Despite Federation in 1901, the state governments retained dominant control over land and continue to influence its allocation and use to this present day. Considerable political power was vested in state lands departments stemming from the responsibility of granting tenure, transfering titles, promoting new agricultural projects (particularly irrigation) and, as difficulties arose with drought, weeds, pests and economic adversity, they administered the (subsidised) drought loans, debt reconstruction schemes etc to prevent dislocation of society and industry. The central role and responsibility of government becomes apparent when we ask the simple question — who owns most Australian land? Most is 'owned' by state governments. Of course most urban and agricultural land is held under freehold title, but the much larger remaining areas are held under leasehold (the pastoral areas), or are unalienated (the unoccupied areas of arid lands).

The state governments are thus amongst the largest landowners and land managers in the world. Responsibility for past and present degradation is therefore unavoidably laid at the doorstep of government and, because of their assumed role in all matters relating to land, they are expected to take the initiative in land (and water) conservation. Individual initiatives have been taken, but our historical origins suggest the majority will abrogate their personal commitment or involvement back to government.

Historically, governments have attended to land degradation only when forced by public opinion and then only in times of perceived catastrophe. Management of land degradation by crisis is inappropriate because the problem is chronic rather than acute and this approach ignores the non-critical, but nevertheless important, events and processes. Science is now contributing positively to inform governments and individuals that prevention of degradation requires management appropriate to whole systems, be they natural or modified agro-ecosystems. In Australia, a pervasive feature of ecosystem management is uncertainty. Given complete understanding we would still be required to deal with the unpredictability of the forcing function, climate.

There are at least two contrasting approaches to the management of ecosystems that have been formalised by Walters and Hilborn (1978). The first is deferred action which presumes that ecosystems cannot be correctly managed until they are completely understood. This approach, which is widely advocated for rare or endangered ecosystems, is both inapplicable and unattainable for the majority of Australia's extensive, varied and frequently degraded ecosystems. The second approach is the use of passive adaptive management which is based on the best understanding (model) available but uses management strategies that are conservative given that mistakes can be expected. Importantly the mistakes, or deviations from the expected, are used to adjust the model as represented diagramatically in Figure 10. An essential component of this system is the presence of a monitoring capability. Feedback from monitoring the effects of a change in land use is needed to avoid adopting inadvisable management policies.

The prolonged attrition of land and water resources in Australia is explicable in terms of Figure 10, since monitoring is either piecemeal or episodic, resorted to only in times of crisis and then allowed to lapse. Very little of our land legislation incorporates adequate resource evaluation or monitoring procedures. This topic will be explored further in the following chapter by Wasson.

In summary, there remain three limiting factors to the effective prevention of land degradation in Australia. The first is insufficient monitoring of degradation processes and inadequate inventory of land condition. This often arises because of the long-term commitment required to characterise properly spasmodic and sometimes initially unrecognised events or processes. Nevertheless, scientific research is making positive contributions towards understanding natural processes and finding the methodology needed to study them. The second, far more important, factor is inadequate application of existing information in an immediate effort by individuals and government to control current degradation. Unfortunately, lack of monitoring also conceals many of the situations where immediate achievements might be attainable using existing information and technologies. The final factor, frequently ignored because it is politically or socially unpalatable, is that the true costs of land degradation have not been fully appreciated, especially the long-term economic and social-welfare costs and instances of irreversible degradation that permanently mar our countryside and limit Australia's productive potential.

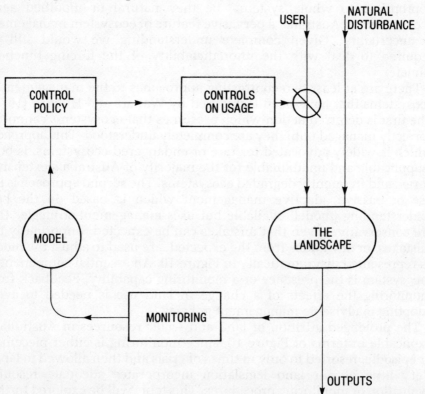

Figure 10 Components of a passive adaptive management system (after Walters and Hilborn 1978).

3 Detection and measurement of land degradation processes

Robert Wasson

Introduction

Detection and measurement are necessary if land degradation is to be accurately located, its past and future trends estimated, and its impact on human welfare assessed. Detection and measurement both depend on and contribute to an understanding of the processes and phenomena of degradation, and it is only by explanation that prediction of the future course of degradation is possible.

This chapter includes a discussion of what needs to be detected and measured, considers the role of natural science in that choice, discusses the nature of explanation and prediction, and attempts to establish a framework for a discussion of processes and the space and time scales of reference to degradation. Wherever appropriate, stress is laid upon the degree to which a process is understood. The intention is to emphasise processes by which we can better understand the phenomena discussed in the first chapter.

What is to be detected and measured?

Land degradation is a change to land that makes it less useful for human beings. The concept of land is broad and, after J G Speight (nd), is the extensive system of physical and biological materials and processes associated with the interface of the solid earth, terrestrial water bodies and the air, and the works of human beings.

A change has occurred in studies of land degradation which places emphasis on the complex links between economic, social, physical and biological processes and effects, rather than identifying specific topics such as soil acidification or tree decline as the only matters of concern

---

This chapter has been read by Mick Fleming, Peter Laut, Ian Moore and Gordon Burch, all of whom contributed valuable ideas and discussion. I also thank Garry Speight and Chris Margules for allowing me to use unpublished material.

in studies of land degradation. In contrast, the first Australia-wide study of land degradation by the Department of Environment, Housing and Community Development (Aust DEHCD 1978a) and its review by Woods (1983) identified examples of degradation in erosion, salinisation, waterlogging, vegetation depletion, fertility loss, soil-structure change and pollution of soil. In each of these examples, the focus is on the physical or biological effects, with land use methods seen as the ultimate causes of degradation. For example, salinisation may happen as a result of ' . . . a rising saline watertable due to irrigation or to dryland agricultural development and practices' (Aust DEHCD 1978a, p15).

While land remains the focus, the perception of causality must not be too simply stated. In a recent account of natural disasters, Hewitt (1983) notes that a hazard does not spring solely from the extremes of geophysical processes. A hazard ' . . . refers to the *potential* for damage that exists only in the presence of a *vulnerable* human community' (p5, emphasis is that of Hewitt). He goes on to observe that actual usage (or the 'dominant view') centres on hazards, for example, hurricanes or frosts, so that the perception of causality or the direction of explanation is from the physical environment to its social consequences.

There are parallels here with land degradation. The similarity between what Hewitt sees as the 'dominant view' of natural hazards (or calamities if human beings are affected) and studies of land degradation is that the causality is complicated. There is a human cause in economic processes. The processes of salinisation involve hydrologic and chemical phenomena, the study of which strictly belongs to the realms of the natural sciences, but there are social consequences of the biological changes brought about by salinisation. These consequences may feed back to land use practices in either the same or different areas. In the case of a natural hazard or calamity, the construction of a village in an alpine valley makes an extreme, but predictable mudflow, a geophysical event with consequences for human society. If human habitation at such a site survives long enough, as in the mountains of Pakistan or Japan, houses and fields may be located with increasingly greater care, thereby reducing the impact of mudflows. Through time, the statistical probability of a mudflow will not change, assuming that sediment supply and climate remain constant, but because of human strategies the probability of a calamity is reduced. In this example, there is no simple relationship between the probability of a geophysical event and a calamity.

In the the book edited by Hewitt (1983) there are two viewpoints which are alternatives to the dominant view. First, to restate the example just given, natural hazard is neither explained by nor uniquely dependent on geophysical processes that may cause damage to human beings or their works. Second, human awareness of and responses to natural hazards are not dependent on geophysical conditions. Means of avoiding or reducing risk depend most on the organisation and values of society and its institutions.

In the case of land degradation, the physical or biological phenomena that are of interest to natural science contain only part of the chain of causality or explanation. In the first two chapters of this volume it has been shown that at least the outline is known of the physics, chemistry and biology of land-degradation phenomena, and in some cases detailed accounts can be given from the perspective of the natural sciences. The failure by individuals and/or governments to 'tackle the land degradation problem' and 'apply existing knowledge' (eg Hudson 1982) often is interpreted by natural scientists as proof of a lack of foresight, mental torpor, complicity or apathy. An explanation of the apparent lack of response among large numbers of landholders and government agencies to widespread erosion or tree death really lies in the same category as the alternative view of natural hazards. Human awareness and responses are not solely dependent upon the existence of salinisation or erosion, a theme explored elsewhere in this volume.

The role of natural science

Throughout this discussion, the focus has been on land not on human society. If it were otherwise we might be considering a topic such as social and economic change, not land degradation. As seen earlier, degradation is a change that makes land less useful for human beings. By concentrating on the natural science of the identified phenomena we cannot specify when a change to the environment has an important economic or social impact.

This formidable problem exists because, as seen earlier, there is no simple relationship between the physical phenomena and the perceptions of land by human beings. From an economist's standpoint, land can be viewed as becoming cheaper and more plentiful as technology allows cultivation, for example, to proceed more efficiently (Blaikie 1984), rather than as a finite indivisible resource (Meadows et al 1972). These two economic views of land can lead to opposing answers to the question: does land degradation matter? The substitution of agricultural technology for the resources of soil, for example, by using the soil as a sponge for chemical fertiliser, reduces the relative value of soil. Once the substitution is made, higher soil losses can be tolerated without fear of productivity decline (Blaikie 1984).

Another way of emphasising the problem of different perspectives of land degradation is to consider conflicts between potential or actual land users. Recalling the definition of land degradation, the specification of a degraded state is dependent on what is considered useful. Conflict between preservationists, conservationists, farmers and miners revolves around this very difficult matter. If the aim of nature conservation is to maintain maximum possible genetic diversity (C R Margules, pers comm), then leaving an area alone may be useful in the long term as a source of, for example, pharmaceuticals. To many conservationists the clearing of forest for farming destroys genetic diversity but represents improvement to the farmer. If there is conflict over what is useful, can we specify what is degraded?

The way out of this conceptual tangle seems to be to accept that the natural and human sciences both have a role in identifying degraded land. In some cases, the natural sciences can make this identification unaided, as in nature conservation where populations of organisms other than humans need to be monitored so that extinctions can be predicted, or where future impacts can be predicted purely from technical information. In many cases, however, the identification of land degradation requires information from the natural sciences on change in the environment with the input of the social sciences aimed at specifying conditions or criteria under which land is said to be degraded. Some examples of these criteria are given in Table 6.

Scales of measurement

For complete detection and measurement of land degradation, a system is needed for monitoring change in physical, biological and social phenomena. In what follows attention will be directed to the physical and biological, with greatest emphasis on physical phenomena, reflecting the author's expertise. When concentrating on change and its explanation, attention must be paid to scale, both in space and time. The detection and measurement of phenomena in different parts of the space-time field require different techniques, and extrapolation from one scale to another requires careful consideration of locations in that field and the methods by which extrapolation is effected.

Attention to scale has also been shown to aid the integration of the efforts of the natural and human sciences. Clark W C (1985) has assembled characteristic length and time scales for phenomena as

Table 6: Examples of degradation criteria

Economic/social	Physical/biological
Farm productivity decline	Reduced genetic diversity
Increased water treatment cost	Species extinction
Need for salinisation amelioration	Soil erosion much greater than formation rate
Navigation problems	Pollution
Stock death from toxic organisms	Erosion

diverse as, for example, crop cycles, animal reproduction, human population growth, droughts and violent weather. These scales extend over more than seven orders of magnitude and imply that explanations, variables and generalisations relevant to one scale are unlikely to apply to others. To paraphrase Clark, the challenge presented by this approach is to pay explicit attention to scale when choosing the physical, biological or social processes to include in any study of land degradation. If the amelioration of a long-term problem like salinity is the target, there is little point in concentrating social analysis just on farm-scale activities. Significant changes to the focus of legislative power and the world's need for food products bear on solutions to the problem of irrigation salinity. Herein lies a large gap in the study of land degradation, identifying both the characteristic scales of process and the important cross-scale influences. To tackle this problem we need careful case studies of land degradation, especially historical studies of ecological and social processes and their interactions with considerable concern for attitudes and action toward land (Nash 1972).

Description, explanation and prediction

Attention to scale demands that phenomena are documented over appropriate areas and time periods, especially if we wish to explain the behaviour of the phenomena as a basis for predictions. Explaining the localisation of severe salinisation near Kerang in northern Victoria, for example, requires a description and understanding of the dynamics of the piezometric surface over an area much larger than the salinised land. Explaining change to vegetation in which individual organisms live for hundreds of years is unlikely to be successful relying solely on observations over a few years, so dendrochronology and pollen analysis of nearby lake or swamp sediments is essential (Wasson and Clark 1985). The uniformitarian principle that the present is the key to the past can be inverted so that the past is the key to explaining the present and also the future (Doe 1983). That is, explanation and prediction use the same rules (Watson 1969).

Trends in time are the result of chemical, physical and biological processes being activated, de-activated and modulated by forcing factors such as land use, rainfall and the introduction of pests. Factors like rainfall, overland flow, streamflow and windspeed have frequency distributions the tails of which contain extreme events like droughts and floods. The frequency of an event is, in general, inversely related to its magnitude, and the interplay of frequency and magnitude produces a frequency distribution of effects (Figure 11) such as sheet erosion, gully extension or dune encroachment. Trends in physical and biological phenomena, therefore, are the sum over time of the effects

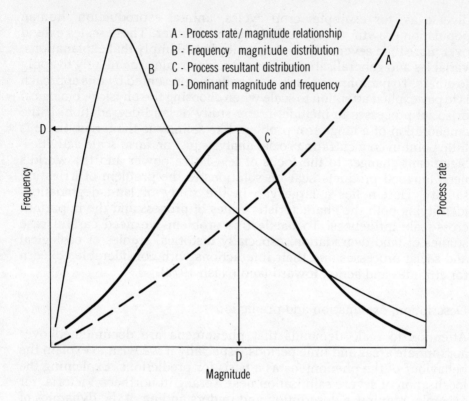

Figure 11 Magnitude, frequency and rate of a process combined to produce a process resultant distribution.

of magnitude/frequency distributions for each of the processes of concern.

We can now list the requirements for a complete description and explanation of physical or biological degradation. They are that the physics, chemistry and biology of the process are adequately understood; the spatial scale over which the processes operate is defined; the magnitude/frequency distributions of the processes are known, including as much detail as possible in the tails of the distributions; and an historical record of change, on appropriate scales of time and space, has been compiled to show by what steps the existing landscape reached its present state.

These attributes can be acquired by a combination of short-term monitoring of processes and effects both in the field and the laboratory, mapping of phenomena such as salt-affected areas, historical analyses (using documents, photographs) and stratigraphic analysis using geologic and palaeoecologic techniques. Such analyses allow some understanding of the past and present, and by quantitative modelling, experimentation and the substitution of space for time (Wasson and Clark 1985) a basis for prediction is formed.

The quantitative and qualitative understanding of the phenomena of land degradation can be incorporated in models to assess risk for future land uses. The simplest case assumes that all economic, social and behavioural factors remain constant and the trend of, for example, the reduction of soil organic matter is predicted. The next step superimposes a change in one of the boundary conditions, like cultivation practice, and a trend is predicted. A well-known example of this second stage is the analysis of the effect of rising atmospheric CO_2 on farm productivity, a method which has been resoundingly criticised by those who see society changing faster than CO_2 (Clark W C 1985, Parry 1985).

In summary, the four attributes listed above for a thorough description and explanation of physical and biological land degradation phenomena are also the attributes needed for prediction of risk. To identify risk, as defined by FAO (1984) for studies of desertification, requires a description of land status (which is really the difference of the area or volume of a phenomenon at different times), and an estimate of the rate of change of area or volume. Finally, to make assessments of hazard, consistent with the definitions given in the earlier part of this paper, we must add to assessments of risk the social and economic factors which lead to changes of land use and land value.

A framework for description, detection, measurement, explanation and prediction both of risk and hazard is presented in Figure 12. Within this framework, the remainder of the chapter examines studies of soil erosion with conclusions concerning gaps in our understanding and information. While the example is specific, the scheme in Figure 12 is thought to apply to all forms of land degradation.

Processes and their magnitude/frequency distribution

Processes

Erosion and transportation by water is best considered within the bounds of catchments because this allows joint consideration of onsite and offsite effects. An understanding of the latter effects requires knowledge of the former so, for example, observations need to be made over the catchment of a reservoir which is accumulating sediment. The onsite effects can be considered separately, however, and because most research effort has been directed to these effects in isolation, we have a much greater understanding of erosion and transportation by overland

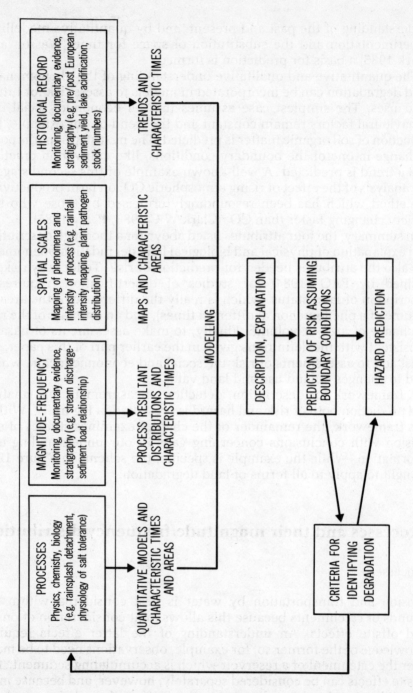

Figure 12 Framework for the study and implementation of land degradation detection and monitoring.

flow and in rills than routing of sediment through catchments by streams. Wind erosion presents severe problems because its area of operation and impact cannot be accurately determined, and so has received less attention.

Erosion occurs when the forces of entrainment (wind, rain and running water) overcome the forces of resistance (cohesion and roughness). The most convenient place to start our consideration of processes is on a hill slope where sheet erosion by overland flow is likely to occur. Rainfall striking the surface of the soil will detach particles and also seal the surface. Infiltration of water below the surface is controlled by properties of the soil and rainfall intensity, so that the prediction of runoff and therefore overland flow is not a straightforward matter (O'Loughlin 1986). Once overland flow has begun, a complex interaction occurs between raindrop detachment and the transportation of particles both detached and eroded by the shearing force of the moving water. As the depth of flow increases, the effect of raindrops decreases (Walker et al 1978) but the respective quantitative contribution by raindrops and overland flow to the movement of particles is far from clear (Kirkby 1980).

As water concentrates, perhaps as a result of randomly distributed roughness elements, flow rate increases and shear stress at the base of the flow rises. Rills and broad shallow channels develop and sediment concentration increases (Loch and Donnollan 1983). Detachment and transport in rills is by running water, a process that can be inhibited by intense rainfall (Moss et al 1979). Rills carry sediment to large channels or to deposition zones, where the deposition rate is proportional to the product of particle-settling velocity and sediment concentration (Rose 1984a). Not all sediment eroded from a hill slope reaches a channel in a valley bottom in a given period of time. Erosion and deposition occur at all scales and all locations from hill crests to the sea.

Open channel flow in gullies and rivers is well understood and sediment transport in channels has been the subject of intensive research (eg Graf 1971). Much less understood are processes of flow within soils and its role in piping, disaggregation, gully initiation and channel bank collapse (Milton 1971). Equally, the prediction of bedload transport rate is still crude, despite considerable advances in understanding of bedform development. Careful monitoring of suspended load concentrations throughout floods in various countries has shown that the peak of sediment concentration often precedes the peak of discharge, producing hysteretic loops on plots of discharge against concentration (Walling and Webb 1982). One explanation of those patterns is that sediment becomes less available as a flood proceeds, but there is little understanding of how sediment exhaustion occurs. Examples of the processes involved may be bank collapse as water level falls so providing sediment for the next flood; animal activity between floods; cultivation between floods.

Sediment transported into lakes, reservoirs or the ocean forms deltas, bodies of laminated muds which have settled from suspension clouds and density-current deposits produced by vertical gradients of sediment concentration. The processes of sorting and settling under varying limnological conditions are reasonably well understood in simple cases (Hakanson and Jansson 1983), but the diversity of water body shape and thermal and density stratification makes monitoring of a representative sample of lakes and reservoirs difficult and the testing of models reliant on only a few examples.

The processes of wind erosion are a little easier to comprehend than those of water erosion because the shearing force of moving air is imposed upon the soil and there is no problem comparable to that of predicting runoff. The foundations to wind erosion research were laid by Bagnold (1941) but extensive review and refinement of the physical concepts have occurred since (see Chepil and Woodruff 1963; Wilson and Cooke 1980). Resistance to wind erosion is not just from frictional and inertial effects of cohesionless grains but, in soils of widely varying particle sizes, cohesion plays a major role. The mechanisms of bonding relevant to water erosion are also relevant to wind erosion, although their specification in model formulations usually differ.

Once movement begins, particles quickly become organised into ripples. The airflow becomes saturated with particles and net deposition begins somewhere downwind of an area which is steadily being depleted of erodible grains. In the source area, a lag forms and erosion stops while the accumulation area is free to migrate as a dune. The finest particles, dislodged during saltation, are carried higher than the traction load that forms dunes. This dust can be carried into the stratosphere if vertical wind velocities are adequate.

Wind erosion is also resisted by roughness elements. Ripples and dunes are erodible roughness elements but shrubs and some grasses are non-erodible although somewhat deformable roughness elements. The effect of non-erodible and non-deformable shrubs on wind erosion can be viewed as either increasing the threshold velocity for particle movement or decreasing the velocity of winds as they enter the vegetation (Wasson and Nanninga 1986). The partitioning of drag on rough surfaces is a difficult problem and analytic solutions of the role of roughness in erosion rate control are not yet available. Equally, the interactions between erosion rate and bed shear stress defy analysis at the moment as most vegetation deforms as windspeed increases.

Wind can erode upslope as well as downslope, can change its angle of attack quickly and can export fine particles enormous distances in one step. Water erodes downslope only, has more definable transport pathways and usually moves fines long distances only by steps. For these reasons it is very difficult to verify by measurement the predictions of mathematical formulations of the theory of wind erosion processes. On balance, water erosion is far better understood and documented despite the relative ease with which some aspects of wind erosion can be studied.

Mass movement on slopes results from the catastrophic rupturing of the soil mantle as pore-water pressures exceed the cohesive resistance of the soil. This resistance is of a kind similar to that discussed previously but includes tree roots in the case of landslides (Gray and Leiser 1982). A large literature in soil mechanics treats the processes by which slopes fail but rarely is it explicitly stated that soil shear strength varies widely across small areas. Relationships expressing the resisting forces are hard to generalise, therefore, and so must be statistical (Singh 1972, Blong 1985).

Experiments in flumes and wind tunnels allow observations of fundamental processes, testing of hypotheses and measurement of rates in controlled circumstances. Field plots separated from their surroundings by physical barriers allow inputs and outputs to be measured, and experiments can be done under manipulated conditions using, for example, rainfall simulators to induce soil movement (Loch 1984). Large plots are often used to test the effect of various kinds of tillage and stubble retention on infiltration and soil movement (eg Queensland Department of Primary Industries 1984, Packer et al 1984).

Understanding processes of bank collapse, sediment transport in channels and by wind, and deposition in lakes requires more speculation and a considerable increase in monitoring effort. Erosion pins, re-surveyed cross-sections, sediment samplers and bedload traps are examples of the monitoring techniques used - all of them troubled by reproducibility and accuracy.The understanding of the effects of processes operating over a few square metres can be extrapolated to larger areas only by way of models. There are two approaches to mathematical modelling (Rose 1984b). The first is essentially statistical in which a large number of observations is ordered in ways informed by insight into physical processes. The second attempts to describe analytically the major features and interactions of processes, often after clarification of the major features is gained by working with a model of the first type.

The best-known example of a statistical model is the Universal Soil Loss Equation (USLE) of Wischmeier and Smith (1978), based upon a vast amount of data collected from erosion plots in the mid-west of the USA. Criticisms of USLE give insight into its main features. USLE estimates erosion as the product of a series of factors, namely rainfall intensity, slope gradient, slope length, soil properties and cropping practices. Kirkby (1980) notes that the multiplication of the factors disallows non-linear effects between factors. Rainfall and soil factors should not be multiplied because of the subtractive effect of soil infiltration capacity on erosive runoff generation. El-Swaify et al (1982) observe for the tropics that the empirical values for parameters in USLE are restricted to the mid-west of the USA and can be changed only after considerable experimental work — even for other areas of the USA. In addition, USLE does not explicitly account for deposition and gives only annual soil losses (Rose 1984a).

Despite these problems, Rose (1984a,b) and Blong (1985) make a case for continuing the effort to make USLE a tool for land management in areas other than the USA. The model currently identifies the factors important to erosion and it is very well documented. There seems little doubt that the type of approach embodied in USLE is the only practical one for mapping erosion risk over large areas.

The difficulties presented by USLE have encouraged interest in models of the second type. Processes are universal and a successful model of them should be more readily transferred than statistically based models. Considerable effort is now being directed to physical models of sheet and rill erosion (eg Rose 1984a,b, Foster and Meyer 1975), while models of gullying processes, stream sediment transport and deposition, and lacustrine sedimentation are highly variable in their accuracy and in most cases are in their infancy (see Graf 1971, Richards 1982, Hakanson and Jansson 1983).

The preceding account of processes of erosion, transportation and deposition of soil materials is brief relative to the available literature but long in the context of this chapter. However, it has been deliberately included to give some insight into the complexity of processes, their interactions and their modelling. The account implies that we may never know as much as we would like because each time a layer of ignorance is removed, a new layer appears. It is disquieting to think that nature might be infinitely divisible but the suggestion indicates that practical solutions require a halt to reductionism. This point will be returned to below.

Magnitude/frequency distributions

Process rates such as wind velocity, rainfall intensity and stream discharge increase with the magnitude of the events. For example, the non-linear increase of sand transport as wind speed increases is a process rate/magnitude relationship (Figure 11). The magnitudes of events are, however, often inversely related to frequency of occurrence or are highly skewed to the right (Figure 11). The product of frequency and rate is the work achieved over a particular period of time, and in

Figure 11 is shown as the process resultant distribution. The mode of this distribution is the dominant magnitude and frequency of a process, or the maximum of total work. The magnitude of this mode is dependent on both the skewness of the frequency of occurrences and shifts of the process rate curve with higher or lower process thresholds. For suspended sediment loads in some rivers the mode is of intermediate magnitude (Wolman and Miller 1960) but is displaced towards higher magnitude events in the case of bedload transport (Richards 1982) because of the high shear stress threshold of bedload particles.

To determine the curves in Figure 11 for a particular process, measurements must be made of the occurrence of, for example, windspeed at a known height over a long enough period to describe its frequency distribution adequately. In addition, the physical relationships between a velocity and its potential for transportation must be known quantitatively so that the process rate-magnitude curve can be drawn. For a magnitude-frequency analysis, the process must be both well understood and well monitored.

In establishing the frequency distribution of events, particular attention must be paid to the low-frequency tail where extreme events are recorded. In the case of windspeed and rainfall intensity, reliance must be placed on instrumental monitoring. For stream discharge, instrumental records can be augmented by stratigraphic study of slackwater deposits which record the height of some extreme floods (Baker et al 1985). By hydraulic analysis, these flood heights can be converted to discharges and, by radiocarbon dating of organic debris which records each flood, the instrumental record can be extended by many centuries or even millennia.

Herein lies a problem. The unique statistical description of a frequency distribution of the magnitudes of many events relies on drawing the events from the same population. If a trend or jump exists in a time-series of, for example, stream discharge records then the series is non-stationary and separate frequency distributions should be constructed for each stationary piece. There are very few records which are truly stationary (Wasson and Clark 1985) and the adequate sampling of extreme events requires long records which are sure to include the results of climatic variation. There seems no alternative, however, to accepting less rigorous statistical procedures in the interests of solving pressing problems.

In principle the magnitude-frequency distribution analysis could be applied to any quantity; for example, the frequency of times when saline water rises into the root zone, where the process rate curve would describe the link between antecedent groundwater levels, annual rainfall and the rate at which the level rises. Equally, a frequency distribution of wheat-disease infestations could be coupled with a process-rate curve where the dependent variable is productivity. Climatic variability and/or change can obviously be analysed in the same way, where the dependent variable in the process-rate magnitude relationship could be a measure of agricultural risk.

Climatic change might make a particular land use a danger to soil resources because of the physical and economic marginality of the enterprise. So climatic variability and/or change can be an agent of degradation if land use cannot be changed. If the mean rainfall in a region has changed then its frequency distribution, especially its tail of extreme events, will be shifted. As shown by Wigley (1985), a change in the mean by only one standard deviation could make an event with a recurrence interval of 20 years become five times more frequently expected. Changes of this order have occurred in Australian rainfall (Pittock 1975) and recent unpublished analyses suggest that windspeeds have also changed. The history of boom and bust on the dry margins of the wheat-belt (Meinig 1962) suggests that the extreme tails of frequency distributions are potent forces in social dislocation. The physical causes of these climatic extremes are poorly understood, obliging prediction to be essentially empirically based.

Space and time scales

The need to identify space and time scales explicitly in the detection and measurement of land degradation has already been argued, and partially discussed earlier in this chapter. It is clear that an understanding of processes operating over a few square metres is useful in land management only if, by means of models, the results can be applied to much larger areas. This procedure has been used to prepare maps of sheet erosion risk and estimate sediment and chemical yield from field-size areas. The USLE has been used in this way by replacing the rainfall energy factor with a runoff factor, producing the modified USLE (or MUSLE, Williams 1975). Another model, CREAMS (Chemicals, Runoff and Erosion from Agricultural Management Systems, Knisel 1980), is different from MUSLE in its formulation but both are applied to areas of similar size. Although claims are made to the contrary (eg Foster and Lane 1982), all these models must be tested against measured soil movement if they are to be relied on, and MUSLE has to be optimised, that is fitted to a substantial body of data, before predictions become reliable.

It is conceivable that MUSLE or CREAMS could be used for areas of 10^3 km^2 or greater by aggregating the results of sediment yield estimates from subcatchments of 10^1 km^2 area. The cumulative errors in that procedure are likely to render the results hard to use.

However, there is a more fundamental reason why models and understanding derived at the scale of 10^1 km^2 cannot be transferred easily to larger areas. The transport of sediment from hill slopes, through drainage lines to a lake, reservoir or the sea involves many steps. An individual particle is never truly lost from the land until it reaches the sea, for it spends much of its life in storage either in colluvium on foot slopes, in river channels, on floodplains and terraces or in lakes. The measured catchment sediment yield will only approximately equal the total erosion rate in a very small catchment because as catchment area increases so too does sediment storage.

Measurement by erosion pins of total sheet erosion in semi-arid catchments of only a few square kilometres shows that up to 45 times more erosion occurs than can be accounted for by channel aggradation and reservoir accumulation (Hadley and Schumm 1961, Hadley and Shown 1976). In humid areas the yield and total erosion may be approximately equal (eg Van der Linden 1983) perhaps because of more efficient transport of sediments from the slope foot in relation to sediment supply, at least in small catchments.

These and other observations are the basis of the sediment delivery concept, expressed as the ratio between sediment yield and the total erosion in a catchment. Total erosion is estimated from a model such as MUSLE and yield is estimated either from sediment load/discharge rating curves in conjunction with flow duration curves for the exiting stream (Vanoni 1975), or from surveys of the amount of sediment trapped in reservoirs (Barfield et al 1979). Of course, sediment yield and delivery ratio have similar functional relationships to catchment area, varying only by a constant.

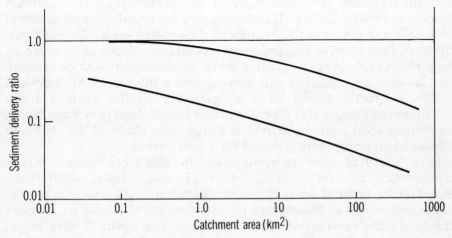

Figure 13 Sediment delivery ratio — catchment area envelope for catchments in the mid-west of the USA. After Vanoni (1975).

An envelope of estimated delivery ratio given by Vanoni (1975) for catchments in the mid-west of the USA is shown in Figure 13, and the ratio is seen to decline as catchment size increases. This is consistent with the direct measurements of both yield and total erosion described above. The interpretation given earlier of this relationship stressed enhanced opportunities for storage in large catchments but Boyce (1975) argued that within a hundred years all the sediment delivered to a river would be carried by it. The delivery ratio is then interpreted as a measure of the efficiency of overland flow transport, largely reflecting mean gradients of catchment slopes. Small catchments have steeper slopes than large catchments, therefore, sediment is delivered more directly to streams per unit of time and area in small catchments. Boyce's interpretation is contradicted by recent studies of sediment budgets (Trimble 1975, 1983, Meade 1982) which show that in the Atlantic Drainage of the USA about 90 per cent of the 25 km³ of soil eroded from the uplands during the last two hundred years has been stored on hillslopes and in valleys. In a smaller catchment further north, Costa (1975) estimated that 52 per cent of sediment eroded from the uplands since 1700 AD is stored as colluvium, 14 per cent is on floodplains and 34 per cent has left the basin in rivers.

These sediment budgets are valuable guides to both space and time scales of relevance to the problem of modelling sediment routing through large catchments. Because no comparable studies exist for Australia, only a diagramatic illustration of what might have happened is given in Figure 14. Clearing of forests was one of the first activities by European settlers in eastern Australia (Powell 1976). Exposure of slopes where pasture grasses were sparse most likely led to severe sheet and gully erosion. The testimony of the grazier John G Robertson (quoted in Powell 1976, p31) demonstrates the speed of degradation of both soils and plants in the Wimmera. Depending on local conditions, the rate of transport of fine sediment eroded from slopes and carried by large rivers probably peaked after the initial maximum of slope erosion, but the coarse bedload is still moving down the rivers. Wasson and Galloway (forthcoming) have suggested a similar pattern for a catchment at Broken Hill where much of the bedload may have begun its journey soon after the arrival of Europeans. Much of the sediment deposited on floodplains is stored for a long period.

Boyce's (1975) view is appropriate in small catchments but is demonstrably false for larger catchments. To quote Meade (1982): 'Time scales on the order of years to centuries are too long for us to apply the predictive power of Newtonian physics and too short for us to make the comforting assumption of a steady state'. And again: 'The principal impediments to modelling the movement of sediments through river basins are the lack of quantitative data on these processes and a lack of understanding of the time scales over which they operate'.

Vigon (1985) has expressed serious doubts about currently available methods of assessing the sources of non-point pollution, mostly suspended sediment and its attached chemical and biological pollutants. The modelling of sediment routing through catchments of 10^3 km^2 is clearly limited not only by the accuracy of the models applied to upland areas but also, as Meade has pointed out, by the time lags involved. Wilson et al (1984) have constructed a model which simulates the erosion, transportation, routing and deposition of sediment in catchments. The routing procedure is that of Williams (1975) in which sediment yield is calculated from each subcatchment using MUSLE, adopting either a single representative particle size or particle size classes. The travel time of sediment is estimated from the hydrologic routing component of the model, so lags can be no greater than hours or days. Gully erosion and flood plain scour can be included in the analysis but the lack of an explicit formulation of floodplain and alluvial fan storage means that the model is only likely to be reliable in small catchments, for which it was designed.

The other major approach to the problem of locating sediment sources and storages has been partially discussed in the case of sediment

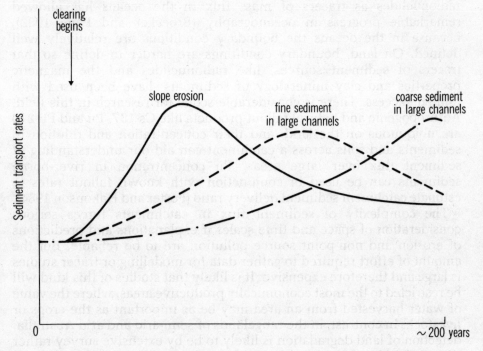

Figure 14 Schematic transport rates of sediment in different parts of catchments in eastern Australia, showing time lags.

budgets. The stratigraphic or geologic basis of the approach is also the basis of sediment-based studies of catchments (Oldfield 1977). The idea is that sediments transported to a lake or reservoir can be interpreted to provide sedimentation rate/depth curves, thereby extending the record of instrumentally recorded stream sediment loads, and can also be used to identify sediment sources by means of the natural and artificial tracers that the lake sediments contain. The remains of organisms and chemical compounds within the lake sediments can also be used to reconstruct, for example, pollution and regional vegetation histories.

Sedimentation rate/depth curves can be constructed only if the sediments are accurately dated. Radiometric methods using the decay of Pb-210 (Appleby and Oldfield 1978), the first appearance of the bomb-produced Cs-137 (Longmore 1985) and Pu isotopes (Koide et al 1985), and the bomb-enhanced production of C–14 (Baker et al 1985) can all be used with the first appearance of weeds, correlations between charcoal laminae and bushfires, and other similar techniques (Wasson and Clark 1985). In some cases, sedimentary laminae are deposited annually so that good resolution can be obtained.

The measurement and modelling of natural and anthropogenic radionuclides as tracers of mass flux in the oceans has allowed remarkable progress in oceanography (Broecker and Peng 1982), because in the oceans the boundary conditions are relatively well defined. On land, boundary conditions are harder to define so that tracers of sediment sources, like radionuclides and the magnetic properties and clay mineralogy of sediments, have been used with limited success. There is considerable scope for research in this field. Anthropogenic and natural fallout products like Cs-137, Pu and Pb-210 are ubiquitous on the land, and their concentration and dilution in sediments and soils across a catchment can aid our understanding of sediment flux over large areas. Pu concentration in river-borne sediments can be used in conjunction with known fallout rates to estimate catchment sediment delivery ratio (Foster and Hakonson 1984).

The complexity of sediment flux in catchments forces serious consideration of space and time scales if explanations and predictions of erosion and non-point source pollution are to be reliable. But the amount of effort required to gather data for modelling or tracer studies is large and therefore expensive. It is likely that studies of this kind will be restricted to the most economically productive areas, where the value of water harvested from an area may be as important as the crops or livestock. In contrast, in the rangelands of semi-arid and arid Australia, detection of land degradation is likely to be by extensive survey rather than intensive methods, although here detailed spatial models using Landsat data are potentially very useful (Pickup, in press).

The reports of land degradation in Australia (Woods 1983, Kaleski 1945) contain maps of current status. Salt at the ground surface in those maps is presumably considered to reflect a degraded state because salt is likely to have been in the subsurface before settlement. In other cases, for example gully erosion, the pre-existing state is much harder to specify and so the maps are only assessments of the spatial distribution of phenomena. Despite these limitations, the problem of comparability of data drawn from different organisations and the unknown theoretical bases of the assessments, this approach is likely to be developed rather than radically changed for areas of the order 10^6 km². Landsat imagery can help the mapping, and act as a tool for generalisation and regionalisation especially when it is linked to other attributes such as vegetation type, elevation, slope and tenure in a computer-compatible form (Graetz et al 1986).

As noted earlier, the USLE approach can be used to map risk and so inform land use planning. The precise formulation of the model is not critical if ordinal results are sufficient. Starting with erosivity maps for both water and wind, it is possible to approach reality by including factors representing erodibility, roughness and eventually management practice. This approach can also be applied to very large areas, but it is not yet clear how all forms of potential degradation can be assessed in this way. Dryland salting is probably the least manageable (Working Party on Dryland Salting in Australia 1982).

The techniques mentioned for mapping status and risk over large areas can be used to estimate trends if status is mapped at various times over a long period, usually from aerial photographs with assistance from satellite imagery, ground surveys, topographic and cadastral maps.

Emphasis has been given to the cost of detailed studies of sediment flux over large areas. Similar remarks can be made about detailed historical accounts of catchments. But there are some important problems which require this approach, namely the estimation of tolerable soil losses, the role of rainfall change in erosion and the decline of the land's resource base. In the first case, rates of soil formation must be known before tolerable loss rates can be determined, requiring well-dated stratigraphically-based pedological studies. Stratigraphy and long rainfall records are needed in the second and third cases, with special attention to the effects of European settlement on the resource base.

Future research and action

In this paper, consideration has been given to the degree of understanding of particular processes and their results, the efficiency of detection and methods of measurement, relative to an ideal set of attributes (Figure 12). Little attention has been paid to the results of past research and measurement but it must be noted that few if any studies of important aspects of land degradation match the ideal. In fact, very few studies include historical records commensurate with the characteristic time scales of the phenomena, there is little explicit attention to spatial scales, magnitude/frequency distributions are poorly known and processes are understood with variable acuity. If explanation is rudimentary, predictions can only be speculative.

Research into processes, characteristic scales and methods of detection and measurement is well advanced and this is likely to continue by virtue of the scientific imperative for knowledge. But the major concern is whether this research and activity is well directed either scientifically or within the framework discussed in the introduction.

In each of the three cases where long-term trends need to be identified, namely tolerable soil loss, role of rainfall change and the decline of the land's resource base, the natural scientist has the main role in deciding the nature of the problem. Farmers and politicians are unlikely to recognise a long-term threat to agriculture if the rate of erosion of a soil is not directly affecting the productivity of the farming sector. The result is that the study of rates of soil formation is separated from issues of productivity and considered 'academic', a convenient label for any study that is not seen as having immediate impact.

The isolation of scientists is resisted at least in part in Figure 12, where the arrows are drawn to show inputs to the establishment of degradation criteria from all interested parties. At the moment there is no methodology available for assessing and integrating the inputs and their interactions for management purposes, no way of adequately dealing with the questions raised in the introduction, and very little recognition for professional researchers who attempt such work (cf Miller 1985). More research into erosion mechanics or the salt tolerance of plants will be of little value unless the multifaceted nature of land degradation is embodied in policy, research and extension. Chamala (1985) gives a way forward in this complex problem but we must heed the warnings of Miller (1985) that often when the methods of the natural sciences are applied to modelling human society and its individual and collective perceptions, it is only trivia that can be adequately included.

This conclusion suggests that carefully interpreted case studies of land use and its physical, biological and social consequences are urgently needed. The land use and environmental history of Australia after the arrival of Europeans has been a series of unique events beginning with the cataclysm of hard-footed animals, going on to the introduction of rabbits, and now the continuation of practices in the 'frontier' areas of the continent already shown to have disastrous consequences. Our history must be told and we must be guided by this history in our choice of detection and monitoring methods.

Two additional approaches to detection and measurement suggest themselves. The first is the 'environmental audit' (Schaeffer et al 1985) where the goals of detection and monitoring are paramount and routine methods treated as ways of improving the audit rather than an end in themselves. The second is to use statistical procedures to tease out possible controlling variables in the time series of, for example, wheat yields (Williams O B 1979, Wynen 1984). By focusing on the goals and the concerns of managers and users respectively, in conjunction with careful historical case studies, it may be possible to detect and establish the real causes of land degradation.

Whichever route is adopted in the detection and monitoring of land degradation, it is clear that little progress can be made unless goals are clearly identified and priorities assigned to research and implementation using the expertise of all interested parties. Monitoring without clear goals is largely a waste of time.

Commentary

Adrian Webb

Introduction

In the statement of problem and objectives circulated before the workshop, land degradation was taken to include 'all those adverse effects that land uses may have on services provided by land'. A similar but more explicit definition has been developed as part of a project within the Rural Development Centre of the University of New England (I. Reeves, pers comm). 'Land degradation is a change in the state of the terrestrial component of global ecosystems that threatens human welfare now and/or in the future'. The important aspect to note is ' . . . a change . . . that threatens . . .'. Hence change per se is not necessarily considered as degradation.

In essence the three chapters presented in this section attempt to provide a sound basis for later chapters through statements on:
• the types, extent and rate of change of land degradation in Australia;
• what we know about it — causes, effects, processes; and
• implications for future action.

In reviewing the three chapters I have highlighted some of the main points and attempted to indicate some broad areas for future action.

Types and extent of land degradation

In the first paper, Chartres focuses on the major land degradation issues of erosion, salinisation and waterlogging, while recognising that aspects such as loss of genetic diversity are expressions also of land degradation. Degradation in Australia can be considered under the broad categories of geological, historical, current and potential. Evidence of geologic degradation is obvious in the landscape features and from the studies of geomorphologists and others. Historical degradation, largely through man's management (mismanagement) interacting with natural factors is well documented for southern parts of Australia. It is generally accepted that land degradation effects in the first 150 years, particularly in the period from the 1850s to the 1930s, were extremely severe in many parts of southern Australia, leading to severe erosion and flooding, and predisposing some areas to secondary salinisation.

Chartres has placed most emphasis on degradation in the grazing lands, particularly in the arid and semi-arid regions. The major external changes were in the stocking pressures (including effects of rabbits) and the use of fire. I can accept the comment by Chartres on the early build-up of stock numbers and degrading effects on pastures for many of the grazing lands. However, I doubt that irretrievable damage occurred across all rangelands. He may be correct for rangelands with less than 500 mm annual rainfall in the southern half of the continent, but in much of the northern pastoral lands, stocking pressures have remained low because in the dry season nutritional quality of the pasture is very poor. This has served as an inherent protection against overgrazing (Mott and Tothill 1984). Stock have usually been removed or have died from malnutrition before they greatly affected the pasture.

The extent of degradation associated with soil erosion is well documented, at least qualitatively, as a result of the collaborative soil conservation study (Aust DEHCD 1978a). This study indicated that water erosion is the most common form of degradation and is particularly important in the cropping lands of New South Wales and Queensland. Wind erosion is of most concern in Western Australia and parts of South Australia and Victoria. Chartres refers to the report by Woods (1983) who extracted much of the interpreted information to widen the level of awareness in the community. Additional to this is the report on salting of non-irrigated land in Australia compiled for the Standing Committee on Soil Conservation (Working Party on Dryland Salting in Australia 1982). This report indicates that 4.2 million ha (0.56 per cent of Australia) are affected by scalding or secondary salinisation as a result of land management since European settlement. The large majority is scalding in arid and semi-arid pasture lands. Seepage salting of approximately 0.5 million ha occurs mainly in Western Australia, Victoria and South Australia and affects twice the area affected by secondary salinisation in irrigated lands. Chartres places most of the severe degradation in irrigation lands in the riverine plains of the Murray and Murrumbidgee rivers.

Chartres gives a broad description of Australian soils and some indication of their erodibility. He makes the point that low fertility, old highly weathered soils, salinity, topographic and climatic factors all combine to render much of the land resource susceptible to degradation following land use changes. However, he understates the actual and potential water erosion on soils in the semi-arid tropics and the sub-humid environments. Erosion is a very real threat to the red earths of the Northern Territory where considerable agricultural development is in hand. The combination of disturbed non-cohesive soils, high rainfall intensities and long slopes leads to serious erosion in a very short time. Similarly, in Queensland and northern New South Wales serious erosion occurs on sloping texture contrast and clay soils where they are cultivated and left unprotected by crop residues in periods of high rainfall.

Chartres raises the point of differing levels of management expertise among land users. Although management techniques may be developed to reduce substantially the risk of erosion in cropping lands, the degree of management expertise required is considerably higher than for the traditional methods.

There have been persistent calls recently for a national survey to assess the extent and degrees of land degradation in Australia. Some have asked for it as an update on the 1978 collaborative study; others have asked because they consider that the early data were not sufficiently objective or that methodologies were inconsistent across states. On evaluation of the current status of degradation, Chartres concluded that, in general, existing land resource maps are not suitable, and no suitable strategy or methodologies have been developed to allow a national assessment. This is now widely accepted and some moves are under way to evaluate various methods of assessing degradation. However, given the different forms and importance of degradation across Australia, it will be no mean task to get agreement on a common methodology, particularly if it is to satisfy all of the requirements listed by Chartres.

Biological and physical factors

In their paper, Burch, Graetz and Noble provide a comprehensive description of the causes of land degradation in Australia and relate it to:

- intensification of land use
- the degree of land clearing
- tillage practices
- soil acidification of ley pastures
- irrigation pollution
- chemical pollution
- fire management
- stock management

The authors highlight the importance of land degradation by reference to the statistics of Woods (1983), which indicate that no more than ten per cent of Australia is arable.

They put the view that economic problems result in the intensive irrigation industries having little flexibility or incentive to adjust their land use enterprises, and this leads to continuing degradation of land and water resources.

Burch, Graetz and Noble present interesting data to illustrate their point that clearing of perennial vegetation in southern and western Australia has been severe. More than half of the south-east region has been totally modified and comprises a high proportion of the land used for intensive agricultural production. Most forms of land degradation due to clearing have caused increased runoff and consequent erosion, as well as secondary salinisation and waterlogging. The continuing widespread clearing of woodlands and forests in parts of Queensland and Western Australia is viewed with more than a little concern by those with experience or understanding of rural decline or dieback.

Burch, Graetz and Noble are critical of fire management policies and suggest that degradation of forests result from these policies. I gather from their statements that there is a reasonably good understanding of the effects of fire and how it should be used to manage our forests. The authors indicate that although there are gaps, there is a good scientific understanding of the biological and physical processes of land degradation in Australia.

In the major cropping areas, recent advances in cropping practices have enabled more stable production systems to be developed. These practices are aimed at minimising soil disturbance, maximising soil cover through retention of crop residues, and the use of rotations. Although there are particular problems associated with various production systems, substantial progress is being made in finding solutions to them. The philosophy has been to develop cost efficient practices, recognising that if conservative farming can be made economically attractive, then adoption rates will be improved.

Detection and measurement

Wasson's paper provides an excellent statement on matters of measurement and data interpretation in land degradation studies. He introduces his paper with an interesting philosophical treatment of 'what is degradation?' He provides a very detailed discussion on detection and measurement, with emphasis on scales, and magnitude and frequency of processes. He comments on the importance of using techniques suited to the scales of the processes and emphasises that extrapolation from one scale to another requires, among other things, careful consideration of locations for data collection. An understanding of processes operating over small areas is useful in land management only if results can be applied to much larger areas using models.

Wasson indicates there is a considerable body of information on the processes of wind and water erosion and of mass movement on slopes. Certain processes such as stream bank collapse and sedimentation in lakes are less well understood and require much more monitoring effort. Techniques for doing this have limitations in reproducibility and accuracy.

Information on rates of change in various forms of land degradation is very limited. In recent years there has been quite an emphasis on assessing the effects of land management practices such as stubble and pasture management on soil loss and runoff and soil water in farmlands. Data have been collected on erosion rates for a number of soils and agro-climatic situations in Australia. Some of the data are event-orientated and this provides an opportunity for the use of climatic records to consider likely frequencies of the results. There is a deliberate attempt being made in several states to develop predictive models for land management decisions using these data. As part of this development, models developed overseas are being assessed for their relevance to Australian conditions.

An important point made in the three papers is that the rates of change induced by the land uses may be so slow that severe effects on the land or water may not be evident for many years. Salinisation from rising groundwater in southern areas is a good example.

Despite the empiricism of the Universal Soil Loss Equation, Wasson suggests that this approach is the only practical one for mapping erosion over large areas. He suggests that where precision is not essential, survey techniques may be useful for estimating trends in land degradation by regular surveying of key areas, and comments also on the development of satellite imagery for this purpose.

One of the most interesting approaches for obtaining information on rates of erosion or sedimentation is the use of natural fallout products such as Cs-137. This technique is likely to be used more widely as research scientists recognise the potential uses. A major limitation in the short term is the limited analytical facilities available.

Conclusions and future action

There are serious deficiencies in information on the extent and rate of land degradation in Australia. However, sufficient is known to indicate what the major problems are and where they exist. One of the limitations in the statistics that are available is the inability to separate historical from current degradation. Methodologies for assessment of land degradation need to be developed to provide a more objective base than currently exists.

Many of the processes involved in land degradation are understood to a level where preventative or remedial strategies can be suggested. There is a strong case for more application of this knowledge allied with suitable monitoring programs to allow better quantification of some of the rates of various processes.

Southern forests and woodlands have been severely over-cleared with resultant problems of salinisation and rural decline. More remedial action seems warranted, but this may be difficult to accomplish without considerable attention to the effects on communities in the affected areas. Land degradation should not be considered solely in biological and physical terms, a point emphasised by Wasson, and Burch, Graetz and Noble.

Modelling of systems appears to be an excellent tool for assessing the long-term effects of different land management practices provided sufficient attention is given to process and scale factors. An immediate need is the integration of information and ideas across the various groups with common goals in Australia and internationally. Management of the land and water resources is a complex responsibility which must be shared across the community. One valuable result of model development is likely to be the enhanced level of communication and interaction which would occur between the various disciplines involved.

II

Social costs of land degradation

4 Onsite costs of land degradation in agriculture and forestry

Michael Blyth and Andrew McCallum

Introduction

The impact of land degradation on the productivity of the land resource has been claimed to be a serious physical problem in agricultural and forestry land uses (see, for example, Balderstone et al 1982, p135 and Schuster 1979, p1). However, the physical existence of land degradation is not necessarily evidence of an economic problem. To establish whether or not an economic problem exists requires knowledge of the costs and benefits of reducing or preventing degradation, and/or the costs and benefits of reclaiming degraded land.

An individual land user considers the costs of land degradation and other costs of production in a similar fashion. The land user will avoid or prevent land degradation to the point at which private benefits from additional prevention, equal to extra profit generated from increased production, are equal to private costs. However, the full social costs of land degradation may involve external costs, for example, in the form of damage to roads or siltation of waterways. There may also be external benefits in that soil lost from one property may enhance the soil productivity of the property on which it is deposited. There is no general incentive for landowners to consider external costs and benefits. The results of private decision making may thus diverge from those which would occur if full social costs were taken into account.

Public policy response to the existence of land degradation is generally aimed at reducing the external or offsite costs and aligning the private and social costs of production or land use. Analysis of policy measures to reduce the external costs of land degradation associated with agricultural and forestry land use requires estimates of the private and external costs, which sum to the social costs, and the social benefits of production. The private costs include the cost of land degradation borne by the individual land user, while the external cost is the cost of land degradation borne by others beyond the source of production. The social benefits are equivalent to the net revenue generated from production.

In this chapter, the focus is on the private costs of land degradation and their relevance to efficient public policy.

The existence of land degradation is explained in terms of the difference between private and social costs of production. As well, a framework is presented for examining private land use decisions and the divergence between private and social optima. This framework is based on an optimal control model.

One possible approach to the measurement of land degradation costs where it is not possible to derive cost estimates directly from the market is the opportunity cost approach. Using some empirical data on soil loss and wheat yield decline, estimates of the wheat income forgone as a result of soil loss for a number of locations throughout Australia are presented.

Several previous studies of land degradation have included estimates of its costs. These estimates are evaluated with reference to a theoretical exposition of the costs and benefits of preventing or restoring land degradation. These previously estimated costs are compared with the cost measures required for efficient public policy to reduce the external effects of agricultural and forestry production activities.

Problems involved in the estimation of private and social costs of land degradation are discussed in the concluding comments. In addition, data deficiencies are identified and suggestions made for further research.

As there is little information available on the impact of land degradation associated with forestry land use, and as the general discussion of land use management is as relevant to forestry as it is to agriculture, discussion of the costs of land degradation under forestry is contained in an endnote to the chapter.

The costs of land degradation

Consider the case of a land user whose resources are devoted to the production of wheat. The wheat production activity also generates soil erosion, the total costs of which are not necessarily accounted for by the individual producer. If the existence of erosion has no noticeable effect on any other individual, then the private and social costs of wheat production are identical and the land user's private decision is socially optimal. On the other hand, if the erosion results in additional costs to others beyond the wheat grower's property, such as damage to roads or siltation of waterways, then the social costs of wheat production exceed the private costs. These additional costs are the external or offsite costs of wheat production. This situation can be illustrated with the aid of Figure 15.

In Figure 15, the marginal private costs of wheat production are

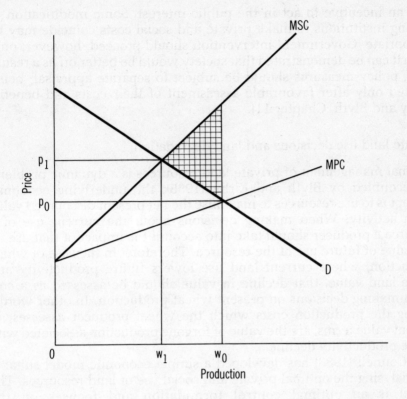

Figure 15 Divergence between private and social costs of wheat production.

represented by MPC, which is equivalent to the competitive land user's supply curve. The marginal social costs of wheat production are represented by MSC, which is the sum of MPC and the marginal external costs of production. D represents the demand for wheat and is equivalent to the marginal social benefits of wheat production. If the price of wheat is p_0 then w_0 tonnes of wheat are produced in the presence of the externality. This is the private optimum level of output. The social optimum level of wheat production is w_1 tonnes, which is less than the private optimum. The price of wheat paid by consumers is p_0 which is less than that at the socially optimum level. Consequently, in the social sense, wheat is overproduced. The net welfare loss or cost to society of the divergence between the private and social optima, because of external costs, is illustrated by the shaded area in Figure 15.

The wheat producer in this situation is unable to discern the relative social demands for wheat and soil erosion. Without inclusion of the external costs in either input or output prices, the producer does not have an incentive to act in the public interest. Some modification of existing institutions to make private and social costs coincide may be appropriate. Government intervention should proceed, however, only when it can be demonstrated that society would be better off as a result. Thus, policy measures should be subject to separate appraisal, being adopted only after favourable assessment of their costs and benefits (Kirby and Blyth, Chapter 11).

Private land use decisions and land degradation

Optimal management of private land resources is a dynamic problem. As recounted by Blyth and Kirby (1985), the underlying economic concept is to use resources to maximise the net present discounted value of an activity. When making decisions about the current use of a resource a producer should take into account the impact of that use on the value of future use of the resource. Therefore, in the case of wheat production, where current land use lowers future productivity and hence land value, that decline in value should be assessed as a cost when making decisions on present wheat production. In other words, among the production costs which the wheat producer assesses, in present value terms, are the value of forgone production associated with future productivity decline.

  McConnell (1983) has developed a simple economic model suitable for analysing the optimal private and social use of land resources. The model is an optimal control formulation and focuses on the intertemporal path of land use, including the conditions under which private and social optima diverge. The components of the model of private land use decisions are a profit function and an expression for the value of farm land. The profit function comprises a single-crop production function and includes as inputs measures of soil depth and soil loss. It is assumed that the land user aims to maximise the present value of the stream of profits from the land and the value of the land at the end of the planning period. Therefore, the land user will degrade the land resource to the point where the return from additional soil erosion is equal to the cost, which is the profit forgone as a result of the loss of future productivity and the decline in the terminal land value as a consequence of soil loss. Furthermore, it is assumed that there are no pollution externalities associated with private land use decisions. Thus, if the private and social discount rates are the same, then the private and social optima are assumed to be equal. However, the model may be specified or constrained to reflect the existence of externalities and, therefore, can be used to analyse the divergence between private and social optima. McConnell (1983) presents the optimal control model of land use decisions in detailed mathematical terms.

Features of an optimal control model which distinguish it as a useful framework for analysing the dynamic optimisation of private and social land use decisions are, first, that differences in time preferences between private individuals and society can be examined and, second, that trade-offs between institutional variables (the control variables) and their interactions with endogenous variables (production processes) can be analysed. Therefore, an optimal control framework can be used in policy analysis to examine the impact of differences in the private and social discount rates; the impact on private land use decisions of regulations and/or taxes on soil loss; the production of particular commodities or the use of particular practices; and the impact of imposing a concept such as maximum sustainable yield on land users. The chief advantage of the control model is that it can be used to determine optimum land use decisions over a specified period of time.

Land values and land degradation

One of the assumptions of McConnell's model is that land users are fully aware of the soil's contribution to the farm's resale value. Therefore, if the land market operates efficiently or perfectly (as McConnell assumes), it should be possible to determine the impact of land degradation by comparing land values among properties which are similar in all respects other than the quality of the land resource base. Such an analysis would provide an indication of the land purchaser's (consumer's) willingness to pay for reduced land degradation.

To test whether or not land values reflect the existence of degradation, a substantial number of market transactions would be needed to obtain a suitable sample for analysis. This may be a difficult requirement to meet, in view of the spatial and temporal variability in Australian agriculture. Furthermore, in keeping with McConnell's assumption, degradation should be clearly visible or readily measurable. However, the presence of some forms of land degradation, such as sheet erosion, may not be obvious. Therefore, it is possible that some land degradation may not be accounted for by individuals in their land purchasing decisions, even though it may have an impact on productivity and returns from the land. Measurement of the severity or degree of land degradation also may be difficult.

There are likely to be several factors involved in the determination of farm land values, in addition to the quality of the land resource base. These include location factors, institutional and government policy factors and personal factors associated with particular purchasers and/ or sellers. The relative importance of these factors varies for different units of land and, consequently, the value of land also varies. Therefore, given that land differs in so many respects, it would be difficult to isolate the effects of land degradation. Despite this apparent difficulty, attempts have been made to account for the effects of degradation on land values. For example, a study was undertaken by Molnar (1955) for two locations in Victoria. He found marked differences in land values associated with variation in the land class, which largely reflected variation in soil fertility. However, with respect to soil erosion, he found that only the more 'spectacular' types of erosion had an impact on land values. In one of the districts he studied, where only sheet erosion and gullies occur, land values were found to be unaffected by erosion. An implicit, or hedonic, price approach was used by Miranowski and Hammes (1984) to analyse Iowa land transactions data. They found that 'differences in soil characteristics are reflected in farm land prices', but they were less confident in ascertaining whether 'the market is discounting the value of farmland sufficiently to account for loss of productive capacity' (Miranowski and Hammes 1984, p748).

Further research into land price formation in Australia would be of benefit. In particular, it is suggested that the hypothesis, that the onsite effects or costs of land degradation are reflected in land values, be tested. At present, there is uncertainty about the impact of land degradation on land values.

An indirect measure of the cost of land degradation

When it is not possible to measure directly the private costs of land degradation, indirect measures may be used to estimate values. The opportunity cost approach is one such measure. Measures of the opportunity cost of land degradation have been included in a number of former studies of land degradation in Australia (for example, see Peck et al 1983 and the Working Party on Dryland Salting in Australia 1982). The opportunity cost approach to valuing land degradation is an indirect valuation technique, in the sense that values are not derived directly from market transactions, as in the case of the land market values approach.

The opportunity cost approach includes a measure of the income forgone as a result of displaced opportunities. Therefore, in the case of wheat production, for example, the cost of soil loss is the value of wheat income forgone. However, this is only part of the necessary valuation information. Equally important are measures of the cost of repair of degraded land and measures of the cost of avoidance or prevention of degradation. Therefore, the opportunity cost of land degradation is equal to the income from production on non-degraded land, less the costs of repairing and preventing land degradation. Together these cost measures provide the vital information to assist private and public decision making.

The information provided by the opportunity cost approach is the same as that which the individual land user considers and, therefore, that required for an optimal control model of land use decisions. Each case requires a measure of land degradation and a measure of the relationship between degradation and plant productivity. On the first requirement, the greatest research effort into the measurement of land degradation in Australia has been in the area of soil erosion. Measurements of soil loss have been made at various locations throughout the country. A summary of these soil loss measurements is provided in Table 7. The various approaches to soil loss measurement include experimental plot and field measurements, actual catchment measurements and measurements using the Universal Soil Loss Equation (USLE).

The USLE, which has not been widely used in Australia, was developed in the United States. It predicts soil loss (A) as a product of six factors, namely rainfall and runoff (R), soil erodibility (K), slope length (L), slope steepness (S), cropping management (C) and erosion control practices (P). Thus, the equation is A = RKLSCP (Meyer 1984). However, according to Edwards and Charman (1980), applicability of USLE under Australian conditions is limited. In particular, they refer to the dearth of experimental data in Australia to validate factor values of the equation. They note that 'the equation was developed largely on soil loss data from medium-textured soils' and that 'there is some doubt as to its applicability to soils on either end of the texture scale, particularly our wheat land soils' (Edwards and Charman 1980, p217). These limitations give rise to scepticism when considering the results of those Australian studies which employ USLE.

The second key requirement for land use decisions and the valuation of land degradation is a measure of the relationship between plant productivity and land degradation. There have been few studies of this relationship for Australian farming systems. In the majority of cases, measures have been made of the impact of soil loss on wheat yields. A summary of the findings of these soil-loss/wheat-yield studies is presented in Table 8.

In each of these cases of the measurement of yield decline, there is no mention of soil depth, only soil loss or soil erosion. However, as indicated by the relationship expressed in Figure 16, it is soil depth rather than soil loss which is the critical factor influencing crop yields.

For example, consider a soil loss event from x_0 to x_1. In this situation the yield effect is zero. Consider an equivalent soil loss event from x_0' to x_1'. In this case, wheat yield response is negative, since the original soil depth, x_0' (soil depth before the loss event), is closer to that depth at

Table 7: Summary of Australian soil loss measurements

Location	Period observation	Crop land[a] t/ha/yr	Pasture land[a] t/ha/yr	Slope[b] %	Soil type	Reference
NEW SOUTH WALES						
Wagga Wagga	1947-76	2.1	0.03	8 (crop)	Red/contrast texture	Edwards (1980)
	1947-76	0.01				Aveyard (1983)
	1948-59	0.30-1.06	0.01			Jones (1961)
	1950-75		0.02-0.21			Adamson (1976a)
	1952-74		0.03-2.02(d)	6-12		Adamson (1976b)
	2 years	3.44				Adamson (1978)
	50 years	21.7				Ring (1982)
Cowra	1943-74	1.8-3.1	0.3		Gradational texture/ mineral fraction	Packer (1981)
	1943-78	0.96				Aveyard (1983)
Wellington	1948-57	1.0-1.2	0.03-0.15		Gradational/mineral fraction	Logan (1960)
Gunnedah	1949-74	0.8-7.4	0.04-0.6		Dark/mineral fraction/ uniform texture	Thompson (1982)
	1949-74	7.7	0.03	8-9 (crop)		Edwards (1980)
	1949-74	3.41				Aveyard (1983)
Namoi Valley		12.5				Junor et al (1979)
		60-125|e				
Inverell	1948-74	1.3-7.0	1.3-2.0		Dark/mineral fraction/ uniform texture	Armstrong (1981)
		150-450|e				Enright (1983)
Canberra	1966-71		0.18(d)			Costin (1980)
QUEENSLAND						
Darling Downs	na	40	0.18		Black deep Basalt/cracking clays	Aust DEHCD (1978c)

Location					Reference
		29	1-8		Alcock (1980)
		7.6-22.6			De Boer and Gaffney (1976)
South Burnett		12.2(c)		Brigalow	Aust DEHCD (1978c)
		75-200(e)			Mullins (1981)
		24(c)		Structured reds	Mullins (1981)
Central Highlands		20.5-23.0(c)		Basalt/Brigalow	
Greenwood		1-36	5		D. Freebairn (pers comm 1985)
Greenmount	na	1-53	6-7		D. Freebairn (pers com 1985)
Nambour	na	1-148	17		D. Freebairn (pers comm 1985)
Mackay	na	42-227	2-11		D. Freebairn (pers comm 1985)
Innisfail	na	80-380			D. Freebairn (pers comm 1985)
WESTERN AUSTRALIA					
Wheat/Sheep Belt	na		61.9		Marsh (1979)
	na		27-200		Marsh (1983)
	na	25			Marsh (1979)
VICTORIA					
Rochester		0.98-34.80(c)			Arch and Dumsday (1981)

(a) Unless indicated, soil loss is estimated from experimental plots. (b) Where not indicated, slope is generally less than 10 per cent. (c) Soil loss estimated using the Universal Soil Loss Equation. (d) Losses measured from catchment experiments. (e) High figures indicate soil loss from a severe climatic event (storm). na, Not available.

which soil loss affects wheat yield. Thus, in addition to soil loss, the critical factor affecting the decline in wheat yield is the original soil depth.

For the wheat grower to make efficient land use decisions, knowledge is required of the relationship between soil depth and yield as well as the volume of soil loss. The soil loss measurement is useful only if it can be related to soil depth and yield by means of a relationship such as that depicted in Figure 16. Unfortunately, there does not appear to be any continuous measurement of this relationship. Previous research has produced discrete measures only, few of which can be used in land management decisions.

The relationship between soil depth and wheat yield is likely to vary for different locations, reflecting different soil types, slope and climate. Furthermore, the relationship will vary for different technologies. For example, the development of a higher yielding wheat variety will result in a new curve, above that of the original relationship, such as that represented by the broken line in Figure 16. Alternatively, a wheat grower may be able to offset yield decline by changing management practices associated with existing technologies, such as the application of additional fertiliser inputs or the introduction of a new cultivation practice. The effect of these changes would be to maintain wheat yield beyond the original soil depth at which yield declined, which is approximately x_0' in Figure 16.

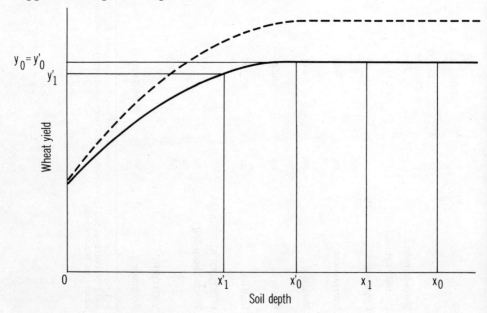

Figure 16 Relationship between soil depth and wheat yields. Source: Adapted from Crosson and Stout (1983).

In view of the variability among the estimates of soil loss and the measurements of the relationships between soil loss and yield and the failure to relate soil loss to soil depth, it is of little meaning to place values on all the measured soil losses. However, a comparison of the costs of equal volumes of soil loss (75 mm/ha and 150 mm/ha) at five different locations in New South Wales provides some interest and value. The cost comparison, presented in Table 9, reveals that the costs of both a 75 mm/ha and a 150 mm/ha soil loss event decrease from the south to the north of New South Wales. This most likely reflects variations in soil quality, including depth, from south to north.

Table 8: Measures of wheat yield decline due to soil loss (several locations).

Location	Soil loss	Yield decline (a)	Reference
NEW SOUTH WALES			
Wagga Wagga	75mm	45.8	Hamilton (1970)
	150 mm	51.4	
	0.3-6.4 mm	50	Aveyard (1981)
	30.3 t/ha	28	Aveyard (1983)
	10 mm	6	Aveyard (1983)
Cowra	75 mm	27.9	Hamilton (1970)
	150 mm	51.9	
	34.9 t/ha	22	Aveyard (1983)
	2-10 mm	4-15	Aveyard (1983)
Wellington	75 mm	21.3	Hamilton (1970)
	150 mm	30.8	
Gunnedah	75 mm	9.5	Hamilton (1970)
	150 mm	29.1	
	88.6 t/ha	17	Aveyard (1983)
	10-30 mm	6-13	Aveyard (1983)
Inverell	75 mm	6.1	Hamilton (1970)
	150 mm	19.1	Daniel (1969)
QUEENSLAND			
Darling Downs	0.08 mm	0.01	Aust DEHCD (1978b) Alcock (1980)
Wyreema-Cambooya			Cummins et al (1973)
VICTORIA			
Mallee	150 mm	6	Hore and Sims (1954)
Walpeup	150 mm	8	Molnar (1964)
WESTERN AUSTRALIA			
Wheat Belt	5 mm	25	Marsh (1979)

(a) Yield decline expressed as a percentage of the yield from a nil erosion treatment.

Table 9: Estimated costs of soil erosion in wheat growing areas

Location	Soil loss	Initial wheat yield (a)	Yield decline	Income forgone (b)
	mm	t/ha	%	$/ha
Wagga Wagga	75	2.2	45.8	138
	150		51.4	155
Cowra	75	1.75	27.9	67
	150		51.9	124
Wellington	75	1.85	21.3	54
	150		30.8	78
Gunnedah	75	1.6	9.5	21
	150		29.1	64
Inverell	75	1.5	6.1	13
	150		19.1	39

(a) Average of ten years to 1983–4 (ABS). (b) Based on price of $137/t for 1984–5 (BAE estimate, 1985)

Note: Soil loss/yield relationships are taken from Table 8.

This review of previous studies of soil loss and the information requirements of land use decision makers reveals some gaps. It is suggested that there would be some value to land users and public policy makers from further measurements of soil loss and soil depth and of the relationship between soil depth and plant productivity. In addition, research should be directed to evaluation of USLE for Australian conditions, and perhaps to consideration of the costs and benefits of similar simulation methods of relevance to this area.

As noted previously, estimates of the income forgone owing to soil loss provide only part of the calculus of decision making on land use. In addition, measures of the costs of repair and prevention are required. Examples of these costs are presented in Table 10. However, since land degradation is likely to be the result of several interacting factors, the solution to the problem is likely to be complex, involving changes in more than one area. A solution may be achieved by different combinations of measures. Choice of the appropriate control measures or the appropriate level of repair is an economic problem. The optimal level of land degradation repair or prevention is that level at which the marginal cost of repair or prevention is equal to the marginal benefit, which is measured as the income forgone. However, the absence of information on soil depth, soil loss and the relationship between soil depth and plant productivity suggests that less than optimal private land use decisions are being made in the production of agricultural commodities. Given that this kind of information is of a public-good nature, there may be an economic case for the government to increase its involvement in the provision of technical information. However, before that action occurs the costs and benefits of increased government involvement should be evaluated.

Distinguishing previous estimates of the costs of land degradation: a theoretical exposition

Several studies of land degradation in Australia have presented estimates of its costs. However, it is often unclear what these cost estimates refer to. The most common estimates are the costs of repairing all degraded land, the value of forgone production owing to the existence of degradation and the current level of expenditure on repair and/or prevention. Efficient private land use decisions and efficient public policy require particular cost information, as indicated earlier in this chapter. In this section a comparison is made between cost information that is available to land users and policy makers and that which is desirable for efficient land management decisions. To facilitate the distinction between costs, reference is made to Figure 17.

In Figure 17, MB represents the marginal benefit curve, which measures the marginal returns to agriculture associated with land degradation repair at a given point in time. Marginal benefits fall with increases in the repair of degraded land, becoming zero beyond the level of repair x_m. This is the level at which all degraded land is repaired to its original productivity which is that level of productivity existing before agricultural development of the land. The curve MC is the marginal cost of repair curve for agriculture. This is the least cost of land degradation repair curve. The optimal level of land degradation repair occurs at the point where the marginal cost of repair is equal to the marginal benefit from repair. This is the profit-maximising level of repair and occurs at x_0 units. The actual level of land degradation repair may also be marked on the diagram. For example, assume that the actual level of repair is x_p units, which is less than the optimal level. This may be the result of private land users underestimating the income forgone because of land degradation and, therefore, undervaluing the benefits of repair. The cost to agriculture of this level of repair may be expressed as the difference in profits at the optimal level and the actual level, or the profits forgone as a result of operating at a lower level of repair. Therefore the cost is the profits associated with x_0 units of repair less the profits associated with x_p units. This is equivalent to the shaded area in Figure 17.

Knowledge of the level of land degradation repair, x_m, where the land's original productivity is restored, is important as it represents the upper limit of the benefits of repair. Furthermore, when assessing technology changes when the costs of repair decline, the new total cost curve can be compared over the full length of the benefits curve.

Table 10: Selected costs of repair and prevention

Type of cost	Location	Cost (a)	Slope	Reference
		$/ha	%	
CAPITAL COSTS				
Contour banks	Darling Downs	40	5	Alcock (1980)
	Southern Queensland/			
	Northern New South Wales	54		Haynes and Sutton (1985)
	Wheat–Sheep Zone NSW	75	3–5	Watkins (Soil Conservation Service of NSW, pers comm March 1985)
	High Rainfall Zone NSW	100	7–10	" "
Reclamation of gullies	Queensland	50	4	van Beer and Steentsma (1981)
	Western Australia	25		Lemon (1983)
ANNUAL COSTS				
Maintenance of	Darling Downs	4–8		Alcock (1980)
structures	Southern Queensland/			
	Northern New South Wales	8		Haynes and Sutton (1985)
	Southern Australia	6		Haynes and Sutton (1985)
Loss of income	Darling Downs	3–7		Alcock (1980)
(contour bank area)				
Stubble mulching	Southern Queensland/	12		Haynes and Sutton (1985)
	Northern New South Wales			
Prevention of damage	Southern Queensland/	1		Haynes and Sutton (1985)
to crops and structure	Northern New South Wales			

(a) Cost relates to year of study

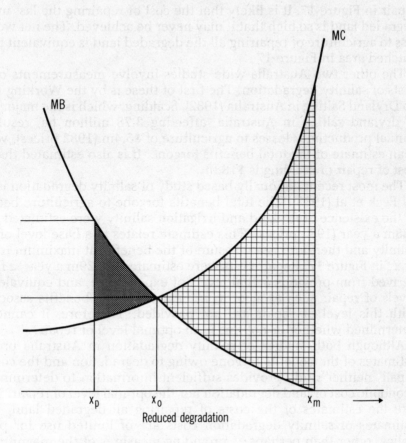

Figure 17 Costs of reducing land degradation.

There have been four large-scale studies of the extent and impact of land degradation in Australia which have included an estimate of the costs of the problem. The first was a study by the Australian Standing Committee on Soil Conservation (1971), in which it was estimated that the cost of controlling soil erosion with structural measures in non-arid regions of Australia was $350m (1970 prices). The second major study was the Commonwealth-State Collaborative Soil Conservation Study (Aust DEHCD 1978a). The often-quoted figure from that study of $675m (1975 prices) refers to the costs of treating all the degraded land in Australia. Like the Standing Committee on Soil Conservation's estimate, this estimate is likely to be an overstatement of the costs of the problems since it could not be expected that all degraded land could be economically treated. These estimates are equivalent to x_m units of repair in Figure 17. It is likely that the cost of repairing the last unit of degraded land is so high that it may never be achieved. The net welfare loss to agriculture of repairing all the degraded land is equivalent to the hatched area in Figure 17.

The other two Australia-wide studies involve measurements of the costs of salinity degradation. The first of these is by the Working Party on Dryland Salting in Australia (1982). Scalding, which is the major form of dryland salting in Australia, affecting 3.78 million ha, results in annual productivity losses to agriculture of $5.4m (1982 prices), which is an estimate of the total benefits forgone. It is also estimated that the cost of repair of scalding is $18.lm.

The most recent nationally based study of salinity degradation is that by Peck et al (1983). The total benefits forgone to agriculture because of the existence of dryland and irrigation salinity were estimated to be $28m a year (1982 prices). This estimate relates to a base level of zero salinity and therefore is a measure of the benefits at maximum repair, or x_m in Figure 17. Repair costs were estimated at $29m a year, a figure derived from previous actual levels of expenditure, and equivalent to levels of repair such as x_p or x_0 in Figure 17. The benefits associated with this level of repair are not provided. Therefore, it cannot be determined whether or not this is an optimal level of repair.

Although both studies of salinity degradation in Australia provide estimates of the benefits forgone owing to degradation and the costs of repair, neither study provides sufficient information to determine the economic cost of land degradation nor the optimal level of repair. Thus, like the estimates of the costs of repairing all degraded land, these estimates of salinity degradation costs are of limited use for public policy, other than perhaps as providing measures of the magnitude of the degradation. Efficient land use decisions and efficient public policy decisions require estimates of the marginal costs of degradation repair and/or prevention and the marginal benefits, equal to the extra income from repair and/or prevention. This entails knowledge of the marginal cost and marginal benefit functions, rather than discrete cost estimates. Therefore, the value of these previous cost estimates to private and public decision makers is particularly low. Furthermore, the shortcomings associated with previous estimates of the onsite costs of land degradation indicate that estimates of the offsite costs, which are also provided, would be of little use to policy makers.

It is unlikely that large-scale aggregative studies can provide the necessary information for efficient public policy and for decisions on private land use. Information for land use decision making may be more appropriately provided at the regional catchment and/or farm levels. There has been a number of small-scale studies of the impacts of land degradation and soil conservation on agricultural land use in Australia.

These include Dumsday (1971), Department of Environment, Housing and Community Development (Aust DEHCD 1978a, b), Junor et al (1979), Alcock (1980), Arch and Dumsday (1981), Bennett and Thomas (1982), Goddard et al (1982), Cooke (1985), and Pearse and Cowie (1985). Each study refers to a particular location and a particular form of land degradation. In all cases, these studies rely on the availability of specific physical data such as the level of soil loss or damage, rainfall and its intensity, slope of the land, length of the slope, soil type, soil depth, typical land use practices and so on. In some cases, USLE has been applied to determine the level of soil loss (for example, Dumsday 1971, Arch and Dumsday 1981). Others employ the results from plot measurements of soil loss and resultant plant productivity decline (for example, Alcock 1980).

The studies by Dumsday (1971), Arch and Dumsday (1981) and Cooke (1985) include evaluation of alternative farming systems, each with a different impact on soil conservation. The objective in each case is to determine that farming system which returns the maximum discounted net revenue. Therefore, the results of these studies are equivalent to the optimal level of repair, x_0 in Figure 17, albeit roughly, given the various assumptions of the models, including perfect knowledge and profit-maximising behaviour on behalf of land users and specification of a sufficiently broad cross-section of practices and measures to ensure the least cost of soil conservation is obtainable. Micro-level models such as these, and the optimal control model discussed earlier, provide the greatest potential value to land users and to policy makers. However, as discussed earlier, any modelling approach is dependent on a measure of land degradation and a measure of the impact of that degradation on plant productivity.

Concluding comments

The availability of suitable data has been the major determinant of the number and distribution of studies on the impact of land degradation on production costs and income in Australian agriculture. There have been few studies of the impact of land degradation on Australian forestry production (see endnote). The availability of data in Australia is in stark contrast to that in the United States, as indicated in a recent review of the effects of erosion of the productivity of crop land by Crosson and Stout (1983). The differences arise as a result of the 1977 National Resources Inventory (NRI), which is regarded as 'the most comprehensive survey of soil erosion in the United States' (Crosson and Stout 1983, p93). As Crosson and Stout recount, 'developed by the Soil Conservation Service, the NRI documents soil types, climatic conditions, land uses and management practices on over 200 000 sample points across the country and generates soil erosion estimates using the USLE and the Wind Erosion Equation'. While such an exercise may seem daunting for Australia, there appear to be some information deficiencies which may be contributing to suboptimal decisions on private land use. It is suggested that these deficiencies be further explored and that the costs and benefits of further public involvement in the provision of technical information on land degradation be assessed.

An optimal control model of land use decisions provides an appropriate framework within which to assess the intertemporal path of land use, including the conditions under which private and social optima diverge. In addition, an optimal control model may be employed to provide insights into effective land degradation control measures and to evaluate the impact of variations in individual rates of time preference on the optimal level of land degradation. The impacts of different land tenure arrangements may also be investigated with an optimal control model. Among the model's basic components is a relation expressing land values as a function of land degradation. This relation approximates the willingness-to-pay approach to the valuation of land degradation, given that it is assumed that land purchasers account for the impact of land degradation in the sale price of the land. This assumption or hypothesis is yet to be statistically proved.

The alternative opportunity-cost approach to the valuation of land degradation is subject to a number of practical difficulties. However, these limitations are equally relevant to the optimal control model and the land market values approach and to actual decisions on private land use. The difficulties lie in the availability of measures of the relationship between soil depth and plant productivity and measures of soil loss. It is suggested that further research be undertaken to provide more of this information. However, the benefits of greater government involvement in this area should be assessed before there is any expansion of current activities.

The information vital to efficient land use decisions, as specified by the optimal control model, contrasts with that provided by previous studies of land degradation in Australia. Previous estimates of the aggregate costs of repair provide little value to land users and to public policy makers, except perhaps as indications of the magnitude of land degradation. It is suggested that micro-level (farm, regional or catchment) modelling approaches may be of greater potential value to land users and policy makers than the aggregative studies of the past. However, these sophisticated models are not necessarily superior to simple budgeting models, as both are dependent on the availability of reliable physical data and knowledge of the costs of repair by various measures. Furthermore, modelling is not without cost and the costs of providing additional information must be accounted for. Some prior analysis of the costs involved, including the need for additional data must be carried out.

To employ an economic model successfully, there needs to be further study of the relationship between soil loss and plant productivity and further validation or modification of USLE for local Australian conditions. The availability of this information may assist the soil conservation services in their extension advice to land users and help authorities devise appropriate incentives to deal with land degradation. In the absence of this information, suboptimal decisions on private land use may be made. Consequently, public policy measures to remove the external effects of land use actions may also be inefficient.

Endnote

The impact of land degradation in forestry land use

Measurement of the onsite costs of degradation on land used for forestry cannot be estimated readily because of the dearth of scientific studies of land degradation processes on forest land and the impacts of degradation on forest productivity (B. Kerruish, Division of Forestry Research, CSIRO, pers comm, April 1985). However, it is possible to delineate the major forms of land degradation to which forestry land use is susceptible as well as the key periods and physical conditions under which land degradation is likely to impose additional costs on foresters. Discussions in the text of this paper concerning the efficiency of private and public land use decisions for agricultural production are equally relevant to land management decisions for forest production.

Forestry land use in Australia can be distinguished as either plantation forest which is intensive land use based on softwood species, or native forests which is extensive land use based on native eucalypt species. There also are some eucalypt plantations, although their total area is small (approximately three per cent in 1983—Aust Bureau of Agricultural Economics 1984). The potential for land degradation is higher on plantation forests because of more intensive land use. Costs or damages are likely to occur at forest establishment, thinning and logging. Soil type, slope and rainfall are the most important physical factors affecting land degradation on forested lands.

The major forms of land degradation which have the potential to increase the costs of forest production are compaction of soils on roads and tracks, the loss of organic matter and soil nutrients owing to water erosion, gullying associated with road and track construction and structural and chemical changes to the soil such as podsolisation and salinisation because of changes in the natural vegetation cover.

While the potential for substantial degradation of forested lands is high for brief periods, particularly in the high rainfall areas on sloping lands and on granite soils, in practice the total level is low, since the land remains reasonably undisturbed for long periods. This is especially so for native forests. 'Access roads are considered to be the cause of much of the land degradation that occurs within commercial forests' (Woods 1983, p22). Roads concentrate runoff which may lead to gullying and increase the chance of land slips. Careful siting of roads and design of drainage can reduce this potential damage (Marshall 1977).

The impacts of land degradation in forestry often appear in the second and later rotations. Productivity declines in the second and later rotations have been linked, albeit tentatively, to the loss of organic matter and soil nutrients, particularly phosphorus (Crane 1983). Crane and others conclude that further research is needed on the dynamics of soil organic matter and nitrogen before appropriate management practices can be recommended for the maintenance of soil productivity. There have only been a few studies of the effects of forestry practices on land degradation in Australia. One is that by Schuster (1979), which concerns the impacts of logging on soils and their rehabilitation in Western Australia. Schuster (p2) notes that logging 'degrades soil structure, altering it physically and hindering normal plant growth'. This leads to reduced root growth and reduced water infiltration. Up to 20-35 per cent of an area being logged may be disturbed. According to research results from the United States referred to by Schuster (p2) 5-10 per cent of the soils in a logged area 'undergo changes that could affect tree growth'. Preventative measures appear to be less practical and effective than restoration. Schuster (pp6-7) concludes that:

'Although improved logging techniques may reduce the problem to some extent, logging on wet podsolic soils will always result in some deterioration of site quality unless suitable rehabilitation techniques are used. These trials indicate that successful rehabilitation of damaged soils is possible, at least in the short term (to two years of age)... The most successful rehabilitation technique is the application to damaged areas of a combination of soil ripping, fertilising and bark mulching, or if possible of soil ripping and the redistribution and burning of logging debris.'

Unlike most agricultural land use, forestry land use is subject to specific guidelines and controls directed at minimising the damages which soil degradation may impose. This is particularly so for forests located within the water-supply catchments of population centres where the potential for offsite damage is high. These regulations and guidelines operate in all states and may cover harvesting operations, road construction and revegetation following logging (Aust Senate Standing Committee on Science and the Environment 1978). For example, in Western Australia, there are now strict operational guidelines which require suspension of all logging operations when soil moisture levels exceed a certain point on a measure known as the Soil Dryness Index. Operational guidelines also prescribe measures to prevent soil erosion, such as provision of bars across snig tracks and filter strips along all permanent or large ephemeral streams. Careful monitoring of field operations is carried out in cooperation with Western Australian Water Authority staff and it is believed that soil erosion is insignificant in forest areas. (F H McKinnell, Department of Conservation and Land Management, Western Australia, pers comm, May 1985).

5 Offsite costs of land degradation

Garrett Upstill and Timothy Yapp

'Soil erosion is the single most important environmental problem facing Australia today'

The quote is from the Hon John Kerin, Commonwealth Minister for Primary Industry, in Toowoomba in July 1983. It reflects an increasing and widespread recognition of land degradation as a national environmental problem of major dimensions. Indeed, from an economic perspective, land degradation shares a number of analytical similarities with traditional environmental problems such as pollution, but there are also some important differences. Land degradation is a complex phenomenon taking a variety of forms. It has effects which are often difficult to trace, measure and evaluate, and it has significance not just for individual landholders but for the community as a whole.

Other chapters in this monograph examine the physical aspects of land degradation, the legal and institutional setting and the contribution of different disciplines to formulating government policies. This chapter argues that an economic-based environmental perspective can help illuminate the problem and contribute to the development of practical and effective responses. It offers a framework for analysis of the social costs of land degradation and some thoughts on the way in which an awareness of these costs can assist government policy making.

This chapter focuses primarily on the offsite effects of land degradation, that is those which occur or are experienced away from the site of the action causing degradation. Examples include the costs of downstream salinity or turbidity. In economic terms these are *externalities* and they comprise a large part of the social costs of land degradation.

For the sake of convenience the scope of this chapter is limited to land degradation associated with primary production. Although other activities such as mining, roadworks and construction also contribute, primary production is the dominant source of offsite problems attributable to land degradation. These are largely non-point source in

---

* The authors wish to acknowledge the helpful suggestions of Dr J A Sinden. Responsibility for errors and omissions remains with the authors.

character, and include pollution caused by the erosion and deposition of soil, associated impacts including salinity, and accumulation of environmental contaminants such as fertilisers, pesticides and biochemical oxygen demand (BOD).

The costs of land degradation: a framework for analysis

There has been a number of surveys of land degradation in Australia which have helped illuminate the physical nature, extent and location of the problem (Woods 1983). However there have been relatively few studies of the economic costs of land degradation. This is disappointing as knowledge of these costs can help:
• decide appropriate levels of expenditure on control of degradation, and
• show the relative significance of land degradation as a regional or national problem.

A simple model

Land degradation is a complex set of processes and the rate of land degradation will vary from place to place and from time to time. Nevertheless, it is convenient in the first instance to posit a simple model of land degradation as a single process whose rate may be measured. There are then four reference points:
• a zero rate: there is no land degradation
• a natural rate: the rate in the absence of human intervention
• the actual rate: the rate given current land management practices
• an optimal rate: corresponding to socially optimal resource allocation and management.
The optimal rate is a useful concept but one which is difficult to measure. It is useful because it explicitly incorporates both physical and economic considerations into a pattern of resource allocation and management which results in maximisation of net social benefit. There is no *a priori* reason why the optimal rate should be below (or above) the actual rate of degradation. In some cases the current rate of degradation may not have high social or economic costs, and a rate of degradation higher than the current rate may be justified. Alternatively the optimal rate may be close to zero, or even negative in the case where optimal allocation of resources involves rehabilitation of already degraded land. The optimal rate of degradation would only fortuitously by zero (Blyth and Kirby 1985).

In practice decision makers will rarely be able to identify the optimal rate of land degradation with any degree of precision. However, it will frequently be possible to identify the direction in which the optimal rate lies.

Economic theory offers a number of reasons why current rates of land degradation may be higher than desirable. These relate primarily to instances of market failure and include information deficiencies, land tenure provisions and externalities. The use of natural resources by the rural sector — resources such as soil, water and vegetation — can generate a number of offsite effects which are not reflected in market prices. These offsite costs must be added to onsite costs such as direct productivity losses to obtain a measure of the full social costs of land degradation.

Drawing on this classification of rates of degradation there are two concepts of the costs associated with degradation which may be particularly useful:

- the *net* costs of degradation corresponding to the difference between the optimal and actual rates; and
- the *gross* costs of degradation corresponding to the difference between the actual and zero rates.

Both concepts refer to the annual or flow costs of degradation rather than to the cost of redressing an accumulated stock of damage.

The first provides the 'ideal' measure of the economic significance of land degradation in terms of lost net social benefit, that is, the net opportunity costs associated with the current rate of degradation. The second is a more accessible measure of the damages associated with land degradation and as such provides a valuable input to analysis and decision making regarding control of degradation. In cases where the optimal rate is near zero and significantly different from the actual rate the gross measure of damage costs can serve as a proxy for the net measure, that is the net social costs of allowing land degradation to continue at the current rate, or in other words the benefits forgone because of degradation.

Offsite effects

The physical effects of land degradation are varied and pervasive. Apart from productivity and aesthetic losses onsite there is a range of offsite effects. Thus the loss of soil from farms has as its corollary a range of offsite (downstream) effects resulting from dispersion of the lost soil. These effects include, for example:

- loss of water storage capacity caused by sedimentation;
- increased turbidity of waterways or eutrophication of lakes leading to losses in recreation and tourism;
- soil-polluted air or water leading to reduced life and increased costs for machinery and appliances;

- damage to roads resulting in higher repair and maintenance costs; and
- loss of native vegetation and wildlife.

The identification and measurement of these effects provides a basis for expressing them in monetary terms. The list may be extended to include less direct effects such as increased swimming and boating accidents associated with excess turbidity (Clark et al 1985).

Table 11 offers a taxonomy of the major elements of onsite and offsite costs as a basis for systematic analysis of the social costs of land degradation. A number of the cost categories are ranked according to measurability; some externalities represent a direct financial burden to the community, for example removing sediment from roads or building extra dam capacity, but others are less tangible and longer term in nature, for example the loss of visual amenity or the loss of wildlife species.

Table 11 does not distinguish between the costs attributable to allowing degradation to continue at present (or increased) rates and those attributable to failing to undertake optimal rehabilitation of

Table 11: Costs of land degradation	
Onsite costs	
1. Productivity costs	
–output	costs associated with loss of product attributable to changes in the quality or quantity of soil.
–input	effects of onsite land degradation on cost of production, eg increased costs of fencing or dams, replacement of machinery, or indirect cost of compensating for ecological changes, eg loss of tree cover for natural predators.
Offsite costs/externalities	
2. Productivity costs	
	loss of product or increased costs resulting from external economic activities, eg downstream salinity effects, air pollution, regional effects associated with loss of flora and fauna.
3. Other costs	
	Increased costs or lost benefits attributable to:
	—damages to water conveyance, irrigation or water treatment facilities
	—flood damage (including silting of roads)
	—decreased capacity of water storages
	—soiling and materials damage
	—increased hazards of navigation
	—reduced availability or quality of natural recreation facilities
	—impaired aesthetic characteristics
	—lost ecological diversity

already degraded land. Both these elements, plus the cost of optimal repair of offsite damages should be included in an assessment of the social costs of degradation. This is not to presume that in every case it will be appropriate to reduce the rate of degradation, rehabilitate degraded land or repair offsite damages. These decisions require additional information on the costs of halting degradation, or of taking remedial action.

The amount or stock of degraded land is of some interest as an indicator of the results of past land management practices within the prevailing social and institutional framework. Nevertheless, neither the stock of degraded land nor the cost of restoring it to its previous productive state is very relevant so far as making decisions about changes to land allocation or management are concerned. What is relevant in a policy framework is identifying the course of action which will result in the land being put to its best social use — that is, identifying decisions which will reduce the continuing losses from degradation and maximise the net present value of the land.

Estimation of offsite costs

Categories of costs vary in measurability as indicated in Table 11, and in immediate relevance to public policy making. Some more measurable impacts represent a direct burden to be borne by the community in the short term, for example maintaining of the quality and quantity of available water, or removal of sediment. Others have impacts which are more widely spread across the community and have longer term implications for sustainable growth, for example loss of flora and fauna. Furthermore, the actual incidence of the various costs within the community will determine to some extent the priority accorded to them in policy matters.

The calculation of costs requires a good background understanding of the physical and behavioural processes of land degradation in order to model the economic relationships. Once the major physical effects have been identified they may be evaluated using a variety of economic techniques for measuring offsite and non-market effects (Sinden and Worrell 1979). Many of these have been developed in recent years to deal with environmental questions and their application is well covered in environmental economics literature (eg Freeman 1979). The techniques fall into two basic categories:

- Use of market data for private goods to develop monetary values for non-market environment goods. This could involve development of cost estimates based on changes in the output of goods and services which result from changes in the quality of environmental inputs.

Costs of substitution, or costs reflected in price differentials for residential locations may also provide suitable indicators.
• Use of non-market data from surveys, questionnaires and bidding games to determine public willingness to pay for particular items. This is the preferred means for obtaining approximate values for less-measurable items such as aesthetic characteristics or ecological diversity.

In most cases measurement is complicated by difficulties involved in establishing accurate relationships between soil loss, physical impacts and behavioural responses, the high costs of data collection and analysis, and the error margins inherent in techniques for estimating monetary values. The practical difficulties are most acute in respect of environmental goods such as visual aesthetics and diversity. Freeman (1979), Clark et al (1985) and others have made the case for providing physical listings rather than cost estimates for such items.

Empirical evidence

Some data on the social costs of land degradation in the USA are available from recent studies conducted by Resources for the Future and the Conservation Foundation. Crosson (1984) estimated the costs of crop land productivity losses (for corn and soy beans) associated with soil erosion to be US$40m per year.

The study by Clark et al (1985) was restricted to analysis of the problems caused by sediment and associated agricultural pollutants entering waterways. The results are summarised in Table 12. The single-value estimate for the damage cost of in-stream and off-stream problems attributable to land degradation was US$6bn per year in 1980 of which crop land contributed US$2.2bn. These figures refer to total annual current costs and are gross rather than net figures, that is no deduction has been made for the investment or other losses necessarily incurred in reducing these damages. Clark did not provide estimates of the costs of biological damages, but noted that they may be very significant. The relative sizes of the estimates from the studies by Crosson and Clark suggest that off-farm impacts have not received the attention they deserve relative to on-farm impacts.

On the basis of the USA estimates, corresponding land degradation costs in Australia could amount to several hundreds of millions of dollars per year. This is indicative at the broadest level only, as a translation of the USA data to the Australian case would require consideration of differences in variables such as the physical condition of soils, the extent and intensity of agriculture, the nature of waterways, and many other factors.

In Australia there has been limited systematic compilation of damage cost figures (Sinden 1984). The figures which are available refer to differing situations and have been developed on the basis of different methodologies. Notwithstanding their lack of consistency some of these are listed in Table 13 as indicative of the magnitude of some of the costs involved.

Policy implications

Land degradation comprises a diverse set of complex physical and biological processes. The development of efficient coordinated policy responses will require the collaborative effort of a number of disciplines. It is suggested that an economic approach based on a broad environmental perspective can contribute to this process by helping illuminate the problem and by pointing to opportunities for effective action and cost savings for the community at large.

Table 12: Summary of off-farm damage costs of land degradation in the USA (US$m, 1980)

Type of impact	Range of estimates	Single-value estimate	Crop land's share
In-stream effects			
Biological impacts		no estimate	
Recreational[1]	950–5 600	2 000	830
Water-storage facilities[2]	310–1 600	690	220
Navigation[3]	420– 800	560	180
Other in-stream uses[4]	460–2 500	900	320
Sub-total in-stream (rounded)	2 100–10 000	4 200	1 600
Off-stream effects			
Flood damages[5]	440–1 300	770	250
Water-conveyance facilities[6]	140– 300	200	100
Water-treatment facilities[7]	50– 500	100	30
Other off-stream uses[8]	400– 920	800	280
Sub-total off-stream (rounded)	1 100–3 100	1 900	660
Total—all effects (rounded)	3 200–13 000	6 100	2 200

Source: Clark et al (1985)

Notes to Table 2:
1. Includes freshwater and marine fishing, boating and swimming.
2. Includes dredging and excavating replacement capacity.
3. Includes delays to shipping, accidents, dredging.
4. Includes commercial fisheries.
5. Includes damage due to increased flood height, direct sediment damage, reduced agricultural productivity.
6. Includes sediment removal from drainage ditches, pumping costs, weed control and sediment removal from irrigation canals.
7. For industrial and management use.
8. Extra costs for use of affected water.

Table 13: Illustrative Australian costs for selected aspects of land degradation.

Annual Cost	Description	Source
$15m (1983)	Cost to Victoria of man-induced salinity	Read (1984)
$5-10m (1983)	Cost to South Australia because of Victoria's salinity	Read (1984)
$94m	Costs of lost production in Western Australia due to land degradation	Robertson (1984)
$7.5m (average 1980-1 to 1982-3)	Cost for repair of damage to road surfaces, removal of water borne sediment, and other erosion damage (figure covers 46 per cent of NSW municipal and shire councils)	NSW Soil Conservation Service
$40m (1985)	Total costs to South Australia of salinity	Easdown and King (1985)
$4m (1985)	Costs of lower production of butterfat and lucerne (Lower Murray Private Irrigation Association)	Easdown and King (1985)
$40 000 (1982)	Cost of keeping sand off roads in Jerramungup district, Western Australia	Carder and Humphry (1983)
$1–2/ha	Costs of siltation and erosion of roads per ha of cultivation in the Darling Downs	Alcock (1980)
$1–2/ha	Costs of sediment in streams and dams per ha of cultivation in the Darling Downs	Alcock (1980)
$4.40/ha	External costs of salinity from further clearing in Victoria's Loddon catchment	Greig and Devonshire (1981)

There appears to be a strong case for a systematic study of the offsite and onsite costs of land degradation in Australia. Present estimates are sparse and inconsistent. The study could trace cause and effect, summarise relevant physical and economic data, evaluate costs for a selected region or regions, examine the incidence of degradation problems and the distribution of associated costs within the community.

A thorough assessment of this type could contribute in the following ways:

• *Establishing the economic significance of land degradation in Australia*

The long-term and insidious nature of land degradation has led to inadequate attention in public policy in the past, despite substantial evidence that it is a major national problem. Aggregate estimates of offsite and onsite costs would provide an important means of showing elected representatives, economic authorities and land management agencies the real costs of failing to act to tackle current problems of land degradation.

• *In priority setting for research and policy action*

A broad picture of the social costs of land degradation would allow assessment of the relative economic importance of different regional problems in land degradation and the extent to which they are essentially offsite or onsite in character. Both sets of information would be relevant to setting priorities for allocation of research funds, for targeting government assistance and for the design of appropriate policies.

At the regional and local level more information on offsite effects and associated costs could assist in ranking alternative soil conservation projects and in identifying areas in greatest need of public intervention. It may also lead to wider recognition that investment in well-designed conservation projects dealing with persistent degradation problems is frequently a cost-effective alternative to major repair works in the future. Although a detailed cost-benefit analysis may be neither feasible nor appropriate in many cases the items listed in Table 11 could serve as a useful check list when confronted by a particular land degradation problem. A judgment could then be made as to the severity of physical degradation, the cost of obtaining economic data, and the extent of analyses to be undertaken.

It may frequently be the case that the environmental costs — the indirect, offsite and longer term costs — will be greater than the direct loss of onsite production. Significant long-term costs to the public sector and the community at large could thus be avoided by appropriate intervention at the farm level based on knowledge of the physical relationships linking onsite practices and offsite effects; for example, the impact of particular agricultural practices on the rate of erosion, the level of turbidity and sedimentation and finally on water quality and storage capacity.

In conclusion, human activities, use of natural resources and environmental quality are inextricably linked, and environmental externalities often transcend local and state boundaries. Responses to land degradation should take account of the opportunities and constraints associated with each of these determinants, as well as the economic and institutional mechanisms available to assist public sector intervention in particular circumstances.

6 Degradation pressures from non-agricultural land uses

Lance Woods

Non-agricultural land uses which may exert pressures on the land resource include forestry, various reservations for special purposes, tourism and recreation, mining, industry, urban and peri-urban development, transport and communications and coastal. These non-agricultural activities do not occupy a substantial proportion of the continent, but they are potentially more important economically and socially than rural land use and are capable of exerting pressures which may lead to intense land degradation. The degrading pressures arising from each of these land uses will now be identified and briefly discussed.

Forestry

Forests are much more than land on which trees are growing. They are remarkably complex ecosystems in which climate, soil and water determine which grasses, herbs, shrubs and trees will develop. The vegetation, in turn, determines the animal life that can exist and the land uses that can be applied by human beings. Forests are the source of many products and services, ranging from timber and paper products to sites for various forms of recreation. They provide clean water supplies, soil and watershed protection, conservation of wildlife and plants and, in some cases, grazing for domestic animals. Multiple land use is common and often encouraged by forestry authorities.

Land degradation may be minimal under established undisturbed forest, but once the forest is put to use there may be an increase in some forms of land degradation. At the extreme, complete clearing of forest without reafforestation usually results in severe degradation of steep land or land otherwise susceptible to erosion.

The obvious use of forests, logging, may range from the selective extraction of a small number of trees through to clear-felling, as with woodchipping. During log extraction soil is compacted not only by the logs themselves but also by the logging machinery. This results in lower

---

This chapter draws on material published in Woods (1984). *Land degradation in Australia.* Department of Home Affairs and Environment, Canberra: AGPS, Chapters 3 and 9.

soil permeability to water. Skid tracks, if they run uphill and cut through the permeable topsoil, may concentrate runoff and, in some soil types, rapidly become gullies before revegetation occurs. Up to 25 per cent of a logged area may be affected by skid tracks and compaction.

Another adverse effect can be increases in the sediment load of streams from catchments which have been logged. Such increases have ranged from negligible to more than a thousand-fold during maximum stream flow. This wide variation can be attributed to differences in soil types, climatic conditions and logging methods. After the completion of logging a few years usually elapse before sediment loads return to the levels which applied before logging began.

In logged areas the residues may sometimes be burned before new trees are planted or before regeneration takes place. If a hot burn occurs, it is likely that most of the organic residues on the soil surface may be burned with the logging residues, leaving the soil unprotected from erosion, which may be substantial before revegetation can exert any effect. The most appropriate residue management, therefore, would be either to leave residues in place or to burn them under such conditions and in such a way that much of the surface organic matter is left.

The establishment of access roads is usually a part of normal forestry operations, both for fire control and for logging. The roads also open up forests for recreational use. Regrettably, those setting out to enjoy the pleasures of the forest can cause fires, leave litter, and cause damage with their off-road vehicles, particularly trail bikes.

In fact, access roads are considered to be the cause of much of the land degradation that occurs within commercial forests by concentrating water flow at particular spots, which may result in gullying. The roads also increase the possibility of landslips on susceptible land. However, these effects can be mitigated if the roads are carefully sited and care taken in dispersing the water collected by them.

Trees are very effective in extracting and recycling plant nutrients, with most of the nutrient pool contained in leaves, bark, surface litter and soil organic matter. When residues are burned nitrogen is lost and other plant nutrients are rendered more mobile and may be lost by erosion or leaching if revegetation is delayed. A small proportion may also be removed from the ecosystems by logging.

The hydrology of a forested catchment is different from that of a cleared catchment. In the former, a substantial proportion (10-30 per cent) of rainfall is intercepted by the tree canopy, where it evaporates without even reaching the ground. Surface litter and soil organic matter encourage high infiltration rates for water that does reach the soil surface, and they slow down the rate of runoff from a wet soil. As a result, flood peaks from a forested catchment are lower and spread over a much longer period than would be the case if the catchment were cleared. Trees usually have deeper roots than the vegetation typical of cleared catchments and transpiration takes place from a greater depth of soil. Consequently, a higher proportion of the rainfall is transpired or evaporated in a forested catchment and less is added to groundwater. The water yield from a forested catchment is thus lower, but the flow is more uniform and the water quality much higher. Permanent clearing, however, may result in salinity problems if salts are present in the subsoil or groundwater and are mobilised by the change in hydrology.

Special purpose areas

Special purpose areas include land reserved for particular purposes, such as flora and fauna conservation, national parks, national heritage preservation, defence and Aboriginal use.

Land use pressures vary greatly, depending on the type of reservation, associated land uses and management, and the type of land. At the one extreme, reserves for flora and fauna conservation should be under no pressure. However, in reserves used for recreational purposes, such as national parks, the soil and vegetation may suffer extensive degradation in spite of efforts at conservation. This is discussed in the next section.

Defence reserves used for training purposes may be degraded by vehicles, equipment, explosives, and other training operations. Examples include soil erosion at Puckapunyal in Victoria following tank exercises and rainforest degradation at Canungra in Queensland following jungle training.

Large areas of land are reserved for use by Aboriginals in the Northern Territory and several other states. Traditional use does not cause land degradation. However, where vehicles and livestock are present they can cause loss of vegetation and soil erosion in the same way and as severely as they can on land used for agricultural or recreational purposes.

Tourism and recreation

The use of land for tourism and recreation is determined by the land's proximity to towns or cities, by scenic values which it may possess and by its unique features. Recreational activities attract large numbers of people to lands which may be small in area. Soil and vegetation may suffer extensive degradation through trampling, the establishment of footpaths, the passage of vehicles, lighting of fires, careless discarding of litter etc and even land and vegetation that are not easily degraded may not be able fully to resist these pressures. Land with scenic values or unique features may, in fact, be quite fragile.

Examples of degradation of land used for recreation include loss of vegetation and soil erosion caused by trampling and vehicles in numerous spots around Ayers Rock; sand dune blowouts along New South Wales and Victorian coasts caused by footpaths and vehicles; and loss of alpine vegetation and soil erosion caused by roads, tracks, and ski-slope development in the snowfield areas.

Recreation pressures may be reduced by careful siting of roads, control of vehicles — particularly off-road vehicles — the establishment of stable footpaths, and fire control.

Mining

Mining is economically very important in Australia but it does not occupy a large area of land. The Bureau of Mineral Resources estimated in 1976 that the actual areas of land disturbed by mining in Australia up to that time excluding areas mined for construction materials and silica was 340 km². The disturbed area was expected to increase about four to five fold by the year 2000.

Mining occupies only a small area of land, but it affects a much larger area through pollution and the development of infrastructure. The location of mining is determined by the location of ore bodies, or by the quantity and purity of the material being mined (ie the likely profitability of the operation). The scope for site selection is far more prescribed than it is for many other land uses. Mines typically have a rather short life, say 10-30 years, which is reflected in minimal infrastructure and lack of concern with long-term effects.

Mining development

The mining operation may be carried out underground or on the surface; processing and storage sites are usually needed on the surface and mine infrastructure has to be developed. These ancillary structures and activities are usually close to the mine site but can also be very distant, according to circumstances and requirements.

Underground mining requires pithead facilities, and sites for dumping wastes (tailings) which may be quite toxic although not large in volume. Underground workings may eventually collapse causing surface subsidence, sometimes as rather spectacular pits.

Surface mining (open cut) usually disturbs much larger areas than underground mining. It may involve total removal of material, as in the case of sand and gravel extraction; or selective removal, as in the case of open cut mining when large amounts of overburden may be shifted and piled up.

Processing and storage, whether at the mine site or elsewhere, involve construction of processing plant and provision for stockpiles, waste disposal, transport and often substantial water supplies.

Other infrastructure may include the construction of towns and recreation facilities, roads, railways, pipelines and ports, dams and water supply facilities, and the provision of energy supplies.

Degradation from mining

Land disturbance is caused by surface mining where overburden is moved, and by extraction of soil, sand or gravel. In both cases, fertile topsoil or soil fertility may be lost, the new surface is very susceptible to erosion, and the surface hydrology is changed. Infiltration and retention of water is usually low. The productive potential of the land for other uses is lowered, eg from crop land to rough grazing land.

Mining wastes include rock or low-grade ore, together with fines from processing. Coarse waste is usually put into waste heaps, or used to construct dams into which the fines are pumped as slurry. These waste heaps or tailings dams are eyesores which tie up valuable land and are sources of dust. Other effects depend on the type of material: for example, coal wastes are liable to spontaneous combustion while seepage through wastes containing sulphides and heavy metals produces very acid and polluted drainage water.

Dust blown from tailings dams and waste dumps and sulphur dioxide and carbon monoxide from burning coal wastes cause air pollution. This is a very real problem where the mine is close to an urban area.

Water pollution exists in two forms; solid particles washed into streams as a result of erosion on mine sites and chemical pollution by acid or saline seepage from waste dumps or tailings dams. The pollution of the Molonglo River by the Captains Flat mine in NSW and the Finniss River by the Rum Jungle mine in the Northern Territory are two well-documented examples.

Some wastes are toxic to plants; pH values as low as 1.6 have been recorded in some waste dumps and surrounding soil. This acidity, as well as directly inhibiting plant growth, releases aluminium, boron and manganese into the soil solution, creating various plant nutrition problems. The red mud from bauxite processing which is strongly alkaline from residual caustic soda, tends to form an impermeable gel resistant to any treatment.

Waste dumps and tailings dams are very susceptible to erosion because of lack of vegetation, steep slopes and the fine dispersed particles of the material. Eroded materials may be deposited on neighbouring land or in streams, with consequent loss of productive potential and pollution. Subsidence occurs as a result of slumping of fill material in waste dumps, compaction of overburden, or the collapse of soil and rock overlying underground workings.

Infrastructure development may cause additional land degradation and other pressures.

Control and rehabilitation

Most new mining development is subject to rather stringent legislative control intended to minimise harmful environmental effects and to provide for rehabilitation of the area affected. It requires extensive collaboration between the mining company and relevant state authorities, eg in New South Wales the Mines Department, the Soil Conservation Service and the Water Resources Commission may be involved.

Mine rehabilitation is expensive, often of the order of $10 000 per hectare, but the high value of mine products allows this cost to be met. Bonds may be required under mining legislation to meet rehabilitation costs.

Rehabilitation usually involves removal of debris, reshaping of waste dumps and tailings dams, addition of large amounts of fertiliser and lime to correct infertility and acidity, revegetation, and structures to control drainage, seepage and erosion.

Industry

The area of land occupied by industrial development is small compared to the total area of Australia and degradation of the land occupied is low, except to the extent that removal from other use, the exclusion of other potential uses and soil pollution may be considered as degradation. The drainage and filling of wetlands to provide sites for industry with easy access to sea transport deserves particular mention.

Industry has important external effects on land. For example, an industry might use raw materials derived from land, such as products from forestry, mining, and agriculture. Sand and gravel extraction and quarrying, in particular, are examples of industrial demand creating land degradation. Any future expansion of agriculture to provide extra crop products for the manufacture of liquid fuels could result in serious erosion, unless carefully managed and planned, as the expansion would be onto marginal land with a higher potential for degradation than existing crop land.

The disposal of industrial wastes can result in serious land and water pollution eg paper mill effluents, smelter emissions at Queenstown and Port Pirie, zinc wastes in the Derwent River and dumping of plastics wastes or used tyres.

Waste disposal is becoming more and more subject to regulation and monitoring. Environmental impact legislation may require that waste disposal be fully considered and satisfactorily provided for when new industrial plants are proposed.

Urban and peri-urban development

Urban use takes up only a small proportion of Australia's land but because initial settlement favoured certain sites it does tend to occupy areas that were once good agricultural lands, particularly on the coast and along rivers.

There are no good reasons, other than location, why prime agricultural land should be selected for urban development instead of less productive areas such as undulating grazing land which may be just as suitable. Urban development policy should be to retain in use as much prime agricultural land as possible even though it may be surrounded by urban development. This would provide greenbelt areas and avoid the problem of loss of prime agricultural land which is beginning to cause concern in Europe and the United States of America. Australia does not have the large untapped resources of prime agricultural land that many people imagine.

Other problems created by urban development include: erosion and dust during the construction phase; increases in the amount and speed of runoff from storm rain which may exceed stream capacity and cause stream erosion; pollution of drainage; and pollution from the disposal of sewage and urban wastes.

Peri-urban development—the establishment of hobby farms or dwellings on large blocks of land around an urban area—creates another set of problems. Land productivity may be reduced, increased demands will be made for infrastructure (water supplies, roads, power supplies etc), and for advisory services which may reduce their availability to commercial agriculture, and pollution from drainage or sewage disposal may be common.

Most of these urban development problems affecting land can be reduced by effective land use planning and zoning based on adequate land survey and evaluation information. This also requires adequate cooperation and communication among the many organisations involved and adequate guidelines for urban and regional development.

Transport and communications

Road and rail transport occupy much more land than do sea and air transport: in 1979 there were 812 000 km of roads open for general traffic in Australia and in 1978 there were 39 700 km of government railways. The actual area occupied by these facilities is relatively small but they have a disproportionately large effect on land and water.

Roads are sources of sediment, both during and after construction, from degrading road surfaces, eroding drains and batters and unstabilised creek crossings. By concentrating runoff at particular spots both roads and railways change surface hydrology which may lead to gully erosion. On the other hand, roads, railways, bridges, culverts and drains may be damaged by erosion, or by sediment and flooding from erosion elsewhere.

Material for road and railway construction and maintenance is taken from gravel and borrow pits and quarries, which erode and change surface flow, resulting in further sedimentation and unsightliness.

Pipelines and powerlines are usually constructed along fairly straight lines with little regard to soil, vegetation or terrain. Tracks cleared along these lines to facilitate maintenance often become gullied because of their siting and lack of provision for drainage.

Coastal land uses

There are many coastal land uses and activities involving this land. These include:

- *Fishing*—dependent on coastal harbours and processing plants, and on breeding grounds in estuaries and mangrove swamps. Fishing may be adversely affected by pollution (eg oyster leases) or from a reduction in the area of mangroves and wetlands.
- *Recreation*—swimming, sunbathing, photography, walking, surfing, boating, waterskiing, fishing, picnicking, and camping may require facilities such as dressing sheds, toilets, marinas, launching ramps, jetties, walkways, caravan parks, campgrounds, motels, car parks, and kiosks. These activities are not all compatible.
- *Urban development*—the wish of many people to live on the coast, whether in cities or holiday areas, leads to linear development along the coastline, which increases the pressure on coastal land particularly if it occupies the fore-dune area; it requires the provision of services such as roads, water supplies, sewerage and garbage disposal; and it may initiate or be affected by erosion, loss of vegetation and pollution (eg Surfers Paradise).
- *Industrial development*—coastal locations may be preferred because

of sea transport, water for industrial cooling, waste disposal at sea, labour supply and infrastructure; it is a source of pollution and may occupy wetland areas to the detriment of fishing and recreation.

- *Minerals*—Australia is a major producer of mineral sands (titanium), and sand extraction for construction is important near urban areas; these uses disrupt dune systems, resulting in dune erosion, dune movement and loss of unique habitats for flora and fauna.
- *Forestry*—coastal forests along the east coast of Australia are important for timber, recreation, honey production and wildlife; clear-felling for woodchips may cause siltation of lakes and estuaries (eg at Eden).
- *Transport*—shipping requires harbours and port facilities and may involve dredging, breakwaters, oil spills etc, which may change coastal erosion and affect living organisms; roads increase pressure through providing access, and may cause and be affected by erosion or landslips if they are not correctly located (eg on top of foredunes).
- *Waste disposal*—urban sewage is often drained into the sea (eg Bondi in Sydney), and other urban and industrial wastes are disposed of on land or find their way into coastal waters; the capacity of the land, sea or coastal waters to assimilate these wastes can be exceeded, or the wastes may be concentrated in marine feed chains (eg zinc in Derwent River oysters).
- *Educational and scientific use*—many organisms and complex ecosystems are unique to the coastal zone; their preservation may necessitate land use controls such as reserves or restriction of access.
- *Water supply*—coastal sands and rivers may be important sources of water for industrial and urban use, as in Perth and the Sydney suburb of Botany; but this water can be easily polluted by nitrate, heavy metals etc, from industrial and urban wastes.

A coastal location is not essential for some land uses or activities that are carried on there; where land is limited, a case can be made for excluding these uses. There are numerous instances of one use or activity being incompatible with another, particularly where a coast-modifying use such as sand mining, or industrial or urban development is involved. Strong competition between land users is a common feature of the coastal zone. The outcome of many land use conflicts may depend more on which pressure group is the most articulate and able to convince local authorities, rather than on national and coordinated land use planning.

Human pressures, such as trampling, use of vehicles, or construction may damage vegetation and this in turn allows erosion to take place. These pressures, together with pollution, may also lead to a reduction in numbers or even the extinction of plants, animals or other living organisms in the coastal zone. These effects can be minimised by restricting the pressures at critical places, eg by establishing walkways across dunes or defining parking areas.

Erosion and build-up of coastal sand dunes by wind and wave action is a natural process which can be substantially changed by loss of vegetation from trampling or overgrazing, vehicle tracks, burning etc. These disturbances may allow the sand to become mobile, destroying any development or vegetation on it or in its path. The problem can be avoided by not building houses, roads etc on coastal dunes, and by minimising disturbances so that the dunes are held by natural vegetation. Treatment includes restriction of access, dune reshaping and replanting of vegetation.

Both land and water may be polluted by industrial and urban wastes. The results include loss of living organisms, loss of amenity values, health problems and increased costs of land and water use.

All of the coastal land degradation problems have inappropriate land use and management as basic causes. Hence they can be minimised by adequate land use planning and management backed up by effective coordination and sufficient data. However, these data generally do not exist, or exist in an unusable form. There are only a few exceptions, such as the CSIRO South Coast study in New South Wales.

Coordination and communication could be improved considerably. Some efforts are being made to do this, an example being an expert Coastal Council set up under the New South Wales Coastal Protection Act of 1979 to advise on the protection and restoration of the coast and on the orderly and balanced use and conservation of the coastal region.

Land use planning and management should be carried out at a regional level if it is to be effective. It is then possible to secure adequate communication, and the people involved are fully conversant with the regional situation and problems. Attempts to direct land use planning and management from a national level, as in England, have tended to fail because the regional situations and problems were not properly understood and the necessary organisation became too cumbersome.

Conclusions

The types of land degradation which may result from these non-rural land uses can be described, and examples and treatment measures given. But the actual national extent of land degradation under each use and the costs of treatment are unknown and this is indeed a very significant gap in environmental statistics in Australia. A number of overseas countries are collecting national data of these types, but no moves have been made to do so in Australia.

During the period 1975-7, the soil conservation authorities in Australia were devoting an average of about 20 per cent of their total effort to non-rural work In one case, nearly 40 per cent of the effort was non-rural. All authorities said that the proportion of non-rural work was increasing. The conclusion from this is that non-rural land degradation is substantial and a significant consumer of conservation resources.

Commentary

Peter Greig

Introduction

If land is degrading, there will certainly be physical consequences, as discussed in the previous section. There may be economic consequences, depending on whether anyone's income or welfare is affected by the physical consequences. The economic consequences, such as they are, may have a bearing on policy, depending on whether the policy maker is 'rational'. What rational means, in shorthand, is that the policy maker chooses policies the economic benefits of which exceed their costs by the maximum amount. So costs have relevance only in a policy context.

Although not all policy makers are rational, the 'maximum net benefit' principle is a convenient starting point for discussion, especially in a forum like this. For the present purposes, therefore, costs are regarded as relevant only so far as they relate to the objective of maximising the net economic benefits to the decision maker, whether freehold farmer or a public agency. A public agency must in theory consider costs impinging on *all* persons affected by degrading land. These are 'social' costs, to distinguish them from 'private' costs, which are those impinging only on the landowner.

The chapters in this section review what is known about the costs of degrading land. As will be seen, the connotation of 'costs' varies between the papers. Blyth and McCallum take costs to mean those that are relevant to decisions based on micro-economic analysis. At the other extreme, Woods uses the term costs as an expression of the physical damage associated with degradation. Between the extremes, Upstill and Yapp define costs as the sum of the productivity lost through land degradation, or the sum of the costs of restoration.

It is also necessary to point out that 'degradation' to one landowner may be 'restoration' to another. Thus, a farm that is reverting to natural bush cover might be 'degrading' from the farm-income viewpoint, but 'rejuvenating' if the owner is a hobby farmer. In what follows, therefore, degradation is defined according to the land use preferred by the current landowner. Thus, degrading land is land that is losing its productive capacity with reference to the current land use. This is the definition that is followed in the rest of this chapter.

When land is degrading, the physical impacts will be occurring both on the site itself, and also off the site. The costs caused by the offsite impacts are discussed in the chapter by Upstill and Yapp. The costs of the onsite impacts are divided between those caused by farming (discussed by Blyth and McCallum) and those caused by non-farm land uses (which are covered in Woods' paper). Forestry is placed in the 'non-farm' category, even though Blyth and McCallum discuss it, albeit in an endnote.

Farm onsite costs

Blyth and McCallum devote their paper largely to an explanation of the most appropriate method for evaluating the farm onsite costs of degrading land, for the purpose of policy making. In their view, all of the four cited national studies into the costs of farmland degradation have not given onsite cost estimates that are useful for policy makers. According to Blyth and McCallum, these studies estimated either the gross costs of restoring land to its pre-degraded condition, or the opportunity costs of the loss in maximum production achievable from the land in its pre-degraded condition. As Blyth and McCallum correctly point out, knowledge of these estimates is hardly useful in guiding policy—if the costs of restoration exceed the value of the additional production thus made obtainable, there will be an economic loss.

Focusing on onsite farm costs, Blyth and McCallum argue that the relevant policy variables are those that impinge on the farmer's income assessed over the long run. Conceptually, the farmer increases productivity, with concomitant increases in the rate of land degradation, to the point where the marginal benefit from further productivity increases is offset by the marginal costs of further land degradation, taking into account the resale value of the land.

The only problem is: no-one knows the relevant functional relationships, either between productivity and the rate of land degradation, or between the rate of land degradation and the resale value of the land (see Powell 1974, on valuing agricultural land). It is not even easy to assess accurately the rate at which any piece of land is physically degrading.

Thus, two points from Blyth and McCallum's chapter are:
- on freehold farmland, the only policy-relevant, onsite costs of degrading land are those that affect the farmer's income; and
- little is known of the physical relationships between land degradation and productivity.

Though Blyth and McCallum focused on the conceptual basis for valuing the costs of degrading land, their examples relate almost solely to soil erosion, with passing reference to dryland salinity on the farm. A comment on the role of farm trees as another integral component of the land would be appropriate.

Agricultural productivity has been increased by the removal of native trees as well as by pasture improvement and soil cultivation for cropping. But productivity gains from clearing may diminish and even be reversed as the point of total tree removal is approached. There is some evidence that long-term farm productivity will benefit by having some tree shelters—or lambing protection, crop protection, insect control, nutrient recycling and for evaporation control around farm dams (see Bird, Lynch and Obst 1984 for a review of the literature). A few trees may add landscape value too, which may be important to the farmer.

An expression of these onsite benefits is that the retention of *some* trees on the farm appears to add to the resale value of the property and this must be regarded as an addition to welfare.

There is a conceptual parallel between the decline or removal of farm trees in the search for increased productivity and Blyth and McCallum's description of the increased rate of soil loss for the same purpose. In both cases, with maximisation of net farm income as the object, it is relevant to consider the onsite impacts and costs of both soil loss and tree loss. Unfortunately, the technical or physical relationships are still unknown, and the best that can be said at this stage is the well-worn remark that more research is needed.

Non-farm onsite costs

Lance Woods' paper gives a review of the types of physical impacts resulting from a number of non-farm land uses, including forestry, parks, recreation, mining, industry, urban and peri-urban development, and transport and communications. Such a taxonomy is a useful starting point for estimating the onsite costs of those impacts in a policy-relevant sense.

As in the case of farmland, the non-farm onsite impacts are expressed most notably in an acceleration in the rate of soil degradation. If the land uses occur in forested landscapes, impacts on plant and wildlife habitats may also occur. However, as pointed out by Blyth and McCallum, it is important for policy purposes to know the magnitude of the physical impacts and their economic impacts on humans. The latter question is not tackled by Woods.

Woods also appears to make the common error of equating 'forestry' with logging, which is just the harvesting phase at the end of the whole biological/managerial cycle known as forestry. That cycle has all the elements, though much more protracted, that are found in agricultural cropping.

It *is* the logging phase of forestry however, that generates the most observable onsite impacts, and these therefore bear some further discussion here. Logging causes potential soil erosion and habitat destruction, but constraints are usually designed and enforced to minimise that potential. For example, logging will be prohibited within a specific distance of a defined stream or water storage. Logging may also be prohibited during periods of heavy rainfall.

The care prescriptions operate much like easements on a property title: they limit the owner's rights over parts of his property. And like easements, the width is determined with much precision but with little science. That is, little is known about, say, the physical impacts on water quality of making the easements larger or smaller. This is an area in which scientists could usefully derive general models of the physical relationships involved. Until this is done, the economic consequences of care prescriptions can only be guessed at.

These and many other constraints are intended to prevent stream sedimentation and to allow restoration of the site and its habitat over time towards its pre-harvesting condition. The importance of time in this process cannot be over emphasised, but it is often overlooked in discussions about the impacts of logging in native forests.

If, as is normally the case, logging recurs at intervals less than the biological climax age of the forest, the site's wildlife habitats will never be restored completely to their pre-harvesting condition. Thus, on that particular site, there may be a permanent loss in habitat for at least part of the continuum of wildlife species. The soil degradation impacts are not likely to be permanent, except where the care prescriptions have not been followed properly.

The costs of these onsite impacts are unknown. As in the case of farming, the temporarily increased rates of soil loss and the effect of this on tree growth and income are unknown, but thought to be small in most cases. The loss of part of the wildlife habitat continuum in local cut-over areas within a regional network of permanently retained habitats is an impact regarded by a section of the community as significant, yet the economic cost is hard to conceptualise and even harder to measure.

Part of the problem is that a large part of the value placed on habitat protection is placed by persons who do not actually reveal their preferences by any expenditures (by, say, visiting the site in question). In such cases, the value they place on the retention of habitat can only be determined by special survey techniques, for example, using a valuation concept called option demand. For an Australian discussion on this topic see Olsen (1975).

Offsite costs

In Chapter 5, Upstill and Yapp stress the need for broad-scale estimates of offsite costs of degrading land as a basis for land use policy making, and give some examples of the methods that have been applied in Australia and overseas.

Examples of offsite impacts abound, but those most useful to the present purposes concern water quality and quantity impacts, and landscape impacts. Stream salination and sedimentation are the most common water quality examples (eg Langford and O'Shaughnessy 1977). Upstill and Yapp quote the work reported in Sinden and Worrell (1979) who show how to estimate the costs of these offsite impacts in both conceptual and practical terms. Again, the conceptual limitation is that the costs must be relevant to policy decisions.

Other practical examples of the estimation of offsite costs are given by Greig and Devonshire (1981) and Reynolds (1978). Greig and Devonshire estimate stream salinity costs caused by tree clearing on farms and Reynolds assesses the landscape impact of removal of farm trees. In both cases, farmers maximising onsite income were causing offsite costs. For the case of stream salinity, the costs were estimated by either the costs incurred by water users, or the costs of the least-cost method of restoring the stream to its previous quality. In the landscape case, the cost of restoring landscapes to previous levels of quality was assessed by the loss in on-farm income caused by restoring more tree cover. The problems of measuring the economic benefits to the viewers of landscape are considerable (but see Greig 1983a or Sinden 1974).

The offsite costs of logging in native forests are similar in some respects to the offsite costs of farming. For example, if an increase in soil erosion does occur, then its effect on streams is similar to that caused by erosion from farming, though less certain and more intermittent. There is, however, one peculiar offsite impact of logging in native forests: a loss in water yield, at least in some forest types.

The loss in water yield has been observed in mountain ash forests in Victoria (Melbourne and Metropolitan Board of Works 1980) and it has a time profile typified by the graph shown in Figure 18. The opportunity cost of the loss in water yield from a catchment can theoretically be measured by the marginal replacement cost of the water. Only *average* costs of water yield are available at present, and these can vary around $100 and $300 per megalitre. A proper evaluation of water yield losses in a given catchment would need to take account of the time profile

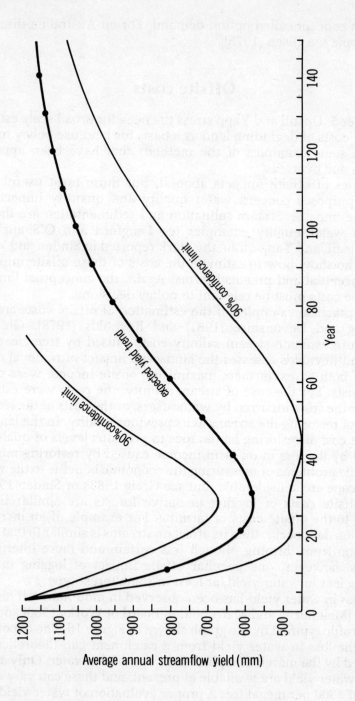

Figure 18 Streamflow-time curve for Mountain Ash.

shown in Figure 18, as well as the seasonal time profile, and a number of other factors. Nonetheless, this is an offsite cost of a legitimate land use, and that cost is likely to be significant and can be measured.

Conclusion

This brief review has shown that the topic of social costs of land degradation is still in its infancy. A few grand scale evaluations of the costs of land degradation have been attempted, yet the results are of questionable significance for policy making. A few studies have attempted to develop the theory of what costs *are* relevant to policy making. Some have been qualititative statements on the *kinds* of physical impacts caused by degrading land, and in some lexicons these are 'costs', though not in the economic sense.

It is important to estimate the policy-relevant costs, both onsite and offsite, of degrading land, but I agree with Blyth and McCallum that the costs ought to be considered on a case-by-case, or catchment-by-catchment basis, beginning with the most rapidly degrading ones. This approach will have a better chance of being policy-relevant than the broad-scale evaluations, which can do little but influence the budget limits for land restoration programs. That may not be an economically efficient approach, given that small-scale information will be needed anyway to help decide where the budget should be allocated.

However, not much can be done to estimate the costs of the impacts caused by degrading land until there are robust models of the physical relationships between land use and site impact. The challenge for scientists is to develop general models of the USLE type. Even if crude, the errors in its predictions are likely to be smaller than those resulting from guesswork, which is the only alternative means for setting land use and land management policy.

The direct relevance of even carefully conceptualised and measured costs of degrading land must be regarded as less than certain. Some of the reasons for this are set out by Greig (1983b), but the principle reasons are that not all policy makers are rational, and even if they were, not all of them consider that only human welfare counts. For all that, the exercise of identifying the physical impacts of degrading land, and the evaluation of their economic costs, is surely a desirable input to policy making on this important topic.

III

Legal, institutional
and sociological
factors in land
degradation

7 Land degradation: legal issues and institutional constraints

John Bradsen and Robert Fowler

Introduction

Regulation and property rights—a philosophical issue

Within modern society, law and the institutions responsible for its administration and enforcement constitute a formidable combination of authority and power. The growth in administrative regulation has been based to a considerable extent on the assumption that the problems confronting urban-industrial communities could not be dealt with adequately through the traditional channels of the common law. Hence, regulation has been accepted as necessary for the public or common good, at least by most sections of the community.

Most regulation has been focused on commercial and industrial activities. However, even where health, building and planning laws have applied to individual home owners, there has been a general understanding and acceptance that such measures have a sound purpose and are of benefit to the community at large. It is interesting to perceive therefore that, in the context of land degradation, a greater degree of resistance has arisen to measures which constrain the use of land for agricultural or pastoral purposes. The hostility displayed recently in South Australia by farmers towards the introduction of vegetation clearance controls demonstrates the point vividly.

At a more general level, there appears to be a view held within some sections of the the farming community that title to land carries with it the individual right to farm the land as one pleases. It is asserted that responsibility should be, and is, assumed voluntarily to manage land carefully because it is in the landholder's own long-term interests so to do. This attitude undoubtedly has influenced the nature and implementation of existing land degradation measures.

The adequacy of self-interest in dealing with land degradation is, however, increasingly being questioned. At the same time there is a resurgent recognition of the legitimate public interest in the problem.

These factors give rise, of course, to consideration of the appropriate means by which to ensure that degradation is dealt with. This paper examines the legal and institutional constraints presently in existence and makes suggestions for change. It should be noted that only legal aspects of the problem are discussed. Consistent with its brief, the paper does not attempt to compare the virtues of regulation and incentives. Indeed it is assumed that both techniques have long coexisted and will continue to do so. Rather the paper is concerned to encourage abandonment of the stereotype assumption that the concept of regulation has a fixed and narrow meaning.

While agriculture and pastoralism are not the only forms of land use which have a connection with land degradation, they are a major contributor. It seems appropriate, therefore, to begin this examination of legal and institutional constraints on land degradation by considering the underlying philosophical issue of the appropriateness of regulation to those forms of land use. The view that property rights should be considered sacrosanct is thought to trace back to common law doctrine on the nature of estates and interests in land, and we turn now to the question of whether the common law did in fact constrain the use of land for agricultural purposes.

Common law and land ownership

It is true that the common law, in evolving the distinctive legal concept of the freehold estate in land, regarded such an estate as embracing an extensive bundle of property rights. Land was regarded as the most important commodity within an agricultural society and hence was accorded a special legal status. The attributes of ownership remain recognisable to this day, as is evidenced by the comments of Blackburn, J. in the *Gove Land Rights* case *(Millurpum v Nabalco Pty Ltd and the Commonwealth* (1971) 17 FLR 141) as recently as 1971:

> I think that property, in its many forms, generally implies *the right to use or enjoy*, the right to exclude others, and the right to alienate. I do not say that all these rights must co-exist before there can be a proprietary interest, or deny that each of them may be subject to qualifications. (Emphasis added)

This definition recognises the broad character of the rights inherent in property or ownership. But it also acknowledges what is often overlooked: namely, that such rights may be accompanied by qualifications. In fact a very long, even ancient, link exists between property rights and social obligations under English law. After the Norman Conquest the common law emerged alongside a feudal system under which, as is still the case, only the Crown could own land absolutely. Under the pyramidal structure of the feudal system of land tenures, all other landholders held an estate or interest in land either directly from the Crown or indirectly through other persons. So fundamental was this system to the social order that land law and constitutional law were virtually merged. Its very essence was that along with rights to land went social obligations in the form of the 'incidents' of tenure, a proposition that was profoundly affirmed in the feudal manifesto, the Magna Carta.

It is an error to see the dismantling of the feudal system and the introduction of freehold tenure between the 13th and the 16th centuries as breaking the link between land use and social obligations. In particular, following the removal of the incidents of tenure, the law of waste emerged and, in the circumstances in which it operated, regulated agricultural land use very effectively in the public interest. Further, the common law developed the law of nuisance to provide a significant additional set of constraints on land use in the public interest. The common law also demonstrated its philosphical position, that ultimately the public interest should prevail, in allowing that in an emergency the Crown could use or even take property, albeit that in some circumstances it must pay compensation. In general the common law has not resisted the acquisition of property in the public interest. The common law recognises, therefore, that land ownership is not absolute.

The common law as an instrument for land management

The common law doctrines of waste and nuisance have long had a regulatory effect in protecting land from onsite and offsite damage respectively.

The law of waste The law of waste had several aspects. For present purposes it is sufficient to point out that it regulated a rural land user in cutting timber and regulated farming to ensure what is best summarised as good husbandry. In the latter respect it precluded in particular the cultivation of certain grasslands, excessive cropping, and, to use modern terminology, certain monoculture. In general it was a system which ensured that farm land was well cared for, and which prompted and lent support to what might be termed a land use ethic within England.

Waste was not developed to restrict holders of the fee simple absolute in using their own land. But there was no need. Even into the 20th century, nearly 90 per cent of all land in England had been farmed by leaseholders. Regulation through waste and its concomitant land ethic had universal impact. The public interest in sustainable agriculture was adequately protected. This remained overwhelmingly the case until at least the Second World War. It is interesting to note that more recently, holders of the fee simple in England have begun to farm land without regard to the traditional land use ethic and unrestrained by waste. Their farming practices revolve around modern technology and are largely dominated by short-term dictates. The common law has found itself unable to respond in a legal system which now is predominantly legislative in nature. The result is that, suddenly, after hundreds of years, Britain has a looming land degradation problem of significant proportions. Legislation is sure to follow.

Australia inherited the British legal system based on the common law and hence the law of waste was received into Australia. But it has been largely ineffectual. There are several reasons. First, it is often pointed out that 90 per cent of alienated land within Australia is held under non-freehold tenure. But agricultural land constitutes, at most, only one-third of the total land mass, and within this region some 50 per cent of the land is held in freehold tenure and is not, therefore, subject to the law of waste. There has been no corresponding experience of widespread private leasing of landholdings for farming use.

Second, governments have not sought to apply the law of waste to lands held from the Crown on leasehold tenures. It may be that waste has been ousted by the legislation providing for Crown leasehold tenures. Such legislation and the covenants incorporated in Crown tenures deal separately with the question of land management. A further consideration, where perpetual leasehold tenure is concerned, is that a very limited concept of the reversion exists. In any event, as discussed below, the thrust of land use covenants has been development not conservation, the exact opposite of the policy behind the law of waste.

Third, the principles which made waste so effective in its birthplace rendered it almost useless in Australia. Waste relied for its substance on common practice based on long experience of what was sustainable agriculture in given conditions. Given circumstances in which unbridled development was accepted practice in the settlement of Australia, and indeed legally required practice where leasehold tenure was concerned, the common law of waste was, in effect, law without substance.

In the arid lands, the picture is somewhat different. The devastation caused by overstocking and clearing led to some attempt from an early date to use covenants to protect the land. To this extent, principles akin to waste may have applied, but pursuant to statute, not the common law.

The law of nuisance The law of nuisance also has a very ancient lineage and for some centuries was a powerful instrument of control over land use practices. The aim, unlike the law of waste, was not to protect land from the actions of its holder, but rather to ensure that these actions did not damage neighbouring land. In other words, whereas waste was concerned with onsite damage, nuisance was concerned with offsite damage.

The basic nuisance test was whether a use of land was reasonable, which tended to mean normal or usual. Given the conservative use of rural land, offsite damage was uncommon. Hence nuisance was not frequently invoked in relation to agricultural activities. Nevertheless, it provided an additional set of safeguards which at its peak constituted a system of virtual strict liability.

In stark contrast, nuisance in the Australian soil conservation context may be a positive hindrance. Where a landholder concentrates runoff as a result of clearing and cultivation, because that is a normal or usual use of land, an action for nuisance will not lie. But where a landholder constructs banks, for example, to protect his land and as a result the channelled water causes offsite damage, an action may lie. This is, of course, because the common law may well take the view that such construction is not a normal or usual use of land. Nuisance may, in the result, prefer damage to two pieces of land rather than one. It is a nice irony that if banks which were not causing damage were removed and the resulting normal broad-scale cultivation did cause damage, again an action in nuisance might not lie.

Conclusion

The common law could function well in a regime of agriculture which was conservative. In Australia where the attitude towards and use of land was, albeit for understandable historical reasons, not conservative but radical, the common law was helpless. It has, therefore, proved to be particularly ill-suited to Australian circumstances.

While doctrines such as waste and nuisance are instructive in demonstrating the obligations that land users traditionally have borne, it has been through legislation that the problems of land degradation have been tackled by the Australian legal system. In this respect, it is significant that a federal constitutional structure was adopted in 1901. As a result, it is necessary to consider not only the efforts of the states to deal with land degradation but also the role and responsibilities of the Commonwealth government.

Instruments of intervention: an overview

Action at state level

An analysis of state laws and institutions which have addressed or influenced land degradation could embrace a very wide range of measures. The earliest measures, dating back to the 19th century, dealt in extremely rudimentary fashion with problems such as weeds. A most recent measure which could be included in a review would be the vegetation clearance controls adopted in 1983 in South Australia. In addition, legislation governing the management of national parks, forest reserves and other public lands could be taken into account.

We have chosen, however, to focus on two particular areas of regulation within the states which we consider have constituted the basic avenues of approach to the problems of land degradation: legislation on soil conservation and the land tenure systems. Some further introduction to these two areas of influence would seem desirable at this point, before proceeding to a more detailed analysis of each.

Soil conservation legislation

Land degradation was first recognised soon after settlement. There was, however, a great reluctance for many years to acknowledge that management practices of settlers were the major cause. In the result, action tended to be taken only in response to what was regarded as an external cause. On this basis, legislation was procured to ensure the control of pests, including rabbits and certain weeds. It was also acceptable to blame drought. In part this explained the *Sand Drift Act* (SA) 1923, which was the first comprehensive legislation enacted in Australia to attempt to deal with the question of soil conservation.

It was, however, the intense worldwide publicity about the dustbowl region in the United States in the mid 1930s which resulted in the widespread recognition that land use management practices were the primary cause of degradation. It is significant to bear in mind that the Australian response was heavily influenced by regular dust storms which literally brought the land degradation problem to the cities. Having said that, it should nevertheless be noted that dust storms had occurred well before the 1930s without much response. Dust helped to produce a response, but there was a real and widespread awakening during the 1930s.

In this climate studies were carried out largely in the arid lands. Their overwhelming conclusion was that the cause of degradation was neither rabbits nor drought but the use and management of the land, in particular excessive clearing, overstocking and unwise cultivation and associated practices. It was even acknowledged at that time that where these matters were a problem, they had to be taken into account in managing the land. The states set up investigating committees during this period and the Commonwealth, given sufficient courage by United States federal legislation, urged the states to act. In the result, NSW enacted its Soil Conservation Act in 1938. SA followed a year later. The first Victorian Act in 1940 was very cautious and it was not until 1947 that the basis for the present Victorian legislation was laid. WA first enacted legislation in 1945 and Queensland in 1951. Tasmania, despite strong recommendations going back over four decades, has still not enacted any soil conservation legislation.

It is one of the remarkable features of the history of soil conservation in Australia that parliaments were so little divided on the need and justification for such legislation. One can read Hansard after Hansard to find bilateral views being expressed. Disagreement was often on insignificant issues. There was, for example, endless debate in Victoria about whether the authority should consist of three or four persons. Even more astonishing is the extent to which speaker after speaker acknowledged the responsibility of landholders to future generations and the right of legislatures to regulate land use in the public interest. It is regrettable that these perceptions did not endure in the years to follow, and that only recently has the cry for action been revived.

Land tenure systems

The role of land tenure legislation, and of the covenants and conditions incorporated within the distinctive forms of land tenure that emerged to facilitate the process of land settlement in Australia, has been paid scant attention. Yet, ultimately, many of the problems of land degradation experienced today have their roots in the policies and obligations imposed through the tenures granted to landholders.

The settlement of land in Australia proceeded on the basis of the proposition that every square mile of the continent belonged to the Crown. Landholders, therefore, had to acquire some form of tenure from the Crown. Free grants of land in freehold began with settlement and continued at a varying pace until about 1836, when grants by way of purchase were introduced under the Wakefieldian theory of land settlement.

A distinctive feature of land settlement in Australia, however, was 'squatting', whereby tracts of unalienated Crown land were simply occupied without any grant from the Crown. Squatters proved to be politically adept and managed to convert their initial trespass through annual licensing into leasehold for a term. This marked the emergence of a unique Australian tenure system. Conflict between aspiring selectors of freehold land and leaseholders prompted increasing regulation of leaseholds and the introduction of the conditional purchase agreement. Ultimately, during the 1890s, the perpetual leasehold emerged as the final, uniquely Antipodean, form of land tenure.

As a consequence, 90 per cent of lands alienated by the Crown in

Australia are subject to tenures other than freehold. Apart from the conditional purchase agreement, a now little-used tenure which enables the purchase of the fee simple by payment of instalments over a period of years, the non-freehold tenures are of a leasehold character, either for a fixed term or in perpetuity.

There is an enormous range of variation in the forms of leasehold tenure. Their essence, however, is the granting of rights to exclusive possession subject to varying conditions imposed by covenants. These arrangements are regarded legally as statutory contracts, since they derive their existence from the relevant land tenure legislation, and they must be distinguished in this respect from private leaseholds which arise by virtue of transactions between citizens.

There is, in the covenants contained in Crown leases, a theoretical potential to ensure the protection of the land. But, as will be seen, this has not occurred, in particular with respect to agricultural land.

The Commonwealth role

From the mid 1930s until the late 1940s there was much urging for the Commonwealth to become actively involved in soil conservation. The national importance of the matter was recognised at the highest level. State Hansards reveal numerous claims that the question of land degradation was of national concern and that Commonwealth assistance was necessary. There was an awareness that since the mid 1930s the United States federal government had led the action in its endeavour to deal with soil conservation. Hence, the pressure for Commonwealth involvement was no federal aberration.

Some limited responses were made at the Commonwealth level. In the ACT, a soil conservation ordinance was enacted in 1947 and replaced in 1960. A limited tax deductibility was granted for soil conservation expenditure. There also was created a Standing Committee on Soil Conservation, which reports through the Australian Agricultural Council. Although there was talk of Commonwealth financial assistance to the states in the 1940s, nothing eventuated until the mid 1970s. Very recently, the National Soil Conservation Program has been established to provide assistance to the states.

With the re-emergence of interest and concern about land degradation, attention has turned once more to the role of the Commonwealth. The states have traditionally legislated to deal with this question and hence the analysis which follows is concerned inevitably with state legislation. But in the final Section of this paper, in which we canvass various options for future action, we will address the issue of Commonwealth constitutional powers.

Soil conservation legislation

Form

One of the most striking features of the existing soil conservation Acts is that the legislation is so different among states. This is typically explained as the result of geographical differences between the states, which may in part be true. But the differences largely reflect accidents of history, quirks of parliamentary debate, drafting styles, ignorance and legislative floundering rather than differences in the needs or circumstances of the land from state to state.

Even where the legislation has certain features in common, history shows that this has tended to result from a copying of other legislation as if its very existence establishes the rightness of its form and content. There is a good deal to be learned from the existing legislation. But a most important lesson is that copying bits and pieces from existing laws will not do. The 1930 model is not suitable in 1980. Soil conservation legislation requires a thorough rethink of its legal techniques based on the philosophy and objectives of soil conservation.

Philosophy

The philosophy underlying the soil conservation Acts shines like a beacon in the Hansard speeches dealing with their introduction. Even though they go back some 45 years, speech after speech acknowledges that degradation is of such enormous importance to the long-term well-being and stability of Australian society that the public interest in protection of the land must prevail over landholders' short-term desire for profit.

When one examines the Acts however, they do not adequately reflect this philosophy in their detailed provisions. The Acts tend to allow that in important respects they may not be put into effect without landholder action, or that their operation can be vetoed by landholders. The Victorian Act, for example, provides at one point that action may be taken by the Soil Conservation Authority 'in the public interest' but can be vetoed by just two landholders. That is a somewhat narrow conception of the public interest.

Admittedly some Acts give greater apparent paramountcy to the public interest. Several Acts, for example, provide for the issue of potent land use orders without even the requirement to consult landholders. But while this appears to be consistent with the philosophy of soil conservation, in practice it is not. These legal weapons are, in their nature, of very limited use. Legislation which relies heavily on these orders has failed, and it would seem must fail, to represent adequately the underlying public interest philosophy. Nor can the legislation do so where an apparent public interest paramountcy is hedged about in legislative complexity and a gauntlet of administrative and appeal procedures.

There can be no doubt that while the interests and cooperation of individuals are vital in a matter of such fundamental long-term importance to society, the public interest must prevail. On the one hand, the very existence of the soil conservation Acts reflects this position. On the other, however, the philosophy underlying the Acts is not adequately represented by the present legislation. For soil conservation legislation to be successful, it is submitted that it must have clear objectives and provide the administrative and legal structures, procedures and, in particular, obligations which are most likely to attain those objectives. An examination of the extent to which these requirements are met under existing soil conservation legislation reveals significant deficiencies.

Objectives

By objectives is meant the end results which the Acts seek to achieve. These are not adequately spelled out. It is not sufficient to call the Acts 'soil conservation Acts' and draft them around terms like 'soil erosion', 'soil conservation' and 'damage' to land. In the 1930s this approach was an understandable start. Everyone knew that erosion meant things like palls of 'dust' with sand drifts covering fences and roads and a landscape scarred by gully erosion. Soil erosion was obvious and soil conservation meant trying to do something about it.

In the 1980s this approach is entirely unsatisfactory for two reasons. First, it is now clear that land degradation is a very much more complex, widespread and insidious problem than was realised when the Acts were first passed. Wind and gully erosion are far from completely controlled. But they are obvious and tend to receive attention. The less obvious but growing problems involve insidious degradation through, for example, sheet erosion, the breakdown in soil structure and salinisation. Second, legal issues, especially in the realm of administrative law, have become far more complex and sophisticated and woolly legislation is increasingly inadequate.

It is far from clear which of the degradation problems are, as a matter of law, embraced by expressions like 'soil erosion' and 'soil conservation'. It is probably the case that, given the history and structure of the soil conservation Acts, some are not covered at all. Indeed it would seem to be clear that where subtle, long-term degradation problems are concerned, the Acts are, with one exception, entirely unenforceable.

The legal answer to this problem is not to draft into the Acts provisions to the effect that their object is to 'control and prevent soil erosion and ensure soil conservation throughout the state'. This may indicate that an Act in some vague sense applies to all land in the state but says very little about the objectives of the Acts in terms of the results which the Acts seek on that land. The legal answer to this problem is to define what is meant by the key terms around which the Acts are drafted.

Only three Acts attempt any definition. The South Australian Act defines 'damage' as including harm, a tautology which is of almost no assistance. The Queensland Act defines both soil conservation and soil erosion. The former is defined as:

> The conservation of soil and the prevention or mitigation of soil erosion.

Apart from the hint that soil conservation means something more than attention to soil erosion, this is unhelpful. Soil erosion is then defined as:

> The natural or accelerated removal or deposition of soil which may be detrimental to agricultural pastoral or forestry activities or engineering or other works.

This definition has two elements. First, it confines soil erosion (and therefore, probably, the scope of the Act) to the movement of soil. A whole range of degradation problems are, therefore, excluded. But it includes, apparently, all such movement. (The inclusion of natural erosion gives rise to some intriguing questions.) Second, the Act applies to such movement which results in, apparently, any detriment. On its face the Act would apply to all detriment whatsoever, regardless of the time frame, but from only one cause. Such a definition of objectives is clearly inadequate.

Only the Western Australian Act, substantially redrafted in 1982, has made a serious attempt to define its objectives. It defines 'land degradation' as including:

> (a) soil erosion,, salinity and flooding; and
> (b) the removal or deterioration of natural or introduced vegetation that may be detrimental to the present or future use of land.

The Act then defines 'salinity' and 'soil conservation'. The latter means:

... the application to land of cultural vegetational and land management measures, either singly or in combination, to attain and maintain an appropriate level of land use and stability of that land in perpetuity and includes the use of measures to prevent or mitigate the effects of land degradation.

These definitions are not entirely satisfactory. The relationship between them, particularly as the terms defined are used in the Act, is problematical. For example 'soil conservation' notices may be issued only where 'land degradation' is occurring. In the result the apparent width of the definition of soil conservation may well hinge, through the definition of degradation, on the meaning of the undefined term 'soil erosion'.

Although a simpler and clearer statement of the objectives of the Acts and use of the defined terms throughout the Acts is necessary, the WA approach is on the right track. Its primary objective is the attainment and maintenance of an appropriate level of land use and the stability of the land in perpetuity.

Obligations

The administrative and legal structures and procedures provided for by the Acts vary considerably. Nevertheless there are some common themes which run through the Acts. One of their most distinctive features is that they do not, with a few exceptions, impose specific obligations on landholders or administrators. They do not provide for a program whereby the objectives of soil conservation will be met. On the contrary, they exist largely in the hope that landholders or administrators or both will act on their own initiative.

The SA Act is something of an exception. This is a legacy of the first comprehensive soil conservation Act in Australia, the *Sand Drift Act* 1923. It was drafted in terms akin to the legislation dealing with certain pests which, in the 1930s, were a species of land management legislation of long standing. For example in South Australia the Thistle and Burr Act was enacted in 1862 and the Destruction of Rabbits Act in 1875. This legacy is most starkly apparent in sec 6j of the Sand Drift Act, which provides that:

a person shall not by cultivation burning off or stock grazing ... create ... conditions as a result of which sand drifts ... and detriment ... to other land ... be caused.

The Act also prohibits certain use of stock routes and clearing until three months after notice has been given.

Other Acts contain certain outright prohibitions on clearing. But these have a limited application, such as, for example, in protected land within NSW catchment areas. The ACT Ordinance prohibits cultivation on certain land having more than a prescribed slope and the Victorian Act prohibits the removal of soil from certain river flats. But generally, obligations imposed directly on landholders by the Acts are rare.

The other obligations which may apply to landholders under the Acts are the result of action taken by landholders or administrators. As noted earlier, some acts rely heavily on voluntary landholder action for the creation of legal obligations. For example, under the Victorian and Queensland group conservation or project area schemes, landholders may initiate soil conservation schemes which, when completed, result in the landholders being obliged to conform to the requirements of the scheme. If, in practice, the initiative for such schemes is taken by the authorities, the landholder's legal power of initiative becomes, in effect, a power of veto. In any event it is hardly surprising that landholders should be reluctant to take voluntary action which results in the imposition of legal obligations particularly when those obligations are linked to the criminal law by being enforced through the criminal process.

Obligations can also be imposed on landholders as a result of action taken by administrators. For example, under several Acts, areas of soil erosion hazard can be created which result in obligations being imposed on landholders in those areas. Most of the Acts provide for the issue of soil conservation or other like orders which, by definition, impose obligations on landholders. The tendency is, however, for administrators to be extremely reluctant to take action which results in the imposition of obligations. Again this is understandable given that the obligations which arise under the Acts are enforced through and operate as part of the criminal law.

The Acts impose very few obligations on administrators. They are drafted in terms of administrative powers, functions and duties, with which some Acts are replete. But, on closer analysis, the Acts tend to provide little or no machinery or accountability requirements to ensure that these capacities are exercised. Many of the so-called duties amount to no such thing. The Queensland Soil Conservation Act provides for example, that one of the 'duties' of the Authority 'shall be':

the prevention or mitigation of soil erosion in all areas of the state.

Such so-called duties are without meaning, given that the Act otherwise specifies that the authority has a limited role which makes the attainment of that duty legally impossible. Indeed, for an extensive period, the Queensland department would appear to have abandoned any real attempt to carry the Act into effect. The Act does not prevent that from occurring.

Another supposed obligation imposed on administrators is to educate landholders and, in the case of the Western Australian Act, the public. Indeed, it is said that the Acts rely heavily on education. But the Acts make no provision to ensure that systematic effective education occurs. If none occurred, no legal consequences would follow.

In general the obligations imposed on administrators are incidental ones, such as apportioning costs for works as between themselves and landholders.

Crisis and crime

It is apparent from the characteristics of the soil conservation Acts that they operate as a species of crisis legislation destined largely to shut the farm gate after the soil has bolted. The Acts are replete with general phrases like 'prevention of soil erosion', which typically have no more meaning than the hollow duty imposed upon the Queensland authority to 'prevent soil erosion throughout the state'. It is the specific administrative and legal structures and procedures which count, and they point to crisis management.

An obvious illustration is the provision made in most Acts for action in areas of soil erosion hazard. Judging by the extent of degradation throughout Australia, very extensive areas should fall into this category. In practice, however, these measures are applicable only to disaster areas. Even then, the constitution of these areas tends to be so hedged about with administrative complexity that in a real emergency the procedure is, to all intents and purposes, useless. Indeed WA has abandoned the concept.

A further illustration of the crisis orientation of the Acts is the concept of the soil conservation order provided by each of them. It has been argued that their infrequent use indicates their effectiveness as a threat. This is fanciful. There is no evidence that, whatever their role in the odd case, they have had any significant overall effect on the reduction of land degradation. These orders await disaster almost by definition. One does not need to know that they are rarely used. It is sufficient to examine their legislative provision. The objectives of the Acts are so vague and the provision of these orders so unsatisfactory in several respects that they would probably be unenforceable in other than the most glaring cases of degradation.

The crisis management approach is also to be seen in other aspects of the Acts. Where the treatment of areas must await landholder initiative, or allows landholder veto, little happens until landholders and their land are in serious trouble. The same tends to be true where the soil conservation authority may take the initiative because the Acts do not ensure that soil capability studies, which are essential to preventive management, will be undertaken. Without these studies, priorities tend to be set by obvious damage. This is not to say that some excellent research work is not being undertaken. Indeed, the scientific knowledge to control degradation is there. It is the administrative and legal structures, procedures and obligations to carry it into effect which are missing.

It cannot be overemphasised that the Acts rely upon the criminal law for their enforcement and in effect operate as part of the criminal law. It is submitted that it is not possible to design a comprehensive, preventive, educational, cooperative system of land management, both from the point of view of the administrators and landholders, which rests on the criminal law. Land degradation is a complex, multi-faceted question. It involves competing interests which include the needs of individuals, the demands of a degree of risk taking and of the protection necessary to ensure future productivity. It is not possible adequately to resolve these matters through the stigma and rigidities of the criminal law and in terms of 'proof beyond reasonable doubt'. Any soil conservation legislation law which relies on the criminal process can operate only as a species of crisis management legislation.

Land tenure systems

The special forms of Australian land tenure

Brief reference has been made previously to the diverse forms of land tenure which emerged in the course of the settlement of Australia during the 19th century. Of major significance was the departure at a fairly early stage in the land settlement process from the English concept of freehold tenure. Through a multiplicity of Acts, new forms of non-freehold tenure such as the fixed-term pastoral lease or licence, the conditional-purchase agreement and the perpetual lease were developed to meet the particular requirements of the colonies.

The common law courts themselves were quick to recognise the distinctive character of these tenures and that traditional common law principles would have to be abandoned or substantially modified with respect to them. For example, in *Cooper v Stuart* (1889) 14 A.C. 286, the Privy Council concluded that the rule against perpetuities could not apply to a reservation included in a Crown grant because its application would detract from the land settlement policies being pursued within the colonies:

> Assuming next (but for the purpose of this argument only) that the rule has, in England, been extended to the Crown, its suitability, when so applied, to the necessities of a young Colony raises a very different question. The object of the Government, in giving off public lands to settlers, is not so much to dispose of the land to pecuniary profit as to attract other colonists.

It is difficult to generalise on the direction of land policy in the crucial period until 1900 during which the distinctive forms of tenure emerged. Thereafter, variations on the basic forms were achieved by way of differing covenants to meet particular circumstances, but no substantially distinct or innovative forms of tenure were developed. Manning Clark has observed that the 'results of the land policy adopted in the period 1788-1850 were a land monopoly and the use of the land almost exclusively for the pastoral industry'. But by 1900, agriculture had developed substantially, particularly as a result of the Selection Acts of the 1860s, which enabled conditional purchase by instalments of land previously held for pastoral purposes. Subsequent drought turned many of these early attempts at agriculture into failure, incidentally contributing in a significant way to the degradation of the land resource, but a more stable industry had been established by 1900.

The tenures themselves reflected through their covenants the underlying philosophy of land development. The prevailing concern certainly was to secure the economic growth of the nation through the development of its land resource. A most interesting view of these tenures is afforded by the Commonwealth Year Book 1910, which presents a comprehensive classification and description of the land tenure systems in each state at that time. The development ethic is evident from the discussion of conditional purchase agreements, where it is noted that:

> Certain conditions, generally as to residence and improvements, have to be complied with before the freehold is granted, but these conditions are usually of a light nature and are inserted chiefly with the object of guaranteeing that the occupier will become of benefit to the community by making a reasonable effort to render his holding wealth-producing.

No distinction is drawn in the Year Book between perpetual and fixed-term leases, which are classified together under the heading of 'leases and licenses'. Its detailed analysis of leasehold tenures on a state-by-state basis reflects that perpetual tenure was at the time a very recent innovation in just three states: South Australia, Queensland and Victoria. In all states, at that time, the occupation of land for pastoral purposes occurred pursuant either to leases or licences for a fixed term.

During the next 50 years, however, much land was alienated for agricultural purposes through tenures other than the fixed-term lease or licence. All states passed 'closer settlement' Acts during the early part of this century authorising the repurchase of alienated lands for the purpose of dividing them up into blocks of 'suitable size' (to quote the Year Book once more). These lands were thrown open to settlement on easy terms and conditions, normally by way of conditional purchase agreement. This resulted in a substantial proportion of agricultural land eventually becoming freehold, once the purchases by instalment had been completed.

In addition, the perpetual lease gained in popularity during the first half of this century. It allowed the retention by the Crown of a limited residual interest in and influence over the land. The soldier settlement schemes following both World Wars are just one example of government schemes in which large tracts of land were granted under perpetual lease. In the end result, there is today as much diversity in the forms of perpetual leases, as a result of variations in the covenants inserted in the leases, as there was by 1900 in the forms of fixed-term lease.

The special forms of Australian land tenure are relevant to the issue of land degradation in two respects: at a basic level, the capacity of particular forms of tenure other than freehold to assist in the achievement of land management objectives; and second, the role of covenants and conditions incorporated in these tenures in promoting or detracting from these objectives. In both respects, the concerns or issues involved are not merely historical, they bear significantly on the approaches which may be pursued contemporarily to the problems of land degradation.

The relevance of tenure form to land degradation

There are suggestions that particular forms of tenure have by themselves accentuated land degradation. In particular, pastoralists have alleged, with support from some economists, that fixed-term leases have detracted from sound land management. The argument is presented in the following terms by Blyth and Kirby (1985, p 109):

> There are grounds to suggest that Australia's system of land tenure, with its emphasis on leasehold and associated covenants, is often inimical to the efficient allocation of society's land resources. The

system creates incentives for land users to follow practices which accelerate the degradation of the land. Uncertainty regarding the renewal of a lease and the lack of full compensation for the land on termination of the lease creates insecurity for the leaseholder. There is reduced incentive to undertake land uses which improve the land or which generate greater returns in the future, particularly beyond the expiry date of the lease. Therefore, in the absence of lease covenants to the contrary, exploitative . . . practices tend to be favoured over more conservation-oriented methods and land is degraded at a faster rate.

There are several questionable generalisations contained within the above passage. Is it the system of land tenure as a whole which is being faulted? Clearly not, for the particular deficiency which is emphasised is the insecurity felt by leaseholders where there is uncertainty about renewal of a lease. Thus, it would seem to be the *fixed-term* lease which is the real subject of the authors' concern. But, given that most agricultural lands are held under perpetual lease or freehold tenure in Australia, why should the fixed-term lease arouse any concern? The answer is that fixed-term tenure still remains common in relation to pastoral land use. Hence, the assertion seems to be, in more specific terms, that pastoralists will desist to a greater extent from exploitative practices if they have more secure tenure. It is in the pastoral land context therefore that the assertion requires further examination.

Pastoral lands

The form of tenure most appropriate to pastoral activity has been the subject of extensive recent inquiry in several states and the Northern Territory. Underlying the inquiries, and the concerted lobbying efforts of the pastoral industry to secure either perpetual leasehold or freehold tenure, is a concern that alternative land uses such as tourism, recreation and conservation, accompanied by the clear thrust of the Aboriginal land rights movement, present a substantial threat to the future of the industry. These efforts resulted in the introduction of legislation in the Northern Territory (the *Crown Lands Amendment Act* 1982 (NT)) to enable perpetual pastoral leases to be granted in place of existing fixed-term pastoral leases. In South Australia, a Bill to amend the Pastoral Act so as to provide for a disguised form of perpetual lease was defeated in the upper house after intensive opposition from interest groups other than the pastoral industry. Moves also are afoot at present within Western Australia to convert pastoral tenures from fixed-term to perpetual leases.

The political motivations which lay behind the pressure for tenure reform in relation to pastoral lands have been disguised to some extent by the argument that fixed-term tenures have in themselves contributed to land degradation. This view is expressed by Blyth and Kirby in the paper previously referred to and draws in turn on observations put forward by others such as Young. It implies that fixed-term leases give rise to a psychological inducement to lessees to exploit their land and that administrative regulation of such leases under land tenure legislation is incapable of overcoming this problem. It is a proposition which requires rebuttal.

There is no clear evidence that land degradation has occurred at a greater rate overall on pastoral lands held under fixed-term lease than on agricultural lands held under the more secure forms of tenure, ie freehold or perpetual leasehold. The report by Woods (Land Degradation in Australia, AGPS, 1983), which summarises the findings of the 1975-1977 collaborative study on soil conservation in Australia, identifies similar percentages of pastoral and agricultural land as requiring treatment for land degradation.

The only evidence put forward in support of the claim relates to a peculiar situation some years ago where certain fixed-term leases in the Western Division of New South Wales were allowed to expire. The lessees were aware some years in advance that the leases would not be renewed. In such circumstance, exploitative practices toward the end of each lease may have occurred, but, in general, the history of fixed-term pastoral leases has been quite different. Such leases have rarely been allowed to run to expiry on the understanding that they will not be renewed. Rather, the practice has been to ensure their renewal in advance of expiry by statutory provisions amending the relevant state land-tenure legislation, and hence the inducement to exploit the land resource has not been present in the same fashion as in the particular situation cited.

It is submitted, therefore, that the arguments against fixed-term tenure for pastoral lands which are founded on the proposition that less degradation will follow the granting of more secure tenure are misplaced. The question of the most appropriate form of land tenure in the arid zone of Australia concerns a more fundamental issue, ie the resolution of increasing competition between alternative land uses. This issue goes to the very heart of future land management approaches within the arid zone. Central to the whole debate on this issue is the proposition that, in certain circumstances, existing pastoral use of the arid zone may be inappropriate from a land management viewpoint and should be halted. The grant of secure, long-term tenures will significantly undermine such a proposition.

To address the question of appropriate arid land tenures, it is desirable to refer back also to earlier land settlement policy in Australia and ask why it was that pastoral activity traditionally was undertaken on short-term leases, while agriculture was accorded permanent tenures in the form of perpetual lease or freehold.

The answer is not related, at least primarily, to land management concerns, even though protective covenants (eg concerning overstocking) have been commonplace in pastoral leases since last century. Rather, it is submitted that the essential policy underlying land settlement and land tenure in the pastoral regions of Australia has been to maintain options for future alternative land uses by deliberately confining pastoral tenures to a temporary or short-term nature. While the reasons underlying such a policy may have changed over many years, the policy itself has been remarkably consistent until recently.

The policy was motivated originally by a concern to legitimise as little as possible the occupancy of the squatters. Subsequently, as vast tracts of land became susceptible to pastoralism, there was a feeling that such an enormous resource should not be turned over by the Crown irreversibly to a small group within the community - an attitude which is still widely held today. And, consistent with the position of the Crown as the ultimate proprietor of these lands, there has been a long-standing public perception that the outback is a communal or public resource. It is common to find a misunderstanding within the community that pastoral leases provide lessees simply with grazing rights in the form of a licence. There is a failure to comprehend that, subject to any reservations contained in the lease (eg in relation to rights of public access), the lessee has exclusive possession of the land for the period of the lease.

Thus, the arguments in favour of permanent tenure in the arid zone which draw substantially on the land degradation aspect challenge a very long-standing settlement policy in Australia. It is conceded that there have been departures from fixed-term tenure in favour of permanent tenure during this century but, until the very recent reactions by governments to pastoral industry pressures, such variations on the basic approach through fixed-term tenure were exceptional and aberrational. The perpetual leases which have been granted in the Western Division of New South Wales (quite extensively) and in Queensland were designed to promote closer settlement of some pastoral land by reference to concepts such as the 'home maintenance area' or the 'living area' (ie areas sufficient to sustain a family in average seasons and conditions). For the remainder, fixed-term leases have remained the norm.

The reality is that once permanent tenure is granted, there is no practical, political possibility of returning to other, less secure forms of tenure. This is evidenced by recent New South Wales deliberations. The Select Committee Inquiry into the Western Division of New South

Wales, in its Second Report published in 1984, concluded that the appropriate tenure for grazing should remain a lease in perpetuity. The only real alternative addressed by the Select Committee was freehold tenure, which had been advanced by lessees as a necessary reform in order for capital borrowings to be obtained from lending institutions. A return to fixed-term leasehold tenure was not seriously canvassed by the Inquiry.

It is submitted that the objectives of sound land management can be pursued through the tenure system without departing from long-standing policy by extending permanent tenure to the holders of the substantial tracts of arid land used for pastoralism. To extend permanent tenure would be to perpetuate pastoralism unnecessarily and to foreclose alternative future land use options, at least in a political if not also a legal sense. Even if governments could afford to resume lands in the future, it would be a much harder political decision to take where permanent tenure exists. Instead, what is required is the inventive adaptation of the fixed-term lease to enable greater security to be enjoyed by lessees while, at the same time, ensuring that more effective land management practices are pursued by lessees. How this may be done will be canvassed in the final section of this paper.

A final suggestion concerning pastoralism and land tenure in the arid zone is that much more serious examination of the cost-effectiveness of the industry as a whole should be undertaken. If an appraisal of the costs to the community of the industry were to be undertaken, looking at such matters as the provision of infrastructure, community facilities, drought relief and the declining condition of the arid land resource, the alleged economic contribution of the industry to the total economy of each state and to the nation as a whole might be significantly reassessed. Instead of pressing towards more permanent tenure, government policy might be more constructive and beneficial to the community as a whole if it were to be directed towards the removal of pastoralism from existing areas where it cannot be substantiated as productive in the economic sense. It would take a courageous state government to assume such a course, but the proposal is as pertinent to the issue of tenure as are the contrary arguments advanced in favour of permanent tenure.

Agricultural lands

Turning to the forms of tenure with respect to agricultural, or non-arid, lands (including many areas identifiable as 'marginal' lands), it is much more difficult to perceive a future practical role for tenure systems in land management. This matter was discussed recently in a report to the South Australian government by an Interdepartmental Committee on 'Freeholding of land used for primary production'. The committee concluded that 'where community controls over land use are required, these should not be exercised through the land tenure system, but by separate legislation administered by land management authorities to ensure uniform application to all lands irrespective of tenure'. It should be emphasised that the committee was concerned only with the agricultural regions of the state, and it acknowledged that its conclusions may not apply or be relevant to the arid areas of the state held under pastoral lease.

The reasons for abandoning the tenure system to achieve land management goals in agricultural lands are related to the anachronistic character of the tenures concerned. Perpetual leases no longer generate sufficient revenue from rentals to meet the costs of their administration. Also, only a limited number of conditions in perpetual leases relate to land management and there were considerable inconsistencies between leases in relation to the covenants imposed. Another factor which influenced the committee was that almost half the agricultural lands in South Australia were already held on freehold tenure and hence were beyond the reach of the tenure system approach to land management. The committee therefore recommended that all lands held under perpetual lease within the settled agricultural areas should be freeholded but, in the marginal areas of the state, freeholding should be deferred until the state's existing land management legislation was more effectively strengthened. It also recommended that stronger, more specific land management legislation should apply to all lands, irrespective of tenure, and 'should be enacted with all possible haste'.

These recommendations were endorsed by the state government shortly after taking office in 1983. The subsequent disclosure of the freeholding policy has caused some alarm, particularly within the conservation movement, which has urged comprehensive new land management legislation (in particular, the complete revision of the Soil Conservation Act) before the policy is implemented. The government itself has asserted that it considers adequate controls exist already with respect to land use and management practices in the high rainfall agricultural areas, regardless of tenure, and appears committed to the pursuit of its freeholding policy.

Thus, the link between the form of tenure and land management seems likely to be severed completely for the agricultural regions of South Australia. The freeholding move is a fundamental one, for it marks a substantial departure (if implemented) from another long-standing land-settlement policy of granting perpetual leasehold in agricultural areas. Unlike the tenure changes being promoted for pastoral lands, however, the arguments do appear to be more convincing. The distinction between perpetual leasehold and freehold tenure is in many respects conceptual, as Else-Mitchell asserted in his

Land Tenures Inquiry Report in 1976. Furthermore, where land is divided almost equally between perpetual leasehold and freehold tenure, a uniform or concerted approach through the tenure system would seem impossible.

However, a contrary approach may not have been utterly beyond reach, had there been an adequate opportunity afforded for further debate on the matter. In his Land Tenures Inquiry, Else-Mitchell concluded that it would be possible by a system of incentives to encourage the voluntary conversion of existing non-residential freehold estates to Crown leasehold (even of a fixed-term character) by offering taxation concessions such as exemptions from land tax and the amortisation for income tax purposes of existing and future improvements on Crown leaseholds. This radical revision of the tenure system would require Commonwealth action in terms of the amortisation proposal (Else-Mitchell himself was concerned with the most appropriate forms of land tenure within Commonwealth territories), and hence such a proposal is unlikely to be feasible within the states.

It seems probable that future legislative action in relation to land degradation in South Australia's agricultural lands will focus upon the subject of land management legislation rather than land tenure reform. It is clear that a number of administrative authorities and enactments will be involved. In other states also, the clear signal is that new legislative arrangements are required to deal with land degradation. The catch-phrase suddenly is 'land management'. In this respect, the South Australian experience in recent times is illustrative of more general trends and concerns.

So, while the land tenure system could (and should) retain a significant function in future land management of arid lands, it is likely new approaches will be developed for agricultural lands independently of the tenure system.

Public lands

A brief additional comment will be made on the matter of land management for public lands. The main focus of this paper has been activities on private land, but it must be remembered that the Crown and other government authorities in each state also constitute substantial landholders. Historically, public lands have been the responsibility of land agencies in connection with the administration of the tenure system, although specialist jurisdictions have emerged in relation to forests and national parks.

The need for a fresh management approach to public lands appears to have received recognition recently in Western Australia. *The*

Conservation and Land Management Act 1984 (WA) provides for the use, protection and management of public lands, waters etc in the form of state forests, timber reserves, national parks, nature reserves, marine parks, marine nature reserves and any other land reserved under the *Land Act* 1933 (WA). A Department of Conservation and Land Management is established by the Act. The Act also establishes a Lands and Forest Commission, a National Parks and Nature Conservation Authority and a Forest Production Council to perform various specified functions. Part V, headed 'Management of Land', provides for the preparation of management plans for any lands to which the Act applies.

A most interesting aspect of this legislation, and the attendant administrative authorities which it establishes, is that there is an attempt to bring together under the one umbrella various and divergent interests involved with public lands management. Similar objectives appear to have motivated the Victorian government to reorganise existing administrative structures in order to establish in 1983 the Ministry for Lands, Conservation and Forests.

It must be questioned, however, whether the management of public and private lands should be encouraged to proceed through entirely separate administrative channels. A different view is held, for example, in South Australia, where proposals for revision of its Pastoral Act envisage that both lands under pastoral leasehold tenure and vacant Crown lands in the outback of the state will be managed by a single authority, the Outback Land Management Authority. The most appropriate administrative arrangements for management of both public and private lands will have to be tailored therefore to suit the existing circumstances within each state.

Covenants and conditions as a land degradation constraint

No thorough study has ever been undertaken of the 'incidents' of Australian land tenures, that is, those obligations imposed on landholders by way of covenants or conditions inserted in the various forms of tenure. Perhaps the closest attempt is the unpublished doctoral dissertation by Fry, 'Freehold and Leasehold Tenancies of Queensland', written in the 1940s. Nevertheless, general impressions are not difficult to obtain from just a brief study of South Australian legislation.

The standard form perpetual lease (*Crown Lands Act* 1929 (SA), Third Schedule) and conditional purchase agreement (Fifth Schedule) each contain covenants requiring the landholder to 'clear so as to render available for cultivation or so as to improve the grazing capacity thereof' specified areas of land. In neither instance, however, is it envisaged that the whole of the land granted will be required to be cleared; rather, a portion is prescribed. Other clauses in both types of tenure are directed to requiring the landholder to destroy vermin and to keep the land free of noxious weeds; to fence the boundaries of the land; and to keep and maintain all improvements in good repair. Pastoral leases granted under the *Pastoral Act* 1936 include covenants to stock with sheep or cattle at specified rates; to destroy vermin; not to overstock; and not to cut any timber without the licence of the Minister of Lands (First Schedule).

Thus, the covenants generally require development of land for agricultural or grazing purposes, but at the same time provide fairly elementary obligations in relation to such matters as the control of vermin and weeds. In the case of pastoral leases, the concern with stocking rates predates the 1936 legislation and reflects an awareness which had emerged by the end of the 19th century that overstocking constituted a major problem with respect to pastoral lands and required restraint through the tenure system. It is interesting to note that the 1936 Act required a covenant that 'the lessee will not at any time during the term of the lease' overstock the land. Earlier legislation, dating back at least to the *Crown Lands Act Amendment Act* 1890 (SA), which made special provision for pastoral leases, had required a more limited covenant to the effect that 'the lessee will not at any time during the last three years of the term of the lease' overstock the land. The 1936 legislation therefore evidences a growing concern with the problem of overstocking and a desire to employ the tenure system to deal with that problem.

It was at about the same time that several other initiatives occurred through land tenure legislation in South Australia, having as their objective the prevention of unsound land use practices on agricultural lands. Obviously, the concern which was emerging during this period about soil erosion problems influenced these initiatives, and it is particularly interesting to note how a dual approach, involving both new regulatory measures concerning soil conservation and the adoption of existing tenure arrangements, was pursued simultaneously. To some extent, the approaches were interdependent.

In 1939, new provisions were inserted in the *Crown Lands Act* 1929 concerning soil conservation and overstocking. Practically identical provisions in relation to destocking were inserted into the *Pastoral Act* 1936 (SA) in the same year: see *Pastoral Act,* s.44a. In the following year, the *Marginal Lands Act* 1940 (SA) was passed. This particular legislation resulted in government control of farm size in the marginal cereal growing areas of South Australia.

A recent review of the marginal lands legislation by an investigatory group appointed by the state government concluded that, while control of farm size 'has been for some regions an important catalyst in a post-Second World War improvement in the economic and physical condition of the marginal lands', nevertheless general land management measures should replace such controls in the future:

> ... the Investigatory Group believes that as a land management tool, control of farm size is complex and difficult to administer, and lacks flexibility in its approach. It is also doubtful, given prevailing trends in farm size, whether it has any real relevance in the foreseeable future.

In 1967, provisions were inserted in the Crown Lands Act to enable 'special development perpetual leases' to be granted over lands the stability or productivity of which could be affected by unrestricted use (ss.66c-66h). The amendments enable 'excluded areas' to be delineated within such leases, ie areas which cannot be cleared, cultivated or used for grazing. Such areas may be increased or decreased at any time after the grant of the lease. In implementing these provisions, consultation is required with the authorities administering the *Soil Conservation Act 1939.*

Much more recently, when introducing its controversial vegetation clearance controls, the state government amended the *Crown Lands Act 1929* by providing that no covenant in any lease or agreement under the Act which required the clearance of land would be enforceable or binding henceforth. Thus, the covenants concerning clearance referred to earlier that are contained in perpetual lease and conditional purchase agreements no longer have any legal force or effect.

This selective review of South Australian tenures suggests that, theoretically, the land tenure system could have played a significant role in the securing of land management objectives. However, historically, the practical result has been quite the opposite, in that the obligations imposed by way of covenants and conditions have been geared primarily to the economic development of land and have shown only an incidental concern for the protection of the land resource. Apart from rabbits and weeds, South Australian land tenures display little interest in issues of land management. The one important exception is the matter of stocking levels, in particular on pastoral or grazing lands, where there is a long history of some effort being made through covenants to secure a form of control. However, administrative supervision by the Pastoral Board of these covenants was, for many years in South Australia, practically non-existent, and the sanction of forfeiture has virtually never been employed. Stocking limits also may have been too blunt an instrument for land management in the sense of encouraging a belief within lessees that the prescribed limits represented an acceptable level in all circumstances. Nevertheless, a revised covenant system could still be used in pastoral areas to deal with the problem of land degradation.

It must be questioned whether the land tenure system can serve an effective role in the future in relation to land management in the agricultural lands. Previous efforts, with a few exceptions, have largely failed. Alternative land management tools may be preferable to the use of the tenure system to achieve relevant goals. These questions will be considered in the final section of the paper.

Towards effective land management

Our analysis of the legal and administrative instruments concerned with problems of land degradation has been focused on the areas of soil conservation legislation and land tenure because so far these have been the major avenues of approach. Clearly, the renewed concern about these problems is causing interested parties to review the adequacy of these existing approaches. In addition, however, consideration is being given to new measures, both regulatory and administrative, outside the traditional areas—for example, the adoption of controls over vegetation clearance in South Australia and the introduction of land management legislation directed specifically at public lands in Western Australia.

As has been noted, much of the current reaction to land degradation problems has invoked the concept of 'land management', and it seems likely that this term will become extremely commonplace in this context in the future. There would seem to be two broad objectives underlying the concept of land management:

- the retention of maximum flexibility with respect to future land use options in order to accommodate changing conditions from time to time; and
- the maintenance of the land resource in a condition which will ensure its capacity to meet the needs of future generations.

These objectives emerge clearly, for example, from the National Conservation Strategy, which now has been endorsed by the Commonwealth and all states apart from Queensland and Tasmania. Obviously, the manner in which such broad objectives are pursued involves a variety of approaches and we cannot propose an all-embracing strategy. What we will do in this remaining section is suggest improvements in those areas of regulatory activity which we have examined. In order to do so, it is appropriate first to consider the Australian constitutional framework, including the powers which the Commonwealth government may exercise to deal with land degradation.

Constitutional powers

Guarantees of property rights

In discussing constitutional powers, primary emphasis is usually placed on the Commonwealth Constitution which distributes power within the federation. But the state constitutions should not be overlooked. The latter may, for present purposes, be seen as entirely flexible in that they do not provide any constitutional guarantees with respect to property or other rights. That is, putting aside the effect of the Commonwealth constitution, state parliaments may from time to time make any laws whatsoever concerning lands within the respective states. The legislation prevails over the common law, over any other state legislation and typically even over constitutional provisions with which they are inconsistent. Any state law or constitutional provision purporting to guarantee property rights (eg certain compulsory acquisition terms) may, therefore, be freely amended or repealed.

The Commonwealth constitution wears a different aspect. It is inflexible in that any constitutional guarantees which regulate or limit the exercise of Commonwealth legislative power must be adhered to unless the proper procedure to change the constitution is followed. The only constitutional guarantee relevant for present purposes is the provision requiring that if the Commonwealth acquires property from a state or any person the Commonwealth must pay just terms. Putting aside constitutional limitations, however, the Commonwealth has, within its powers, the same legislative capacity as the states, including the inability to limit its future law-making power. It is also the case, of course, that valid Commonwealth legislation prevails over the common law and also over any state legislation with which it is inconsistent.

Powers of Commonwealth

The Commonwealth should not be seen as having only a federal capacity. In the Commonwealth territories (including the NT until 1978) and within the states, the Commonwealth has a unitary legislative capacity, in that the Commonwealth alone can legislate for its own activities, those of its Crown instrumentalities and any activities within Commonwealth places. This capacity embraces a great deal of land use.

In this unitary capacity, particularly within the states, the record of the Commonwealth is not an impressive one. The Commonwealth would not appear to have any legislation, nor indeed any comprehensive policies, dealing with the question of degradation resulting from such land use. The matter requires attention.

The Commonwealth is primarily seen in terms of its federal capacity; it is often taken that the states have the constitutional power to deal with land use and that if the Commonwealth uses its powers in such a way as to bear on this question it is in some sense subverting the Constitution. This embodies three fundamental misconceptions.

The first is that the Commonwealth Constitution confers power on the states or, in terms, divides power between the Commonwealth and the states. The fact is that it confers power only on the Commonwealth. There is not one word conferring power on the states in relation to land use. This is not a semantic quibble but a matter of real constitutional substance. The states have literally what is left after the full ambit of Commonwealth power has been determined. One cannot ask what constitutional power the states have except by asking what power the Commonwealth has. In short, the Constitution does not confer power on the states with respect to land use. Their laws dealing with the matter will be valid only if there are no Commonwealth laws doing so. The Commonwealth's failure to act in this matter of national concern reflects not so much an absence of constitutional power but a lack of political will.

The second misconception is that the powers given to the Commonwealth are given for particular purposes. This is not so. They are typically given in general terms. For example, the Commonwealth is given power to tax. If it chooses to levy taxes in such a way as to break up large estates, as it did in its first decade, or to ensure soil conservation, that is entirely constitutional. The Constitution does not say the Commonwealth has power to tax but not so as to influence land use. To the contrary, taxation and expenditure have long been used as policy instruments, as the constitutional founders well knew. Where limits on the taxing power are intended, they are clear. The Commonwealth may not, for example, tax state property or impose taxes which discriminate between or destroy states. But any attempt to insist upon further constitutional limitations reflects a refusal to recognise that the Australian founders were aware that they were creating a nation.

The third misconception arises from the failure to appreciate the compelling width of general powers conferred expressly. The Commonwealth is given, for example, power to legislate on trading corporations. There is no such thing as a law which in some pure sense is just about trading corporations. The words used empower the enactment of laws about the activities of trading corporations; and not just about their trading activities. The same section of the Constitution also gives power with respect to foreign corporations. It cannot seriously be suggested that the Commonwealth has power only with respect to the foreign activities of foreign corporations. A corporation which uses land to grow and sell, that is trade in, primary produce is a corporation engaged in trading and hence a trading corporation within the scope of the Commonwealth power. The Commonwealth can, therefore, legislate to regulate the activities of that corporation. This includes by stipulating measures which the corporation must take to prevent land degradation.

This misconception is found in Appendix E of the recent Report on Land Use Policy in Australia 1984 in that it draws a distinction between direct and indirect Commonwealth federal, as distinct from unitary, powers. If the Commonwealth has power with respect to corporations but none with respect to parks, a law which prohibits corporations from mining in parks is not somehow indirect legislation about parks. It is simply one of many laws which could, quite constitutionally, be made about corporations. The Appendix only compounds confusion by classifying the external affairs power as a direct power and all other relevant federal powers (it omits the corporations power) as indirect.

There is a number of other powers which the Commonwealth could use to deal with land degradation. These include the power to legislate with respect to trade and commerce. The Commonwealth could, for example, regulate the export of produce according to environmental considerations. Under other powers it could provide that certain financial institutions should tailor their lending to comply with appropriate land use policies. To the extent to which land degradation is a matter of international concern, the historical and contemporary signficance of which should not be understated, the Commonwealth could legislate under the external affairs power.

The exercise of Commonwealth power

Historically, it is untenable to argue that it is somehow unconstitutional to use Commonwealth powers to influence land use throughout the nation. The fact is that the Commonwealth has promoted, subsidised and guided the development of land and influenced agricultural policies almost since its inception. For example, through its involvement in the soldier settlement and other schemes, and through its taxation policies, it has been a major promoter of the clearance of land. It also provides drought relief, which may have land degradation implications.

There is no logic in the argument that the Commonwealth may use its powers to promote the exploitation of land but not to ensure its protection. Indeed it has been recognised at the highest level for over four decades that land degradation is a matter of national concern.

Despite the views expressed above, it is not suggested at this time that the Commonwealth should legislate to deal with degradation provided that satisfactory steps can be taken within the existing framework. The role of the Commonwealth should be to act in collaboration with the states to identify and monitor the nature and extent of the problem of degradation and formulate land use policies. In particular it should identify the extent to which the problem gives rise to national concern and seek to ensure that appropriate minimum national standards of prevention and control are met.

It may be that the Commonwealth can meet its responsibilities through the provision of funds to the states. It is clear that very considerable funds will be required. The funds should, of course, be provided for specified purposes and within the framework of an appropriate program. In this regard the Commonwealth should take into account the adequacy of state institutions and legislation.

There is a long history of the Commonwealth providing funds to the states, through Section 96 grants, in the national interest. In the present context this began with the collaborative study in the mid 1970s and should be continued. Several purposes to which funds may be directed are mentioned below.

Different considerations arise in the case of the clearing of land whether for agriculture, woodchipping or other purposes. Bearing in mind that clearing is the precursor of degradation and that the history of the Commonwealth as a promoter of land clearance may require to be redressed, the Commonwealth should exercise its constitutional powers as far as possible to ensure that no further clearing takes place without full and adequate consideration of the question of degradation.

Soil conservation legislation

For the present, at least, it may be assumed that enactment of substantive legislation to ensure the prevention and control of land degradation will be left to the states. It is quite clear that some form of legislation is required. Incentives and assistance can be justified although there is no clear evidence that they deal significantly with the problem of degradation. The issues are too long term, too complex and cross too many property boundaries for adequate resolution through landholder self-interest. Throughout the world the facts are overwhelmingly clear. The unregulated, self-interested use of land results in degradation.

It is also clear from the level of degradation in Australia and from the history of the operation of the soil conservation Acts, which are the primary legislative means through which to deal with degradation, that they cannot cope with the problem in their present form. This is not to denigrate the achievements of the soil conservation authorities. The performance of some individuals has been remarkable while some authorites have a reasonably proud record. All may be cautious but the legislation clearly allows, even demands, such an approach.

A continuation of the present legislative policy based largely on voluntary action by landholders, extension and education, with land use orders a remote possibility, is not the answer. The evidence and surveys show that the present approaches have had limited impact.

There is no doubt that education and changed attitudes towards land use are crucial, but they are not being adequately promoted by such things as demonstrations. It is necessary to adopt the principles of adult education, whereby landholders learn through accepting and undertaking the responsibility of developing and implementing management policies for their properties which are consistent with soil conservation.

In short it is time to stop running away from the need to spell out land use obligations. The cracked record of the story of soil conservation in Australia is that obligations will be necessary. 'But', it is always said, 'not yet'. That response has been repeated for 40 years. A 1983 Queensland report, while acknowledging that Queensland soil erosion 'approaches disaster status', repeats it yet again. When is the time ripe? Is it when degradation is not merely disastrous but cataclysmic? Is it when we run out of words or when prevention is no longer possible or when the costs of restoration become prohibitive? What may have been reasonable caution is becoming irresponsible procrastination. Land use gives rise to obligations. A program must be devised to ensure that those obligations are met.

Much of the difficulty arises from an unthinking response to any suggestion that obligations should be provided for and from rigid, if not fearful, attitudes invoked by use of the word 'regulation'. There tend to be assumptions that obligations involve the exercise of police powers, and that soil conservation of necessity means imposed management and reduced production. What is required is a preparedness to reflect positively on the nature and extent of the degradation problem and on the means by which it might be dealt with. For example, as suggested elsewhere in this paper, civil rather than criminal law should be used. Further, obligations may include flexible, negotiated self-regulation which maximises individual initiative and innovation.

It is appropriate to mention at this point that soil conservation Acts should also deal with the obligations of administrators. The legislation should embody targets and a structured program to ensure that, as far as possible, those targets are met. The legislation should also provide for the review of the whole system to ensure that it does not simply slip into a groove and limp along.

Alternative approaches

There are broadly three approaches which legislation could take to ensure that obligations are met. One is the strict liability approach whereby any damage to land or any damage beyond a prescribed level is rendered unlawful per se. This appears to be the direction in which United States law is heading. Soil loss volumes are set. If they are exceeded, the landholder is liable to prosecution. This is similar to the approach taken by Australia's first soil conservation legislation and, as has been the case for over a century, it is also the typical basis for legislation such as that dealing with the control of plant and vertebrate pests. Such laws are, of course, a species of land management legislation.

This approach is, therefore, not out of the question. But, while imposing obligations on landholders, it is not recommended. It cannot adequately take into account the individual circumstances of a particular landholder nor the complex, long-term problems involved in land degradation. It does not place a positive emphasis on land use management nor provide a suitable framework within which to develop an approach to management based on education, negotiation and cooperation. Its reliance on the criminal law is bound to ensure its inefficacy. Furthermore it suffers from failing to impose appropriate obligations on administrators.

A second approach would be to enact legislation which establishes a regime like that which traditionally applied under the common law in England. It would ensure that law akin to the law of waste applied to all onsite damage and that law akin to the law of nuisance applied to all offsite damage. The state or the owner of leased property would have a right of action in the former case while the state or the landholder of affected land or offsite interest would have a right of action in the latter case.

This approach would require legislative definition of the principles and objectives by reference to which good husbandry and reasonable use could be assessed. There is a long and respectable legal history which could aid the development of such legislation. Furthermore it would not rest on the criminal law. Nevertheless this approach would suffer from some of the deficiencies of strict liability law and is not recommended, at least standing alone. However, it has clear merit not least because it would establish the fundamental legal and social proposition that land should be used with reasonable care. Serious consideration should be given to the use of this approach as an adjunct to the third approach, to which we now turn.

This embodies many of the legal techniques found in the present legislation, but it requires that these be provided for in a different way. These techniques comprise a four-stage process which can be summarised as follows. The first stage involves land utilisation determinations which broadly decide the use to which land should be put. The second involves land use planning which identifies areas having common land characteristics or problems and which establishes guidelines as the basis for the third stage, which is land use management. This involves the application of soil conservation principles within a given area to particular land. The fourth and final stage is enforcement. Some further, brief explanation of how these techniques may be embodied in the reform of existing soil conservation Acts follows.

Suggested reforms

It goes without saying that the effectiveness of these techniques turns on an adequate definition of the objectives of the Acts. This cannot be achieved by extensive listing of so-called powers and functions of administrators which characterise some of the present Acts. Many of these are, as indicated above, largely devoid of legal meaning or effect. It requires a careful definition of key phrases, such as soil conservation or land degradation, around which the Acts are drafted and upon which the techniques dealt with below rely for their substantive effect.

Existing soil conservation Acts, generally speaking, do not, and should not, provide for land utilisation determinations. In the case of land used for agriculture, for example, the Acts simply accept a prior 'decision', often in fact the result of ill-considered policies or short-term market forces, that land is appropriately being so used. It may be that the Acts would result in different forms of agriculture being adopted or even, in rare cases, of land being withdrawn altogether. Also, given the link between clearance and degradation, the Acts should provide that no land may be cleared for agriculture or any other purpose until the question of degradation has been adequately considered. (Soil Conservation Acts cannot, of course, embrace the other issues involved in land clearance such as protection of rare fauna etc.) In general, however, soil conservation Acts should accept land utilisation determinations and within that use seek to ensure that degradation does not occur.

The present soil conservation Acts all embody in various ways each of the other three stages. Land use planning is involved, for example, in the creation of soil conservation districts, areas of erosion hazard and project areas. Land use management of particular properties is also provided for in association with some forms of planning such as project areas, while soil conservation orders are, in effect, land use management orders. Criminal enforcement looms as a significant feature of the Acts. Even sanctions such as soil conservation orders rely ultimately for their effect on the criminal law.

An analysis in some detail is required to demonstrate that, with exceptions, the present provision of these techniques is unsatisfactory. But for present purposes several illustrative points may be made. First, where these techniques are dealt with together, the matrix is far too complex. For example, some group or project area schemes embrace land use planning and land use management and in some respects link that with enforcement in one cumbersome process. Second, where land use planning awaits landholder initiative, action tends to be ad hoc, without adequate scientific basis and insufficiently preventive. Third, where land use management is provided for either as a form of enforcement, or without adequate land use planning, it will not result in comprehensive improvement in land use management. This problem is compounded by the use of the criminal process. Fourth, the Acts do not have a clear conception of land use management and place, or allow administrators to place far too much emphasis on works. Fifth, each stage is largely devoid of obligations except, for example, in the case of land use management orders which, as they are presently provided for, verge on the unusable.

The third approach, again stated very briefly, is as follows. Appropriate land use planning must be based upon scientific land-capability studies. The aim should be to identify districts or areas and their needs, essentially on an ecological basis. Bearing in mind economic or other imperatives, priorities should then be established for the sequential application of appropriate land use management practices. The legislation should impose clear obligations on administrators to see that such studies and land use planning occurs within a specified time frame. Commonwealth funding will be essential to undertake the necessary research and to establish and maintain a suitable database.

In accordance with the priorities established and within the land use planning guidelines, landholders should be responsible for the initiation of land use management proposals for their properties which accord with the objectives of the Acts. The assistance of the Soil Conservation Authorities and, of course, private enterprise should be available. It is important that within this framework the maximum possible initiative, consistent with the needs of soil conservation, be left with landholders. Land use management, either on an individual property or group basis, should be carried out as far as possible in a spirit of negotiation, cooperation and education. Again, Commonwealth funding should be available to assist landholders in planning and management of their properties. Soil conservation and productivity are not necessarily incompatible and such assistance should, of course, include management for the latter consistent with the requirements of the former.

Enforcement should be separated as far as possible from land use planning and land use management, and should play a less significant formal role in the structure of the Acts than at present. In any event it cannot be overemphasised that soil conservation legislation should be decriminalised. Where enforcement action is required a civil process such as the use of injunctions is clearly preferable. In general, enforcement should be directed not at the landholder but at the needs of the land.

This whole structure and procedure should not be open ended but should be given a timetable and goals. It should also incorporate a required review at the conclusion of a period or periods. If adequate progress is to be made it is essential to establish a dynamic regime with specific means directed to clear objectives.

Tenure systems

The pursuit of land management objectives through the tenure system will require significant alteration of existing tenure arrangements. It would seem futile to seek to invoke the tenure system where lands have been acquired under permanent tenure such as the perpetual lease, for reasons which we have outlined earlier. Hence, the role of the tenure system in land management is confined essentially to the arid zone, where pastoral leases for fixed terms remain common in many parts of Australia. The real challenge in this context is to establish land management techniques through the arid land tenure system which can achieve the same ends as the reforms that we have advocated with respect to soil conservation legislation for agricultural lands.

A fundamental premise on which we would advocate the retention of the tenure system in the arid zone to achieve land management objectives is that, by continuing fixed-term tenure in preference to perpetual lease or freehold tenure, maximum flexibility is ensured for future land use options. At the same time, decisions can be taken more readily to ensure the maintenance of only long-term, sustainable uses, by the resumption of leases in appropriate cases. These objectives are much more difficult to attain through separate land management legislation where permanent tenure is enjoyed by landholders.

The need for some greater degree of tenure security on the part of lessees could be accommodated under the fixed-term tenure system by providing a right to seek renewal of the lease at a fairly early stage during its term. Proposals in South Australia to amend the existing form of pastoral lease will afford to the lessee a right to apply for renewal of the 42-year term every 14 years. Hence, the shortest period which will remain at any time, assuming renewal occurs, is 28 years. Such an approach would achieve much greater stability and certainty for lessees than the previously widespread practice throughout the states of intensive lobbying by pastoralists to secure statutory renewals of leases before their expiry.

Related to the proposed process of regular renewal of the full term of the lease in South Australia are more specific land management measures. At each renewal, it is intended that covenants will be reviewed and renegotiated. Hence, among other matters, stocking limits can be considered more regularly and routinely through the tenure system. However, a 14-year span may still be too long to respond to immediate land-condition concerns, and other management techniques are also most necessary.

One such technique which has attracted discussion and some support in South Australia is the management plan. There is some resistance to the idea that such plans should be imposed, and it would seem more appropriate to empower the relevant authority to cooperate with lessees who should have the primary obligation for the production and implementation of management plans. In this respect, covenant review may provide a most useful back-up, since authorities could negotiate for the inclusion of a covenant to require the preparation of such a plan, or subsequently to implement aspects of a plan, or to revise the proposals contained in an original plan.

The preparation of the plan should have, as its aim, the development of specific management principles and policies for the leasehold land. To faciliate this task, it is essential that authorities themselves engage in broader surveys on a regional basis and regular monitoring and assessment of the condition of the land. These are tasks which can and should be required to be performed by the relevant administering authority through the terms of the legislation. To date, such tasks have been limited by inadequate funding and personnel, but the statement of a statutory obligation to undertake them may serve to increase the commitment of resources by governments to their performance.

Another aspect of the tenure system which requires enlightened reform is the matter of penalties. The sanction of forfeiture has been largely ineffective since few authorities have had the will to impose it. More imaginative sanctions for breach of lease covenants are desirable, including orders to reduce stock numbers, to desist from grazing on certain areas, or continuing fines for non-compliance (as are provided for in some land use planning Acts in relation to breach of development conditions).

Finally, one further important issue which can be regulated through the tenure system, although attempts to do so to date have been most inadequate, is the exercise of rights of public access to the arid zone and the complex related issue of the liability of lessees as occupiers. Once again, there are proposals in hand in South Australia to deal with this matter through amendment of the Pastoral Act. The proposals themselves bear some reasonable similarity to those put forward recently in New South Wales by the Select Committee which inquired into the Western Division.

Overall, the various reforms suggested herein to pastoral land tenures embody the same broad techniques and approaches to land management as were proposed earlier for soil conservation legislation. The only difference is the vehicle through which they are to be most effectively implemented.

Conclusions

We have made numerous criticisms of existing legal and institutional constraints on land degradation and suggested some fresh directions for the future. At the broadest level, there is a need to rethink traditional legal approaches to the problem of land degradation. We would urge a departure from the long-established model of poorly defined regulatory obligations sanctioned by criminal penalties in favour of a more precise scheme of civil regulation. Such a scheme should be based on clear statements of statutory objectives; the prescription of land management guidelines on a regional basis; cooperative land-management techniques; and civil remedies or restraints where landholders fail to cooperate or to meet basic land care standards established by legislation.

It has also been suggested that the management of arid lands can be undertaken in a similar fashion through revision of the existing tenure system, so as to provide more explicitly for land management objectives. However, current moves to switch from the predominantly fixed-term form of leasehold tenure in the arid lands to more permanent forms of tenure, such as perpetual leasehold or freehold, should be resisted.

We have also sought to challenge conventional assumptions concerning the role of the Commonwealth and its constitutional powers with respect to land management, while recognising the practical necessity for the Commonwealth to have an appropriate political will to act. An important collaborative role exists for the Commonwealth in this area, through assistance (particularly of a financial nature) in the identification, monitoring and treatment of land degradation problems and in the formulation of appropriate land use policies at the national level.

Clearly, problems of the magnitude that confront the Australian community with respect to land degradation require fresh outlooks and approaches. It is possible, through the particular perspective of the legal system, to contribute significantly to such developments.

Endnote

In its soil conservation aspects, this paper builds upon an earlier paper entitled, 'Soil Conservation: Legislative Measures', presented by Bradsen to the Third Environmental Law Symposium held in Adelaide in August 1984. This paper also reflects ideas and analyses which are developed more fully in a report on Australian soil conservation legislation being prepared by Bradsen as part of the National Soil Conservation Program. That report, which is nearing completion, will contain the references to Parliamentary debates and other historical material which is absent from the present paper.

Commentary
Michael Barker

Preliminary Comments

I should first like to commend the authors on their most thorough chapter dealing especially with soil conservation laws and the Australian land tenure system, fields in which contemporary Australian lawyers do not often tread.

I immediately wish to highlight the context in which the chapter appears to have been written. First, it assumes the desirability of government intervention in those areas of activity leading to land degradation. Kirby and Blyth (Chapter 11) suggest that this assumption should not unquestioningly be made. Secondly, the chapter perceives a need for a rigorous legal response to land degradation, especially through the creation of land use obligations. Chisholm (Chapter 12) highlights the availability of economic techniques of land use control rather than traditional legal ones of command-and-control.

The chapter therefore provides a distinctly legal view of the lie of the land. No apology need be made for this (nor the pun). However, we must ask ourselves whether the traditional legal way is the only way, or the best way to deal with land degradation.

My comments will largely remain within these disciplinary boundaries, assuming that law will continue to play an important, even if different, role in responding to land degradation.

Introduction

Law, as a discipline, is often depicted as cautious, conservative, and generally an obstacle to progressive forces in society; in short, a guardian of established traditions and interests. A case can be made out to support this contention. Law is, however, capable of being the vehicle of societal change; it is capable of taking on a dynamic, progressive form. The chapter by Bradsen and Fowler clearly and unashamedly adopts the dynamic, progressive view of law. In their paper they argue that, ultimately, law created by the people, in Parliament, can introduce a new land ethic and, by implication, require its acceptance. They also contend that the deliberate choice of one form of land tenure, a legal concept, over another has significant ramifications for the acceptance of a new land ethic. In my comments I will briefly observe that, in one way, too much has been made of too little in the prolonged Australian debate on land tenure. The tenure debate arises again in Chapter 8 (Young). I will then spend a little time to consider the need for lawyers, and others concerned with policy reponses to the phenomenon of land degradation, to take cognisance of the historical development of laws which may generally be characterised as pertaining to environmental resources. Implicitly the authors adopt something of the approach I would advocate; I believe, however, that they should be explicit in their acceptance or rejection of the analysis I shall put forward.

Land Tenure

The debate concerning the value of freehold or leasehold land tenures has bedevilled Australian land policy since the European settlement of the continent in 1788. It is not new, although some of the reasons for the debate may be, or at least may be reasons not so strongly emphasised in years gone by.
Consider this assessment of the debate:

Save in the Northern Territory, where conditions are exceptional, the difference between the two tenures has narrowed until it has become infinitesimal. Australian tenures have all accepted the idea of *limited freehold*, that is, a title dependent on the performances of conditions such as residence or improvements, these conditions applying to the transferee as to the original holder.
'It was a very short step', said a Labor Premier, 'from the perpetual lease with its nominal rental which the Labor Party called a leasehold to the fee-simple which the National Party called a freehold'. The difference was largely one of nomenclature; the imposition of improvements had bridged the gap; and the point was whether purchase money should be paid at once or spread over a number of years. Provided that the conditions suffice, aggregation is not more possible under the one than the other, while the large resumption powers of Australian Governments still further narrows the gap. The issue ....has assumed a changed form.

Those words were written by the historian Sir Stephen Roberts in 1924 (Roberts 1924). I would suggest that over 60 years later in 1985, the issue has assumed not only a changed form but lingers only for reasons which are broadly conservationist in nature.

As Roberts' study shows and also that of others (see eg Powell 1970) the political conflict over land policy historically pitted those who would aspire to accumulate large landholdings (the squatters) against those who would like to see the land resource kept in the public domain to satisfy, over time, the land needs and aspirations of the 'ordinary' citizen. Before the gold rushes, governors like Gipps strove to protect the interests of the Crown over these vast tracts of land; whereas after the gold rushes the broad egalitarian leanings of the populace (affected by the teachings of the Chartists and the later land tenure reform movement in England and North America represented by the likes of Henry George) suggested that all members of the community should have an equal opportunity to own land. Consistent with this was the view that the government should not finally alienate land to private owners, as in a freehold grant.

As Roberts suggests, over time the differences between the leasehold system and the freehold system were barely discernible in practice, and the point has been made over and again in more recent times. Bradsen and Fowler acknowledge as much by conceding that in agricultural, non-arid zones, (but not the arid zones) perhaps freeholding of the lands presently subject to forms of leasehold tenure, could proceed. I agree, for two reasons.

First, I believe that the political concern can no longer be substantiated that modern squatters will monopolise the land resource in these areas. Accordingly, leasehold tenure offers no real advantages over freehold. Indeed, freehold tenure may, if anything, promote a greater respect for the land by the land user. Second, it has been demonstrated, on the balance of probabilities if not by a higher burden of proof, that in practice the leasehold system has not been effective in controlling the incidence of land degradation in this country. While there may be institutional explanations for weak administration in the past, I would suggest that the imposition through legislative means, of any form of land use control to ameliorate land degradation in the future, is likely to be more politically acceptable, where the *primary* resource ownership issue has already been resolved in favour of the land user. Additionally, I believe a system of statutory land use control, which is open to public scrutiny and influence, is administratively preferable to that of the more subtle, more private, leasehold land use control system. Finally, on a more complex philosophical level, such an approach would largely place urban, and non-urban lands on exactly the same footing, that is, subject to direct statutory controls without discrimination. Urban lands, of course, have long been subject to direct regulation by land use planning controls.

For historical reasons, and by reason of their more ecologically fragile condition I would support the distinction made by Bradsen and Fowler that lands in the arid regions should remain in the public domain. In relation to such lands, a debate will ensue on whether a short-term lease of land is a satisfactory form of tenure. In this respect the carefully considered views of the paper writers warrant close examination.

It is appropriate to mention, before leaving the tenure discussion, the nature of land use controls over Aboriginal land in Australia. In recent years, large tracts of land in the Northern Territory and South Australia have been conveyed to their traditional Aboriginal owners. In the future, Aboriginal ownership of significant areas of land in other states may arise through the operation of either state or Commonwealth land rights legislation. A detailed analysis of current land rights laws (see Barker 1984) discloses that, by and large, Aboriginal land remains subject to relevant state/territory and Commonwealth laws of general application. Of course, in some remote areas relevant planning or environmental law may not presently have application for administrative reasons. Planning and environmental agencies simply may not have extended their controls into such regions.

Although there can be little doubt that such laws of general application are capable of applying on Aboriginal land, there can be no doubt that direct attempts to control land uses over, say, traditional Aboriginal land, have the potential to conflict rather dramatically with the general principle of Aboriginal self-determination — or autonomy — which underlies much recent land rights legislation. The actual outcome of this conflict will be due less to legal theory than to the prevailing political climate. I dare say that where Aboriginal land users are employing satisfactory land use practices, the powers of government will not be used to coerce them to do otherwise. Such a pattern of control is, of course, rather familiar — it reflects the basic relationship between farmers and government today. I believe, however, that there will be a considerable degree of reluctance to intervene directly in land use activity on Aboriginal land, especially on traditional lands in remote areas, at least in the short term. During this period, Aboriginal autonomy will be considered the first priority.

The evolution of environmental resource legislation

I now wish to turn to a more speculative topic. I consider that in 1985, with nearly two decades of experience with legislation and techniques designed to protect the environment behind us, we must make some attempt to analyse the evolution of environmental control generally before we design special laws and administrative structures to deal with land degradation. I make no bones about it, I treat land degradation as an item of environmental distruption (broad though its nature is) just as I do pollution in its various forms, poor land use planning, heritage destruction, and so on.

I would analyse the evolution of environmental resource legislation in the following way. First, historically, a period of virtually uncontrolled physical development of a resource to supply a perceived demand. Second, a period when the imposition of legislative control over exploitation of the resource is considered necessary to ensure its orderly, economic development, with some basic conservation requirements inserted in the legislation consistent with the sustainable yield philosophy inherent in such controls. Third, the later enactment of a specific law, in response to political demand during a period of raised environmental consciousness, designed to alleviate or reduce the disruption being suffered by the resource. This consciousness may arise from either a 'shallow' or a 'deep' ecological perspective. Finally, the integration of various resource specific laws (both those that control resources in economic terms and those that protect environmental quality) in recognition of the complex ecological relationship between all resources and their usage.

I am not prepared to say how far the integration suggested in the final step may go. Conceivably one could devise a supra-environmental resource law, or a supra-environmental resource agency to administer existing laws. At a lower level, and certainly a more realistic response in our present state of development, would be the creation of a consultative process to bring together the various parties engaged in resource planning, allocation and protection. Institutional concerns are highlighted by Junor and Watkins (Chapter 14).

To relate these evolutionary steps to the land degradation issue before us, I would suggest the present state of play falls somewhere between the second and third steps. In the 1930s we appeared to take the second step with the enactment of soil conservation legislation. I do not believe the legislation then passed went so far as actually to *attempt* to alleviate disruption of the land resource so much as to faciliate the more sustainable exploitation of soil. I am not convinced we have yet taken the third step. To me the Bradsen/Fowler proposals for reform of soil conservation law largely fall into the third step, though acknowledging the inevitability of the fourth step. It is my contention that the activities leading to the causes of land degradation have been largely ignored by environmental control until recently and that a major task for policy makers in this area today is to combine the third and fourth steps in the design of new approaches to the problem.

Proposed approach

1   The technique of resource planning should be extended into all areas of land use. At present, it is mainly urban land which is affected by the planning technique. There is no reason why agricultural and pastoral land should not also be subject to formal land use planning. Consider, for example, the potential of the *Environmental Planning and Assessment Act*, 1979 (NSW): section 5(a)(i) provides the first object of the Act:

> to encourage the proper management, development and conservation of natural and man-made resources, including agricultural land, natural areas, forests, minerals, water, cities, towns and villages for the purpose of promoting the social and economic welfare of the community and a better environment.

> In this way the initial use or non-use issue may be determined in a more holistic way than at present, especially if the issue is not the sole responsibility of a single-purpose agency but one upon which other resource planning and using agencies are able to pass an opinion. Ultimately, perhaps, land use questions could be taken by a department of environmental resources.

2   In conjunction with the planning exercise, or rather to come into play once planning is complete, is the need for a workable management system combining various control techniques designed to produce appropriate environmental use of land.

> These would include economic incentives to conserve land (see Chisholm, Chapter 12), direct regulation to restrain proscribed unacceptable land management techniques and some basic legal land use standard to reflect the desire of the community that land users should not abuse the land resource.

It is in relation to the latter possibility that the paper of Bradsen and Fowler makes some particularly interesting suggestions. Shying away from criminal, or quasi-criminal sanctions, it is suggested that there are three possible ways of achieving the desired result: strict liability for any onsite damage to land, or damage beyond a prescribed level; a nuisance-type law for offsite disruption; and the better provision of present discretionary abatement powers in soil conservation services. Chisholm (Chapter 12) suggests economic techniques which may better achieve the objectives of these controls. If it is desired to have an explicit behavioural standard spelt out in legislation (and there are those in the community who would believe legislation should communicate accepted community morality to land users), then may I suggest a possible fourth technique.

Using pollution control legislation as an analogy, and others which I shall shortly mention, it could be provided by legislation that agricultural, pastoral, and forestry land users employ the best available practicable means to prevent or minimise land degradation. To avoid definitional arguments about when 'land degradation' has occurred, it may be preferable to require 'good farming practices', 'good forestry practices', and so on, to be employed. This may be an expression lacking in precision but, nevertheless, one capable of interpretation by a court in the light of evidence presented by experts to show what is, and what is not, acceptable behaviour in that industry. Examples, of sorts, are now to be found on the Statute Books. The *Forestry Act*, 1916 (NSW), s.8A(1)(c) requires the Forestry Commission, inter alia, to preserve and improve, 'in accordance with good forestry practice', the soil resources and water-catchment capabilities of Crown timber lands. (This might be considered an enforceable obligation: see *Evans and Spicer v Forestry Commission of NSW* (1982) 48 Local Government Reports of Australia 266). In a similar vein, the *Petroleum (Submerged Lands) Act* 1967, s.97(1) requires petroleum exploration activities and petroleum operations to be carried out 'in a proper and workmanlike manner and in accordance with good oil-field practice'.

An objection to such an approach may be that 'good practices' are already causing disruption. My response to that is that new practices evolve in the light of environmental disruption, and courts are capable of developing a standard in the light of changing data and evidence. I share the authors' desire, however, to develop a standard which does not depend on administrative definition or explication, but rather depends for its efficacy on the (on the whole) impartial and objective construction of the standard by the judiciary, that branch of government which is well-used to performing such a task in both the common law and statutory arena.

Concluding comments

Space prevents me from touching on other points of legal analysis raised in the most useful paper of Bradsen and Fowler. Some of these matters are of great importance, especially the potential and desirability of Commonwealth intervention in the issue of land degradation, and will undoubtedly be raised in later sessions. I might just note in closing, however, that to the extent that the chapter creates an impression that it would be an easy matter for the Commonwealth to legislate with respect to land degradation, it does a little overstate the case. While the corporations power in particular might enable significant control over land use, it certainly would not support comprehensive regulation. To say that does not, however, deny in any way the broad principles and analysis set out in that part of the chapter.

8 Land tenure: plaything of governments or an effective instrument?

Michael Young

Introduction

The first land use policies to be used by Europeans in Australia were based on the tenure system. Settlement began with leases, licences and the issue of grants of land to people as rewards for achieving certain goals. The feature which differentiates land tenure from most other systems which seek to influence land use is that it mixes incentives, rewards and penalties all into one. Moreover, the system is specific in the way it issues and withholds property rights. Most other systems work solely by withholding property rights. But, as most rights to use land have already been issued in the form of a freehold title to a landholder, the option to return to a land tenure system to control land use throughout Australia's productive lands has largely been forgone. No government is likely to exercise its powers of eminent domain and legislate to acquire and/or resume all these freehold titles and reintroduce a tenure-based land use control system. Throughout most of the less productive rangelands of Australia, however, this opportunity remains.

Recent developments

Earlier this century a cynical NSW politician bemoaned the fact that the Western Lands Act, which establishes the tenure system for the control of grazing in western New South Wales, had become 'the plaything of governments' (King 1957 p163). Since 1979 there have been inquiries into pastoral land management in Western Australia ('Jennings Report', Western Australia 1979); pastoral land tenure in the Northern Territory ('Martin Report', Northern Territory 1980); pastoral land administration management and tenure in South Australia ('Vickery Report', South Australia 1981) and the Western Division of New South Wales (four reports New South Wales 1983, 1984a,b,c). In addition, there has been a series of reviews of pastoral land administration, management and

---

Parts of this chapter are drawn from Young (1985a).

tenure in South Australia which have produced more reports (South Australia 1984a,b).

The complexity of the issues which surround this topic is demonstrated by the simple observation that each of the above inquiries has produced a dissenting report. It will be interesting to see whether or not the latest inquiry, that into pastoral land tenure in Western Australia, will maintain this track record. Earlier this year the Western Australian government, however, was able to commission and receive a report on the Kimberley region which was unanimous; it recommends, among other things, that pastoralists be given more secure tenure (Western Australia 1985). The Queensland government has also reviewed its grazing tenure system in the last five years but has chosen not to release the resultant reports for public examination.

Land use objectives

Historically, government requirements and expectations from pastoral land have changed. For example, in South Australia it was originally hoped that these lands would be farmed (Meinig 1962) while in more recent times some people have expressed concern that these same lands may not be able to sustain pastoral production without degradation (South Australia 1981). The rhetoric which is flowing from the state inquiries and also from documents including the recent National Conservation Strategy (Anon 1983) suggests that Australians want their arid pastoral lands or rangelands to be efficiently used for pastoral production without land degradation and managed in a way which retains all future land use options. It also demonstrates that at least in Parliament's view there are big differences between leasehold and freehold land.

With the exception of some parts of Queensland most of Australia's rangelands are still held under leasehold tenures. These objectives can be achieved by attaching covenants and conditions to leases (Young 1979a). For example, leases in South Australia contain covenants which require lessees to 'not . . . overstock the land . . . which in the opinion of the Minister or the Pastoral Board would have the effect of depreciating the ordinary capacity of the land for depasturing stock' (*Pastoral Act* 1936-1974, Schedule A). But concern for the declining condition of South Australia's pastoral lands is only a fairly recent phenomenon. It was not until 1939 following the receipt of the Ratcliffe Report, and also a report from a committee appointed to inquire into Soil Conservation, that the government passed amendments which enabled the minister to penalise a lessee who overstocked his lease. This was followed in 1960 by amendments which introduced covenants prescribing the maximum number of livestock that could be carried on a lease (Young 1981). The stocking rate maxima were set by adding roughly 25 per cent to recorded average stocking rates over the past 20 years. This policy, in the eyes of lessees, appeared to transfer the onus for deciding the maximum livestock which could be carried on a lease from the lessee to the Pastoral Board. Most landholders responded by increasing stocking rates to the set maxima and subsequently many made requests to the Pastoral Board to exceed the stated maximum number of livestock they could carry on their lease. The recent South Australian inquiry (South Australia 1981) found that over the 15-year period between 1964 and 1979, 33 per cent of lessees on average carried more than their maxima (with the board's consent). Their data imply that the attachment of livestock maxima to South Australian pastoral leases may have increased stocking intensity by as much as 10 per cent, the opposite of the effect which was sought.

Closer settlement objectives are still written into all state Land Acts, but are now pursued only by Queensland and New South Wales where constraints on size, in the form of restrictions on the aggregation of pastoral lands into larger holdings, still apply (Young 1979b). Past closer settlement has led to about one-half of Queensland's sheep properties being less than one living area and over one-half of New South Wales' sheep properties being less than one home maintenance area in size (Young 1980). Stocking intensity on these small substandard holdings is generally higher and, as a consequence, the vegetation on them is generally in poorer condition than those which by good fortune were not reduced to a 1940-1950 living area (Young 1985b). Despite good intentions these governments may have caused more degradation than they have prevented. If left to free market forces, New South Wales' and Queensland's pastoral land, their pastoralists and their consolidated revenue may all be in better condition. The same most probably applies in all other states, but this does not mean that with appropriate modification there are not social gains from the introduction of a pastoral land tenure system which recognises the interdependence of the economic, social, political and ecological system managed by pastoralists. A necessary precondition to the design of an effective system is careful attention to the differing impacts of rewards and penalties on land use.

Rewards and penalties

In the rangelands, ideas about adopting resource pricing policies which seek to achieve optimal land use are usually quickly discounted because, given current technology, the cost of obtaining the information necessary to apply them is too expensive. It can be argued that drought policy provides the exception; however, such policies usually have the perverse effects of rewarding those managers who do not plan for droughts and discouraging investment in drought avoidance strategies (Robinson 1982, Freebairn 1983).

Generally, governments are too inflexible to set rangeland resource pricing policies which match the diversity of environmental conditions on each property. Consequently, they are restricted to setting policies which recognise two or perhaps three physical states of which each is defined by a set of performance standards. For example, if the goal is to maintain land condition then a reward for reclaiming degraded land or maintaining land in good condition (state one) is offered and a penalty is set for degrading the land (state two). Between these two states there is a fuzzy transition area where encouragement and cautioning statements are used to attempt to persuade people to move in the desired direction. Regulatory actions are used to discourage people from managing for state two and, ideally, incentives offered to encourage the pursuit of the first state. Thus a regulatory action is likely to be most effective if it can quickly penalise the breach and then offer a reward which will encourage the offender to change his management strategy to one which takes the land in the preferred direction. On their own, regulations are frequently counterproductive because people tend to avoid them if they can, while incentives which make the adoption of a recommended practice synonymous with self-interest are likely to be much more effective (Young 1984). An interesting recent development in this area is the conscious use of peer group pressures through regional advisory groups etc to increase the stigma of changing land into the (socially undesirable) degraded state. This is a particularly powerful strategy in small rangeland communities where all people are readily identifiable.

The alternative, of course, is to offer such strong financial and social incentives for people to maintain land perpetually in good condition so that penalties are never necessary. Any examination of the popular literature would suggest that, at least for our arid pastoral lands, the public does not believe that the economy provides such an environment for pastoralists. These beliefs arise from the public's observation that pastoralists, like most other people, discount the long-term consequences of their actions and appear to exploit their land to the fullest in good periods and push it to its limit in droughts as they strive to conserve their livestock in the knowledge that when it breaks they will be worth many times their within-drought value. With the tenure system, however, if an attractive reward is offered to maintain land in good condition then movement towards this preferred state, rather than some other state, is more likely. If no rewards are given and a penalty is imposed for not maintaining land in good condition then it is likely that either a loophole will be found to avoid the penalty or, alternatively, the measure used to determine whether or not the undesired state has been reached will be found, in retrospect, to be inappropriate. The South Australian experience with livestock maxima is an example of this latter consequence.

Security of tenure

Within the tenure system there are two penalty-reward system options; those which influence viability and those which influence social status. The first set of options plays on the immediate cash flow received by a pastoralist through the threat of forfeiture, issue of fines, rent concessions, subsidies etc. Generally forfeiture is politically impossible and, as the cause of overgrazing is often the existence of a financial problem, fines are inappropriate. Similarly, rent concessions soon become capitalised into lease values and lose their effect or become political playthings. Moreover subsidies, because of their agricultural industry rather than pastoral land orientation, are usually counterproductive. The alternative approach is to mix a series of incentive and regulatory instruments through the tenure system in a manner which plays on pastoralists' desires for security of tenure and self-esteem. Elements of this second approach exist in many tenure systems but, as yet, they have never been combined into an effective package.

All Acts guarantee pastoralists who hold only term leases the right to apply for a new lease before it expires. History has demonstrated that whenever a substantial number of lessees are within 15 years of lease termination, political pressures emerge and after a period of inquiry most, if not all, leases are reissued for a new term. For example, in South Australia a Royal Commission in 1927 led to an amendment to permit all leases to be reissued in 1929. Similarly, in 1960, when 54 per cent of leases expired between 1971 and 1975, legislation was passed to enable all leases to be surrendered and immediately reissued for a further 42 years. By default and as a consequence of the increasing concern of pastoralists as the remaining term of their lease shortens, most term lease covenants tend to be reviewed simultaneously approximately every 20 to 30 years when a majority of pastoralists become concerned about the security of their investment.

Western Australia is the only state which has formalised this review process. There all leases expire on the same day in 2015 but 20 years before that day, a ministerial review is required and the Minister must report to all lessees before 1998 (Young 1979b).

Term leases guarantee use rights for a specific period of time and enable lease boundaries to be altered or whole leases allocated to another person when the lease expires. A key difference between the term and a perpetual lease is the recognition which a perpetual lease gives to the grazing value of the leased land. On expiry or resumption of a term lease, a term lessee is paid compensation only for improvements placed upon the land and receives the same amount of money irrespective of the way he has cared for, improved or abused the vegetation found on the leased land. In contrast, a perpetual lessee who leaves vegetation in good condition is conceptually entitled to receive more compensation than one who leaves it in poor condition. Moreover, on resumption, a perpetual lessee who improves the condition of vegetation will be paid financial compensation for his efforts while a term lessee is entitled to compensation for only that proportion of the value of the improved condition which could be realised through increased carrying capacity before the end of the term of the lease. Although, conceptually, there is a difference in the amount of compensation due to a perpetual lessee who maintains a lease in good condition, in practice because of the discount rates used by valuers, the real monetary difference on resumption is marginal. Nevertheless, it has an important psychological effect. Under the perpetual system lessees are told that there is an incentive for good management while under the term lease system they are told that there is no reward.

The next aspect of the tenure system to examine is the method by which term leases are usually reissued. If any uncertainty surrounds the reissue of a term lease, lessees can be expected to adopt short-term exploitive management strategies which they themselves would regard as wasteful if their tenure was more certain (Batie 1982; Ciriacy-Wantrup 1968). Conversely, lessees who have security of tenure are given an incentive to avoid land degradation and to focus more of their attention on lease covenants. This will be particularly true if the administrative authority removes this security until the covenant breach is rectified. Benign tolerance of covenant breaches and the existence of inappropriate or obsolete covenants, however, preclude such administrative action. Effective regulation depends on credibility, which flows from the maintenance of consistent regulatory activities and the association of breaches with appropriate action.

Conversion to permanent tenure does not preclude any future land use options providing lease covenants can be periodically changed. Land can still be acquired for other uses and, as a decision to resume and then reallocate a lease to an alternative use has to be made at least 15 to 20 years before expiry, the financial difference between compensation payable under both forms of tenure is usually negligible. Furthermore, when a change in use or allocation is optimal, the administrative and political costs associated with continually having to defend the decision until lease expiry, to prevent degradation until that date, to deal with a lessee's desires to sell etc, usually make it expedient to acquire the lease immediately and not wait until expiry (Young 1983).

The efficiency of lease covenants as a land use control instrument

Covenants are similar in many ways to planning regulations enforced by local government authorities: they are a set of performance standards for land use which lessees are required to meet. They also apply to each and every land title to which they are attached, but it could be reasoned by holders of freehold land that special exceptions should be made for the planning regulations which apply to their land. But when there are many titles within an area, the administrative cost of attaching covenants to each title may outweigh the benefits of doing so. In arid areas, however, properties are large and environmental conditions vary substantially between them. Consequently there is a need to set land use controls which vary within leases, perhaps on a paddock basis, rather than introduce controls which vary only between leases. Thus, it is possible to argue that control of arid land use may be more cost-effectively implemented through lease covenants rather than planning regulations because they make statements which are specific to the rangeland in question.

The simple attachment of covenants to leases does not necessarily mean, however, that they will be effective. All recent state inquiries have found that very few lessees know which covenants are attached to their leases. Moreover, as many land titles tend to be held in bank vaults and solicitors' offices, a substantial proportion of lessees have never seen their lease instrument and a few do not know when their term lease expires. Some incorrectly believe that their perpetual lease runs for only 99 years (for a discussion of this see New South Wales 1983). Consequently, if covenants are to affect land use then administrators must periodically draw the standards they require to the attention of lessees. The annual mailing of rent notices provides one opportunity to do this but there are many other avenues.

Consequence of not reviewing lease covenants

If covenants are intended to ensure that land use standards are met then a necessary precondition is that provision be made for periodical review and adjustment. Unless review enables obsolete or inappropriate covenants to be removed and new ones to be introduced simultaneously across entire regions the tenure system will always appear to be inconsistent and administratively ineffective. Eventually the system becomes irrelevant to contemporary circumstances and the system loses credibility (Young 1984).

Within Australia's rangelands perpetual leases are found in New South Wales, Queensland and South Australia. On the southern fringe of the last state's pastoral lands, there is a narrow strip of perpetual leases which were issued to pastoralists late last century and lie outside any local government area. Pragmatically, and in recognition of the now inappropriate nature of their old covenants, no attempt is made to regulate land use on these perpetual leases as none can be applied in a consistent, legally enforceable manner through the tenure system (South Australia 1981). South Australian land administrators continually express concern about this situation but, as yet, no politically acceptable solution has been found.

Perpetual leases in Queensland have been issued only since 1968. In this state, however, little attempt is being made to control land use by attaching covenants to leases. Consequently the problems caused by the existence of inconsistent covenants between adjoining leases have yet to emerge.

Although most perpetual leases in New South Wales were established between 1934 and 1952, a considerable proportion was established by simply extending term leases issued between 1902 and 1934 to leases in perpetuity with no change in covenants. But successive administrations have required that different covenants be attached to new leases. Properties which originally consisted of a single lease now contain several with varying covenants and the occurrence of conflicting lease covenants continues to increase as properties amalgamate. Administrators now consciously avoid discussing the specific performance standards expressed in covenants and required from lessees. As new covenants cannot be added to perpetual leases under present legislation, methods of overriding lease covenants are being developed. For example, in 1979 a new section was added to the *Western Lands Act* 1901 which enabled recent cropping development to be controlled. This new section (s.18DA) introduces a new licence system with cultivation restrictions which are superior to all lease covenants and in may cases contradict them. The alternative approach is to amend legislation to provide for the periodic review and amendment of perpetual lease covenants. This was recommended in New South Wales (1984) but met strong opposition from graziers who feared it would be used to reduce their rights. In South Australia, however, pastoralists are advocating the adoption of such a tenure system. In recognition of the importance of maintaining security of tenure, any covenant changes which reduce the productive potential of the leased rangeland must also be accompanied by the payment of full compensation to the lessee for his lost rights.

Periodic covenant reviews should be conducted on a regional basis and staggered so that only one region is under review at any time. Land use planning and assessment concepts suggest that any covenant changes should be consistent with the regional land use strategy for the area and made at the same time as those plans are reviewed (Young 1984). The rotation of reviews around regions will reduce any sense of insecurity induced by the process by keeping pastoralists accustomed to the process. It will also increase administrative efficiency as staff become accustomed to conducting them and procedures are formalised.

Consideration can be given to the conduct and frequency of covenant reviews only after the nature of covenants and the need for review has been identified. The adaptive management principles espoused by ecologists such as Holling (1978) suggest that the review process should be continuous. This, however, creates politically unacceptable accusations of inequitable, inconsistent treatment, and feelings of insecurity which in turn stimulate the adoption of short-term exploitative management practices. Traditionally, urban land use planners advocate a review every five years, but while the issues to be considered at each review remain relatively simple and as reviews create insecurity, a longer period may be more appropriate for rangelands. As indicated above, political circumstances usually precipitate a review every 20 years and thus a ten-year to 15-year covenant review period would seem appropriate.

Enforcing compliance with lease conditions

If the prevention of land degradation is a prime objective of covenants then they should explicitly state this and set clear performance standards. A consequence of this necessary precondition to effective land use control is the development of a monitoring system which is capable of determining whether or not the performance standards are being met at minimal cost.

In the past the sole penalty for breaching a condition of lease or covenant — forfeiture — has been rarely used because it is politically and administratively almost impossible to implement. It involves not only the seizure of a lessee's livestock, but also eviction from the family home and acceptance by the bureaucracy of responsibility for continuing management of the lease. It is tantamount to requiring a judge to hang all offenders irrespective of their offence. Threat of forfeiture has been used, however, with varying effects. It creates a hostile environment where the onus is on the lessee to adjust his management to a different position or to prove that no wrong has been done. There are no rewards, nor clear indications of what is required. Moreover, the threatened penalty may seem to be out of all proportion to the severity of the breach and does not remove the cause of the breach.

When covenants are breached, penalties in proportion to the extent of the breach are much more likely to be used by administrators. In addition, to limit insecurity induced by the prosecutions, an equitable appeal system is necessary. As an incentive and reward for good land management, lessees should be guaranteed the right to use an area of land continuously, providing they comply with lease covenants and conditions. Conversely, as an initial, but not necessarily final penalty, those who breach conditions should have the term of their lease reduced to a shorter period of, say, 19 or 20 years and have their management automatically placed under review or probation. When financial pressures from creditors have caused the breach, this 'lease reversion', by threatening the security of creditors, may remove some of that pressure while the more usual penalty — a fine — will only increase it. Furthermore, lease reversion reverses the negotiating framework and provides administrators with the opportunity to reward improved management with the reinstatement of the originally held guarantee of continuous use rights. In practice, lessees would learn that as soon as they appeared to have recognised their error and to have taken the necessary steps to have it rectified, their original lease would be reinstated. Statutory lease reversion is desirable, because in the past, ministers of the crown who have been responsible for penalising covenant breaches have, for political reasons, been reluctant to prosecute breaches. Normally, under the above framework reversion back to a lease with a continuing tenure would occur within one year.

The desirable characteristics of a pastoral lease

Together, the above observations suggest that if leasehold tenure is to remain an effective land use control system throughout arid Australia then provision must be made to ensure that:
- the covenants and conditions attached to leases are periodically reviewed;
- all lease covenants and conditions are simultaneously reviewed on a regional basis;
- the covenants attached to leases within each region are reviewed approximately every ten to 15 years in a manner which does not threaten pastoral security;
- full compensation is paid to the lessee whenever the government

wishes to either (a) reallocate land to another use or person or (b) decrease the permitted intensity of use;
- covenants objectively state that which is required of a lessee;
- administrative actions are taken whenever objectively collected data indicate that a substantial covenant breach has occurred;
- appeals to administrative actions in response to covenant breaches may be made to a court of law to determine whether, beyond reasonable doubt, a breach has occurred;
- compliance with lease conditions guarantees continuous use rights to the lessee, his heirs and successors;
- by statute, non-compliance with lease conditions removes the guarantee of continuous use rights until the situation is rectified.

None of the 27 forms of lease in arid Australia meets more than two of the above criteria. Until new forms of lease which meet these criteria are introduced it is likely that administration within Australia's rangelands will remain ineffective. In recognition of the importance of guaranteeing continuous use rights to pastoralists who comply with lease conditions, it is suggested that leases which incorporate the above characteristics should be called continuous leases. Those who meet lease conditions are guaranteed the right to continue to do this. Lands which cannot be managed within this framework are probably incapable of sustained pastoral land use and, consequently, should be immediately resumed. In every state there is now sufficient information to identify all such lands and this process is an obvious precursor to the implementation of the above system.

Administrative cost

To advocate the adoption of a new tenure system which calls for a new administrative bureaucracy with strong indirect market intervention requires consideration of both ethical and efficiency considerations. In the long-term, effective land use control requires administrators to be aware of the consequences of their recommendations. In an uncertain rangeland environment information and knowledge is uncertain and, hence, there is a danger that administrators will simply use their powers as a shield for poor understanding. This has happened in the past and is likely to happen in the future.

The cost of establishing the above administrative framework may also be prohibitive. For example, the recent Western Australian inquiry into pastoral land management observed that its government was spending in the vicinity of $7000 per annum (per lease) on pastoral administration and extension in return for $1000 per annum in rent. In most states there is now in excess of one bureaucrat trying to prevent rangeland degradation for every ten pastoralists. Such close scrutiny is unheard of in other industries. If all these administrators were employed as extension officers without regulatory power Australia's rangelands may be in better condition. Perhaps if we cannot trust pastoralists, we should have no pastoralists at all.

Social bases of farmers' responses to land degradation

9

Roy Rickson, Paul Saffigna, Frank Vanclay and Grant McTainsh

Introduction

Numerous studies of Australian agriculture have found land degradation (water and wind erosion, salinisation, waterlogging) to be a serious threat to the sustainability of its soils (Chartres Chapter 1, Gasteen et al 1985). There are several different dimensions to the study of land degradation: an understanding of how erosion physically occurs is clearly necessary, but there are technology and management practices which, if used by farmers, would greatly reduce the amount of agricultural land now inadequately protected against erosion.

We argue that studies of the physical and biological nature of land degradation must be complemented by socioeconomic analyses. These studies would include how farmers generally relate to their land, how farmers perceive and define erosion, and what prompts them to invest time, energy and money into learning about erosion and on-farm management practices which may help towards its abatement.

Farmers' responses to soil erosion and other forms of land degradation have several components. Agricultural technologies and practices based on European cultural traditions and experiences have often proved incompatible with Australia's climate and soil; Burch, Graetz and Noble (Chapter 2) state that it has taken Australian farmers over a century to modify the agricultural traditions of Europe to suit Australian conditions.

Modern production processes in agriculture have increased food production and farm earnings, but have caused onsite land degradation and offsite pollution. Douglass (1984) and Schnaiberg (1980) argue that petroleum and capital-intensive agricultural methods of production have stretched natural systems beyond their capacity to sustain production and absorb waste. Farmers, faced with cost-price squeezes, find themselves in a 'treadmill of production' (Schnaiberg 1980)

* The research upon which this chapter is based has been supported by the National Soil Conservation Program and the Australian Research Grants Committee.

requiring intensive efforts to improve crop yields, income and, thereby, personal security, with neither time nor energy left over for conservation. Rapidly increasing crop yields in the United States have been accompanied by certain negative outcomes including widespread onsite erosion and sedimentation of drainage ditches and natural water ways (Forster and Abrahim 1985) and also offsite damages from erosion-salting of public water supplies and deposition of pollutants in streams, lakes, reservoirs and estuaries (Clark, E H 1985). For instance, Heady (1975) contends that farmers' commitment to intensified row-crop farming has been a major cause of externalities such as erosion, salinity and water pollution.

Cultural traditions and production systems affect farmers' perceptions of their land and their decisions on its use. These perceptions and decisions are part of the relationship between farmers and their land, and should be studied if effective land management programs are to be developed. Caldwell's (1970) conception of environmental management is an important point of departure for our approach to the study of socio-economic factors in land degradation. Environmental management is defined as the administration or control of the relationships which people have with the biotic or abiotic environment.

Our analysis concentrates on the following:

- Soil erosion as a perceptual stimulus to farmers; how easily can farmers recognise erosion on their farms? Are farmers reluctant to admit to erosion when they do recognise it?
- Farmers' estimates of the consequences of erosion for sustained crop yields on their own properties; do farmers see a relationship between soil erosion and crop yield?
- Farmers' perceptions of soil erosion as a major problem on their own properties compared with erosion elsewhere: in this research, Queensland's Darling Downs and neighbouring farms. Are farmers more likely to consider erosion a problem for other farmers, but not for themselves?

Soil erosion as a perceptual stimulus

Soil erosion, is commonly perceived (eg in the Australian television documentary 'Heartlands') in rather dramatic terms. Farmers, soil scientists and politicians, when publicly discussing erosion or promoting action programs, usually refer to highly visible forms of degradation such as deep gullies, extreme land slips and large dust clouds of wind-eroded country soil blowing into urban areas. In fact, the principal types of erosion (rill or 'gully' and sheet erosion) which farmers commonly experience are, in their earliest stages, not easy for them, or anyone else, to see.

Flooding is a significant cause of erosion and its effects on farmland are usually quite clear. Farmers' experience of severe flooding can sometimes cause them to take the fatalistic attitude 'that nothing can be done about it, so why try' (Williams, M 1979, Chamala et al 1985, Chamala and Rickson 1985). This attitude predisposes farmers to put soil conservation as a low priority in their future planning. Farmers perhaps fall back on fatalism to contend with the psychological stress caused by personal encounters with severe instances of land degradation (cf Barr 1985, Cunningham and Jenkins 1982).

Research findings show that land degradation, in its most subtle and devastating forms, provokes a variety of responses from farmers, ranging from fatalism, indifference and inertia to active interest and firm commitments to preserving the soil. Rill and sheet erosion develop gradually over time and are difficult to identify in the early stages. Farmer resistance to erosion control, therefore, lies partly in the physical nature of incipient erosion and partly in the reluctance they may have to admit to that there is erosion, or other forms of land degradation, on their farms (Barr 1985, Chamala et al 1985, Nowak 1983).

Farmers' perceptions of erosion

Farmers' experience and perceptions of erosion are similar to their experience with salinity and drought. Their reactions to salinity, studied by Barr (1985) illustrate the point. He describes drought and soil salinity as *pervasive* hazards in comparison with *intensive* hazards such as earthquakes or damaging storms. According to Barr (1985 p20):

Pervasive hazards are often more slowly recognised than intensive hazards. Salinity is a particularly pervasive hazard. In particular, it is characterised by its extremely long duration, very slow speed of onset, and diffuse spatial dispersion. The spread of salinity through an area might even be called insidious. Often the growth of the problem is imperceptible unless viewed through the perspective of several years.

The insidious nature of pervasive hazards, as defined by Barr (1985), is a significant reason why farmers usually misperceive either the existence or extent of erosion on their properties. Miscalculations of erosion range from denying its existence to dismissing reports of erosion and its effects as overstated. For example, Chamala et al (1982), in their study of the Darling Downs found that farmers underrated the prevalence of erosion and that most felt the reported levels of erosion were greatly exaggerated.

North American studies by Hoover and Wiitala (1980), Nowak (1982), and Heffernan and Green (1981) found that farmers misperceive and underestimate erosion conditions. Rickson and Stabler (1985) report data showing that farmers greatly undervalue objective evidence that farm soil erosion is a leading cause of public water pollution. Cary (1982) concludes that over a long-term farmer experience with pervasive hazards such as soil salinity, erosion and drought can lead to psychological adjustment to the condition and to underestimating its seriousness for them, on their own land.

A significant part of farmer responses to erosion is based on their assessment of short-term personal and economic costs against future benefits from investment in soil conservation. It is most probable that farmers use conservation methods when conservation is compatible with their personal economic goals of maintaining or increasing crop yields and income (Chamala et al 1982, Pampel and van Es 1977). Australian farmers respond most favourably to new technology when it is easy to use, inexpensive and promises short-term profits (Donald 1970). Soil conservation, in contrast, requires that farmers set long-term planning horizons. Douglass (1984 p25), after an examination of several recent studies, concludes that '...farmers' plannning horizons would need to be fifty years or more and interest rates for discounting future benefits would have to be very low, or even zero, for them to voluntarily adopt techniques which would indefinitely maintain the productivity of their land'.

Productivity can be maintained, at least in the short-term, by application of fertiliser. However, there are limits to the degree that fertiliser can compensate for eroded topsoil and most farmers' short-term planning horizons will not include that measure. The effect, from Crosson's (1981 p16) study, '...is to weaken the long-term advantage of conservation tillage relative to conventional tillage'.

The problem is further complicated because effective use of many conservation measures requires intense 'site-specific' analyses of their applicability for a particular farm, and their use usually requires farmers to change their land management practices. Conservation tillage and strip cropping, for instance, although potentially promising for sustained production and substantially reduced erosion, require extensive 'onsite' analyses by farmers and professional consultants. Their successful use depends on tillage techniques being accurately matched with the range of crops, soil types, soil depths and slopes which characterise the land. A great investment of time, personal energy and money is necessary if erosion and alternative technologies to control it are to be understood. When conservation measures require changes to land management practices farmers will be correspondingly slow to adopt these measures.[1]

We agree with Blyth and McCallum (Chapter 4) that micro-level economic studies of farms and farmers are necessary to model accurately land users' responses to soil loss and land degradation. We would argue that farm level sociological studies are also necessary. For instance, soil loss is not an economic problem unless farmers consider it to be so. An expert may assess that a farm may have serious erosion, but farmers and professional soil scientists often differ in evaluation of land and its condition (Chamala et al 1985, Earle et al 1981). Private land use decisions associated with erosion are first of all affected by the farmer's decisions on whether erosion is a personal economic problem in either the short or long-term (Cary 1982).[2] We study how farmers see the long-term effects of erosion on crop yield and whether they see it as occurring on their farms or only somewhere else.

Research methodology and findings

Farmers' estimates of erosion and crop yield

A sample of farm owner-operators was drawn from an area on the Darling Downs in Queensland. The region selected for study has had a history of serious soil erosion and is an area designated as a soil erosion hazard zone by the Queensland Department of Primary Industries. The area has a range of physical conditions (soil types and land slope), crops and farm sizes. Ninety-two properties are included in the analysis.[3]

Our study compares farmers' estimates of the relationship between erosion and crop yield, over time, with experimental evidence of the relationship. The experimental plots were at Greenmount on the Darling Downs.[4]

Figure 19 summarises farmers' perceptions of the relationship between crop yield and soil erosion. The data are derived from farmers estimates of the percentage crop yield farmers would expect on their land if they had an annual loss of 5 mm or 1/5 inch of soil loss per year after periods of one, two, five, ten and twenty years.

The median measure of central tendency is used in Figure 19 since it is not affected by extreme estimates. Most farmers expect that erosion will affect crop yields and that the effects on yields are cumulative. For example, at the ten-year point in Figure 19, the median response by farmers is 80 per cent crop yield or a 20 per cent yield decline over ten years if their farms were to lose 5 mm per annum of topsoil over that period. The median response after 20 years of soil erosion was a 50 per cent yield.

There is a large range in the estimates. Farmers whose responses placed them at the 75th percentile said their crop yield would be 80 per cent (20 per cent yield decline) after 20 years of erosion at 5 mm per

LAND DEGRADATION

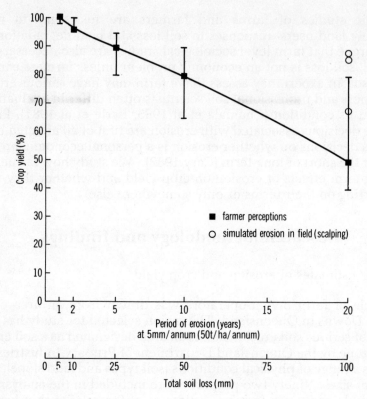

Crop yield (%)

■ farmer perceptions
○ simulated erosion in field (scalping)

Period of erosion (years)
at 5mm/annum (50t/ha/annum)

Total soil loss (mm)

Figure 19 Comparison of farmer perceptions of the effect of soil erosion on crop yield with the results of field experiments.

annum. At the other extreme, the 25th percentile, farmers expected 40 per cent of crop yield (60 per cent or more yield decline) after 20 years of erosion. If we consider farmers above the 75th and below the 25th percentile at the 20-year point in Figure 19, the differences are clear. The 25 per cent of farmers above the 75th percentile expect to have 80 per cent, or more, of crop yields after 20 years of erosion while farmers below the 25th percentile expect dramatic declines as low as five per cent or ten per cent yield.

The data demonstrate that farmers see a relationship between erosion on their farms and crop yield. Significantly, their planning horizons extend further than suggested by other researchers (Douglass 1984, Crosson 1981). This is a critical variable and not before measured by researchers attempting to explain variance in the level of adoption by farmers of soil conservation measures, and their acceptance of public conservation laws and programs. When a farmer sees an association between erosion and the property's crop yield, their long-term private economic self-interest appears to coincide with society's need to preserve farmland and sustain agriculture. Past research has found that a self-interested response by farmers to public programs can occur when there are conflicts between their individual economic interests and society's long-range conservation goals (Rickson and Stabler 1985).

Bultena and his colleagues (1981 p37), in a study of farmers in midwest America, found that the private self-interest of farmers was a factor predisposing them to withhold personal support of soil conservation programs. According to the authors: 'This self-interest can pre-empt the public interest (centralised, comprehensive planning), particularly if the benefits of public programs are not perceived as outweighing the personal costs these programs might impose on landowners'.

Uncertainty about the applicability and cost effectiveness of conservation measures to conditions on their farms, tends to constrain farmers from recognising erosion as a problem for them. 'Self-interest' therefore includes the 'personal' and economic cost of change, embracing the investment of time and energy by farmers to learn about erosion, work with relevant professionals and to decide on what conservation practices and works are suitable.

A basic question, in the analysis of our data, is what 'short term' means to farmers as they respond to land degradation. Production technologies, such as hybrid seed, pesticides, herbicides and other innovations provide short-term gains. However, conservation measures, particularly conservation tillage, have no distinct short-term crop yield advantages over conventional tillage (Crosson 1981). Farmers, in our study, generally do not see significant crop yield declines after one year of 5 mm loss of topsoil, but the median response after two years is about a four per cent decline. Their estimates of crop yield declines thereafter, at five, ten and twenty years, sharply decrease, as is evident in Figure 19.

Is the belief by farmers that there will be a small reduction in crop yield due to soil erosion after one or two years sufficient to motivate them to adopt conservation measures? This is the short-term question for research. On the other hand, is planning sufficiently long-term that estimates of crop yield losses because of erosion after five, ten or twenty years would lead to decisions to use conservation measures? Crosson (1981) notes that little is known about farmers' time horizons, their expectations about future crop prices, and how they discount the future costs and benefits of present investments.

The limited research on planning horizons suggests that farmers are fairly insensitive to the long-term effects of erosion on crop yield. Douglass (1984), cited above, concludes that farmers' planning horizons would have to be 50 years or more before perceived relationships between erosion and crop yield would effectively influence them to use conservation measures. Crosson (1981) has a similar conclusion. Our findings show that farmers see erosion as definitely affecting crop yields after five years of soil loss at 5 mm per year and effects on crop yield increasing sharply from that point on up to 20 years.

Research findings reported by Vanclay and Rickson (1985) lead to a more optimistic conclusion about the relationship between planning horizons and farmers' adoption of soil conservation measures. They found that a planning horizon of 20 years was sufficient to motivate a sample of Australian farmers to use soil conservation measures. A related question is whether farmers think erosion on their properties is a major problem to them and are they willing to accept that erosion at the rate we proposed (5 mm/annum) could actually be happening to them.

Where do farmers think erosion is a problem?

Farmers are very reluctant to accept that erosion is an individual problem on their own farms. North American and Australian research literature shows this quite clearly. As an example, Nowak (1982 p37), in a summary of several studies of North American farmers, says:

> . . . farmers will readily admit there is a very serious problem at national or state levels; it is just a problem at the community or district level. And *that other guy down the road has a problem, but I don't.* The closer one comes to home, the less likely one is to acknowledge soil erosion as a serious problem.

Table 14: Farmers' perceptions of the extent of erosion problems in different locations: Queensland, Darling Downs, local area, their own farms.

	Darling Downs (N = 89)*	Queensland (N = 89)	Local area (nearby farms) (N = 90)	Own farm (N = 90)
	(%)	(%)	(%)	(%)
Not a problem	0.0	0.0	0.0	15.5
A small problem	1.0	3.4	16.6	36.7
A medium problem	11.3	18.0	36.7	36.7
A major problem	87.7	78.6	46.7	11.1
Total	100.0	100.0	100.0	100.0

* The number of observations (N) varies slightly across categories due to missing values. For example, in the question on how much of a problem erosion was on their farms, two farmers did not answer the question. In answer to the above categories, 'don't know' was an alternative. There were only two 'don't know' responses (Darling Downs and Queensland) which indicates that farmers are rather certain of their judgments.

Nowak's (1982) findings are applicable to our research findings (see Table 14). Farmers, in our study, see a relationship between erosion and crop yield, but hesitate to accept erosion as a problem on their own farms. They say, however, that erosion is a major problem elsewhere, including their neighbours' farms. As Nowak (1982) found, the closer to one's own farm, the less likely is a farmer to acknowledge soil erosion as a 'major' problem. When we examine the 'major problem' category, only 11 per cent of farmers see erosion as a major problem on their own farm while 47 per cent consider erosion as a major problem on neighbouring farms, 88 per cent see erosion as a major problem on the Darling Downs and, taking the state as a whole, 79 per cent define erosion as a major problem in Queensland. These results are consistent with Nowak's (1982) findings for North American farmers.

Farmers are concerned about erosion and its potential effects on crop yield, as is evident from the data in Figure 19. However, we may also conclude that most farmers do not accept that erosion is a condition of critical concern for them as individual property owners. Hence, erosion is not accepted as a major problem on their farms by about 89 per cent of the farmers.

Accuracy of farmers' estimates

Most studies, including ours, conclude that farmers misperceive either or both the incidence and extent of erosion on their own farms. Farmers often disagree with the scientific evaluation of erosion conditions by professional soil scientists and agricultural extension agents (Chamala et al 1985, Rickson and Stabler 1985, Nowak 1982). However, we would expect a range of accuracy among farmers' estimates of the extent of erosion and its effects. We attempt to deal with this question by comparing farmers' estimates of the relationship between erosion and crop yield with field experiments.

Figure 19 includes independent data points reporting field experiments designed to measure the effects of erosion on crop yield. There are three data points. The first observation shows that a crop yield of 88 per cent would be the result of a soil loss rate of 5 mm per year for 20 years. The second data point is at 86 per cent crop yield and the third point shows 68 per cent yield after 20 years. When we compare the experimental data points with farmers' estimates, the median response is to *overestimate* the long-term effects of soil loss on crop yield. Interestingly, they overestimate the effects of soil loss on crop yield but, we conclude, *underestimate* the occurrence of erosion as a major problem for them, on their own land.

The variance between the data points in our field studies illustrates the difficulties for anyone judging the relationship between crop yield and erosion. Agricultural scientists establish, on a general basis, that there is a relationship between crop yield and erosion. How experimental data can be translated so that they are applicable to the physical conditions of a particular farm is another consideration (Nowak 1982).

The difficult question remains of how accurate farmers' judgments are about the incidence and extent of their farms' erosion problems and how important these observations are for their use of soil conservation measures. Our study, and those of numerous others in North America and Australia, conclude that farmers misperceive the incidence and extent of erosion on their own land.[5] However, farmers' perceptions of the effects of erosion on crop yield and erosion as a major individual problem for them are associated with use of soil conservation measures.

Vanclay and Rickson (1985), in a further analysis of these data, matched the soil conservation measures used by each farmer with the farm's physical conditions and potential for erosion (eg slope, soil depth and soil type). Farmers were thereby classified as 'protectors' and 'non-protectors'. Farmers defined as 'protectors' were both more likely than 'non-protectors' to see a relationship between erosion and crop yield, and to think erosion was a personal problem for them.[6] Regardless of the accuracy of farmers' observations, when they believed that erosion had an effect on crop yield and that erosion was a particular problem for them, they were likely to have adopted sufficient soil conservation measures to be defined as soil 'protectors'. Nevertheless, we would expect that as the accuracy of farmers' judgments about erosion increased, their use of conservation practices would be more extensive and effective.

Perceptions by farmers of the impact of erosion on crop yield, acceptance or non-acceptance of erosion as a personal problem, and the accuracy of their judgements about the extent and effect of their property's erosion are basic to their voluntary adoption of soil conservation measures. Classical innovation/adoption studies in rural sociology and agricultural economics implicitly assume that voluntary adoption by farmers of soil conservation measures is a primary means for reducing soil erosion (Nowak 1983). Voluntary adoption by farmers is based on their individual assessment that conservation is, on balance, a present and future benefit greater than future benefits from other types of investments. Farmers' evaluation of benefits and costs includes a number of economic and non-economic values (Craig and Phillips 1983).

Governments in Australia have accepted soil conservation as a public benefit and are paying more attention to the offsite or off-farm consequences of erosion. To depend entirely on farmers' voluntary use of soil conservation measures to achieve long-term public benefits recognises neither social nor economic realities (Rickson and Stabler 1985). However, an acceptance by farmers that governments have the right to plan, enact and enforce soil conservation law would be a major step forward. Government enactment and active enforcement of soil conservation law is another important mechanism to persuade or force farmers, no matter how gently, to use appropriate soil conservation measures (van Es 1983). However, farmers must accept that soil conservation is a personal necessity for them and that it is politically legitimate for governments actively to enforce conservation law.

Conclusion

In this paper, we have reported findings from our research on Australian farmers' response to land degradation. The specific focus of the paper has been on farmers' perceptions of the relationship between soil loss and crop yield, and the degree to which they see erosion as a problem on their farm. These two types of perceptions of erosion by farmers are essential for an understanding of farmers' attitudes to erosion generally and of how they respond to erosion on their own farms. Farmers 'see' a relationship between soil loss and crop yield, but have trouble 'seeing' erosion on their land or admitting the fact to themselves and others. Significantly, farmers defined as soil 'protectors' (Vanclay and Rickson 1985) saw a long-term relationship between erosion and crop yield, and accepted erosion as a medium or major problem on their farm. Perhaps, a ten or 20-year planning horizon is sufficient to motivate farmers to use soil conservation methods. Our research leads to a more optimistic conclusion about the length of farmers' planning horizons and associated investment in soil conservation measures than the findings of the North American researchers (Douglass 1984, Crosson 1981).

Assessment of rill and sheet erosion on a given property is a complex task for both farmers and agricultural scientists. Its complexity, as we have discussed, is based partly on the physical nature of erosion and particularly the fact that its first stages are hard to detect and interpret. Because of the insidious nature of the early stages of this type of erosion it is fairly easy for farmers to deny its occurrence on their farms, especially since to admit to erosion and try to reduce it requires substantial investments of time, energy, and money. One might expect that cognitive dissonance and similar theories explaining selective perception by persons could be applied to farmers' perceptions of soil erosion.

A major complication in the management of soil erosion and programs to promote farmer adoption of soil conservation measures is whether farmers see erosion on their farms and can accurately estimate how much of a problem for them it is. They are concerned, at one level, about the effects of erosion on crop yield to the extent that they may be overestimating its effects. However, one of the very great challenges for Australian farmers, agricultural scientists and planners lies not so much in convincing farmers that soil conservation is important or that Australian agriculture is at risk but, on the contrary, understanding how farmers approach the problem of erosion on their farms and respond to its incidence and extent.

Endnotes

1. Crosson (1981 p12) identifies a number of necessary management changes if farmers are to use conservation tillage methods effectively. For example, he says: 'Weed, insect and disease problems likely will be more complex, requiring more knowledge of the properties of a wider variety of pesticides and of how to apply them to get adequate pest control, or of crop rotation sequences and disease- and insect-resistant varieties as substitutes for pesticides. A greater variety of machinery and equipment may be needed with conservation tillage, increasing the need for knowledge of the right combination for the particular circumstances and perhaps also making more demands for machinery and equipment maintenance'. Farmers must have more knowledge and more access to professionals if they are to have the managerial expertise to match conservation tillage methods adequately with the capability and erosion potential of their land. Revision of land management practices is required of farmers when they decide to use soil conservation measures. However, erosion, if not controlled, can, through landslips and gullying, substantially affect how farmers work their land. Some farmers, faced with this event after flooding, decide to use conservation methods even if they may have to change long-accepted ideas about working their properties.

2. Crosson and Miranowski (1982), agricultural economists, argue that farmers recognise erosion when it happens to them, but do not invest in soil conservation because it does not increase their short-term property values. Nowak (1982) says that such assumptions fit abstract theoretical models in economics rather than 'real' farmer reactions to erosion. Other economists studying farmers' soil conservation efforts take farmer recognition of erosion as a variable (rather than as an assumption) along with other economic variables such as farmers' aversion to risk and debt levels (Ervin and Ervin 1982).

    Our data provide a somewhat different interpretation. Farmers accept that there is a relationship between crop yield and erosion. Also, if farmers are more capable of recognising erosion on other farms than they are on their own, it may be that the price they would pay for another property would be affected by their perceptions of its erosion potential and conservation measures already in place. Soil conservation, therefore, could have an indirect rather than a direct affect on property values, increasing the value at point of sale but, perhaps, not figuring in a short or long-term economic calculus by farmers when evaluating their own land without any consideration of selling it.

3. Shire records were used for sampling. The sample was limited to farms of 60 hectares (148 acres) or more in a particular shire. This yielded a total of 281

properties from which successive random samples were drawn. Out of 160 names randomly drawn from the list, 20 owners did not operate their own farm, ten farmers could not be reached by telephone, and five were not farming. Thirty-three farmers refused to be interviewed. Reasons given for refusal were mainly personal illness, bereavement and time pressure. There were 92 completed interviews.

Prospective respondents were first contacted by mail and then called by trained survey interviewers to arrange an appointment for interview. Interview times ranged from one to four hours with an average length of two and a half hours. At the end of the interview, respondents were given a questionnaire to measure selected farmer attitudes about soil erosion and selected environmental attitudes. The questionnaire was completed in the presence of the interviewer. There were several questions in the interview asking farmers to describe their farm-soil types physically, and also the extent of slope on their land. Land classification data such as the land use group, erosion value and soil depth were also recorded for each farm from land inventories for the area (Vandersee 1975).

Soil was removed ('scalped') to simulate the effect of soil erosion over a 20-year period at the locally occurring rate of 50 t per ha per annum (5 mm per annum). The top 100 mm of a Black Earth (Irving soil) at Greenmount on the Darling Downs was removed using a blade mounted on a tractor. The 'scalped' plot (10 m x 3 m) and an adjacent control plot were cultivated and cropped with wheat to grain maturity. Plant tops were removed from three replicate plots in each treatment, dried at 60 degrees C and weighed. The yield of plant tops from the eroded treatment was expressed relative to the mean yield from the control plot. In Figure 19, the crop production data are expressed in terms of dry matter rather than grain yield because the experimental plots received heavy rainfall immediately before harvest which caused significant grain loss. Furthermore, the grain loss was not constant between treatments, the better crop on the control site suffering most. For these reasons, dry matter was used as a measure of crop production as it was unaffected by rainfall.

The methodology employed in drawing the sample reduces the possibility of bias in selecting only farmers with low levels of erosion on their farms. Interviewing was conducted intensively within a relatively small area on the Darling Downs. This procedure allowed the study of neighbourhood interaction, and is less likely to contain sampling errors. It is therefore very unlikely that all farmers would be correct in reporting that they do not have a major erosion problem on their farm while they believe that their neighbours have such a problem. Because we are studying groups of neighbours, any difference between what a farmer reports about his own farm, and what the farmer's neighbour reports, is, technically, due to misperception, either by the farmer or the neighbour, or both. A farmer may misperceive the seriousness of soil erosion on his own farm, while fully appreciating it on his neighbours' farms. Recognition or acceptance of erosion as a major problem on a farmer's own property requires time, energy and money be spent to correct erosion. Farmers generally take a great deal of pride in their own farm and acceptance of erosion as a personal problem is an 'emotional' as well as a rational decision (Barr 1985). Thus, some cognitive dissonance applies in the recognition and acceptance by farmers of erosion on their land.

While farmers, in our study, appreciate that if they had erosion at the rate of 5 mm per annum from their farms, their crop yields would decline, they did not all accept that they were experiencing that level of erosion. Comments from interviewers were that some farmers had difficulty in answering questions about drops in yield and erosion because they could not conceive their farm losing 5 mm of topsoil per year. A number of farmers felt that most of their soil was lost during severe storms and then washed down gullies and not through more gradual processes of erosion. The concept of erosion occuring at a low nominal rate, 5 mm per annum, is not accepted by many farmers. This form of erosion is hard to detect and, unless farmers are actually looking for erosion, in a careful and systematic manner, it will not be obvious to them. Nevertheless, gradual forms of erosion (sheet, for example) contribute significantly to land degradation in terms of removal of topsoil and offsite pollution. Losses of 5 mm per year are fairly common in the area we studied. It would be expected on farms not adequately protected by instalment of soil conservation works and practices.

6. This was based on a discriminant analysis and refers to statistical associations adjusted for effects of other variables in the discriminant function. Other variables included land use group and other physical variables for the individual farm as well as attitudinal variables and socioeconomic characteristics of the farmer.

IV

Behavioural causes, economic issues and policy instruments

10 Land degradation: behavioural causes

John Quiggin

Introduction

Land degradation is essentially the result of human action. Irrigation in unsuitable areas may lead to rising water tables and salinisation, overgrazing and unsuitable cultivation practices promote erosion and excessive tree clearance leads to a variety of ills including dryland salting and dieback problems. While in some cases these outcomes are similar in kind to natural processes (eg erosion) the timescale is usually shortened dramatically. These natural processes have largely determined our existing environment, but they rarely constitute a perceived problem unless they have been accelerated or aggravated by human action.

This fundamental fact has not been adequately incorporated into policy discussion surrounding land degradation. The role of agricultural practices in promoting or arresting land degradation has certainly been recognised. However, this has usually occasioned calls for repentance rather than analysis of why farmers make the decisions they do. In policy analysis, more attention has been paid to treating the symptoms of degradation through ameliorative works, rather than dealing with its causes. Ameliorative works may sometimes be the best response to land degradation problems, but they are likely to be ineffectual, or even counterproductive, if they are introduced without taking account of the underlying behavioural causes of the problem.

In this paper, some of the factors which lead farmers to make decisions tending to promote or restrain land degradation will be discussed. This discussion will, in turn, be used as the basis for a brief discussion of the policy options open to governments. Emphasis will be on economic aspects of the farm management decision. Thus, it will be assumed implicitly that farmers are basically making business decisions on grounds of profit and loss, and that they will take land degradation into account mainly as it impinges on these variables. While this may seem unduly pessimistic in the light of attempts to promote a conservation ethic among farmers and the community in general, it would seem unwise to wait until such an ethic is firmly established before mounting an attack on the problem.

Before examining particular causes of land degradation, it is necessary to give a brief account of the analytical approach which will be taken. This approach differs from the one most commonly used by Australian policy economists, which is based on the concept of market 'failures' or 'imperfections'.

The imperfections approach

One common approach used by economists in assessing problems such as land degradation is to take as a benchmark the rate of land degradation which would be generated by competitive market processes under various ideal conditions, for example, full information, perfectly operating capital markets etc. The actual situation is then appraised in terms of divergences from this ideal, often called 'imperfections', and policies are formulated accordingly. For economists using this framework, the identification of imperfections is both the vital precondition for the justification of any given policy program, referred to in this framework as 'government intervention', and the main basis for selecting the appropriate form for such programs. In some cases, this approach is complicated by the addition of a vaguely defined category of 'government failure', which acts to cast doubt on the validity of recommendations for intervention, without providing a clear-cut alternative framework for analysis.

This approach encounters a number of major difficulties in dealing with complex environmental problems such as land degradation. These problems have been discussed in more detail in Quiggin (1986a), so only a brief outline will be given here. The most important problem with the imperfections approach in general is its incapacity to cope with the way in which a variety of considerations of efficiency and income distribution is intermingled. In the case of environmental issues a second problem arises with nebulous categories such as 'existence' and 'option' values. These are considered to involve market failures in that people set value on the existence of environmental goods, such as whales, and the Tasmanian wilderness, but there is no market in which they can express their demands. Such values undoubtedly exist and are expressed in socio-political decision processes, but economists' attempts to measure them have not been very successful (Quiggin 1986a). In view of these and other problems, an alternative approach will be adopted in this paper. The problem will be attacked by seeking to identify those factors, such as discount rates and information problems, which may influence the farm decisions which determine the rate of land degradation. The question of whether the existing rate of land degradation is excessive will not be tackled, but will be left to political processes to determine. The fact that the critical decisions will ultimately be made in the political arena makes the provision of accurate and timely information all the more important. If there are widespread illusions (whether optimistic or pessimistic) about the rate and extent of land degradation, politicians are likely to respond with inappropriate decisions.

Assuming that current rates of land degradation are considered excessive the next task is to identify cost-effective policies to reduce them. The term cost-effective must be treated with care. It does not refer solely to budgetary costs incurred by government, but to all the costs imposed on members of society by the adoption of particular policies. Moreover, the way in which costs are distributed across society must be taken into account. A policy which is widely perceived as unfair is unlikely to be adopted even if it is very effective in combating land degradation. Furthermore, policies must take account of the behavioural factors which generate the problem in the first place. Otherwise it is quite likely that they will be ineffectual, or even counterproductive in the long term.

Finally, just as it is impossible to determine, on a technical basis, the optimal rate of land degradation, so the determination of the resources, if any, which are to be devoted to the problem, is ultimately political. This decision will depend both on the perceived severity of the problem and on the availability of cost-effective countermeasures. Economists may be helpful in identifying the most cost-effective countermeasures, but they cannot ultimately determine which, if any, forms of government action are socially desirable.

Some of the major factors determining the way in which farm decisions affecting land degradation are made include discount rates, uncertainty, property rights and the side-effects of government policies with objectives other than soil conservation. These factors will be discussed in the remainder of the paper.

Discount rates

Most farm decisions involve some element of choice between present and future income. This is clearly true of decisions affecting land degradation. For example, a higher stocking rate now is likely to reduce the carrying capacity of land in future years. In general, a given future income will be valued lower than the same income today. The simplest way of representing this in models of decision making is by using a discount rate. If the discount rate is r, income received n years from now will be discounted by a factor of $(1+r)^n$. For example, with a discount rate of five per cent, $100 in ten years time would be valued at $100/(1.05)^{10} = \$61.39$ today. Clearly, as the discount rate increases, the present value of future income declines. Because most practices resulting in land degradation are likely to increase present income at the expense of future income, the higher the discount rate used by farmers, the more rapid will be the rate of land degradation. There are, as always, exceptions. For example, livestock industries may require more long-term investment than cropping industries, but be less damaging to land. In this case, some of the effects of increased discount rates may reduce land degradation.

The discount rates used by farmers in decision making depend partly on their preferences for present and future income. However, they are substantially affected by opportunities for borrowing and lending provided by financial institutions. For example, if a given project has an annual yield of seven per cent, and it can be funded by money borrowed at five per cent, it will be worthwhile to undertake, even for a farmer with a strong preference for present incomes. Similarly, in the absence of financial alternatives, an elderly farmer with no heirs might be expected to ignore conservation practices in order to maximise income in the short run. However, it may be more beneficial to adopt conservation practices and borrow money against the (increased) value of the farm.

In the ideal theoretical case of a perfect capital market, all farmers would operate with the same discount rate, which would equal the market rate of interest. Those who were initially impatient would simply borrow until their marginal rate of time preference equalled the interest rate while those who were patient would save. In practice this is unlikely to be the case. Different farmers will face different interest rates (depending, for example, on the accuracy of credit-rating information) and the rate for the same farmer may differ depending on whether a given investment decision is financed by borrowing, by drawing on savings, by diverting money from other farm investments or by reducing present consumption expenditure.

Variations of this kind are likely to provide the most policy-relevant aspects of discount rates. While it is clearly impractical to lower general interest rates in order to promote conservation, it may be possible to find cases in which particular categories of farmers or of investments are subject to high discount rates, and where some form of remedial action is possible. One possible example arises with drought relief. Conservation practices are particularly important during droughts since the danger of degradation is very severe. On the other hand farmers may find it very difficult to borrow, even for conservation projects which are clearly financially sound, because of the danger that the farm enterprise as a whole will go bankrupt. This problem is easier to diagnose than to solve, however. The solution most often advocated, namely the provision of concessional finance, involves many difficulties. In particular, it simply transfers the bankruptcy risks from the banking sector to the public.

It is quite difficult to determine the discount rates which are relevant when farmers make decisions affecting land degradation. It is possible to observe market interest rates, but it is important to remember that these are expressed in nominal terms, while discount rate analysis usually proceeds using real (inflation-adjusted) rates. This is further complicated by the various provisions for the taxation of interest, profits and capital gains. Furthermore, loans at low interest rates are sometimes rationed, while high interest rates may contain a substantial allowance for default risks. After attempting to take these factors into account, Quiggin (1983) estimated that the real pre-tax discount rate for farm investment decisions averaged about five per cent.

For followers of the market failure approach, it is clearly important to ask whether this rate is excessive in comparison to the social rate of time-preference. This is extremely difficult because a wide range of discount rates has been advocated for projects of this kind. Treasury economists have argued for a rate of ten per cent or more, while other economists have argued that the appropriate rate is the social rate of discount which may be as low as three per cent. Many environmentalists have argued (on grounds of intergenerational equity) that environmental assets constitute a special case for which a zero rate of discount is appropriate. A zero rate of discount would justify almost any project which reduced soil degradation, while a rate of ten per cent would almost justify strip-mining the soil. For example, a program costing $50 000 per annum would not be justified even if the alternative was the complete loss of Australia's present agricultural production from the year 2100 onwards. A recent study published by Resources for the Future, Lind et al (1982) examines many of the issues in greater detail.

Even if it is not possible to give a satisfactory answer to the question of an optimal social rate of time-preference, it is important to compare the rate of discount used by farmers for conservation-oriented measures with the rate of return available from public projects in this area and from public investments in general. There are clear general arguments for allocating public funds to the project with the highest yield, whether this involves loans to farmers, public soil conservation projects or other public projects. There are however, important arguments for giving preferential treatment to areas such as soil conservation.

The first of these concerns the potentially irreversible nature of soil degradation. This means that if preventive measures are not taken now, and the costs of soil degradation turn out to be greater than are currently expected, it will be too late to do anything about it. By contrast other public projects can be deferred without irreversible losses. Of course, it is not clear to what extent current land degradation is irreversible. It is clearly not totally irreversible, in the sense that, say, the damming of the Franklin would have been. On the other hand, with present technology, prevention is generally much cheaper than cure, so much degradation may be irreversible in practice.

A second set of arguments concerns inter-generational equity. It may be argued that public investments should be separated into two classes. The first would consist of those with payoffs in the short to medium term (say the next 20 years). Since these mostly benefit the present generation, they should be evaluated using current market interest rates. On the other hand, as the market interest rate takes little account of the needs of future generations, benefits occurring further in the future should be discounted at a rate below that yielded by market processes.

In practice, the question is unlikely to be resolved by arguments of this kind. Discount rates will mainly be important in rationing the amount allocated for soil conservation between competing projects.

Property rights and institutions

A second major factor in determining farm decisions is the distribution of costs and benefits from those decisions. This is largely determined, not by the inherent nature of the problems, but by the set of institutional and property-rights structures within which decisions are made. For example, farmers' water-use decisions will be radically affected by a shift from a situation of open access to a given supply of water (where the farmer receives the benefits of their own use, but costs are borne by all water users) to a situation where the water must be bought from an owner or group of owners.

There are two major sets of property-rights issues affecting land degradation. The first concerns the property rights in land itself. Many patterns of property rights in land are possible including tenant farming, sharecropping, communal ownership and multiple usage. However, the major patterns prevailing in Australia are freehold title and leaseholds on Crown land. Both systems have properties which may accelerate land degradation. In the case of leasehold, the obvious danger is that the lessee has no interest in what happens to the land after the lease has expired. It was said of tenant farmers in 19th century Ireland, where nine-year leases prevailed, that the first three years were spent repairing the damage done by the previous tenant, the next three in farming efficiently and the last three in running the land down as fast as possible.

In general, the adverse consequences of leasehold will be least when the terms of the lease are long and it is possible for the owner to monitor the behaviour of the tenant and control it through lease covenants etc.

The major dangers with freehold title arise in areas which are being newly developed or where new techniques or products are being introduced. Since freehold land goes to the highest bidder, market processes will allocate it to those with the most optimistic expectations of such variables as carrying capacity, since these people will be willing to pay more than the holders of more cautious beliefs. For example, in the 19th century, in South Australia a series of good seasons led to a widespread belief that 'rain follows the plough', so that apparently desert areas would bloom once agriculture began. However numerous the sceptics might have been, their views had little weight since they could not be expressed in the land market.

In view of these competing dangers the author's preference would be for a system based on freehold ownership in well-established agricultural areas with long-term leasehold involving careful monitoring of land use in more marginal areas.

The second major concern is about environmental assets other than land, such as streams and catchment areas. In many cases property rights in those assets are ill-defined and a situation of open access prevails, at least for members of some group such as the riparian landholders for a given stream. In general, the less well-defined are the property rights in a given asset, the more it will be subject to various kinds of degradation.

A number of different property-rights structures may be imposed on environmental assets, and different institutions will be appropriate for different resources. The major categories are private ownership, public ownership and common ownership. Private ownership is most appropriate in cases where the resource is clearly defined and easily divisible. An obvious case is that of water from a stream. While water is plentiful, open access may be the optimal rule. However as water becomes scarce, its use will be improved by dividing water rights among various users. The economic theory of property rights has concentrated heavily (excessively in my view) on this transition from open access to private property.

Where resources are not easily divisible, but their use can be confined to an identifiable group, systems of common ownership are likely to be appropriate. These involve allocating the asset to a group of owners who may either control its use by the imposition of rules governing use, or levy an access charge and divide the proceeds among themselves according to some predefined rule. Although systems of common ownership are rare in Australian agriculture, they are common in other agricultural systems and in other types of property such as strata and company-title flats. In order for the decision procedures required by a common property system to work smoothly, it is generally desirable that the number of owners should not be too large and that the benefits from improved usage should be evenly, or at least unimodally distributed. The possible role of common property institutions in dealing with land degradation problems such as dryland salinity is examined in Quiggin (1986b).

Finally, public ownership is likely to be appropriate in the case of assets with complex and ill-defined patterns of usage. As in the case of common ownership, publicly owned assets can be controlled either by regulations or by pricing measures (in the case of polluting activities which degrade the resource, these are referred to as Pigovian taxes). Economists generally prefer price-based measures although this depends on factors such as monitoring costs and uncertainty. (Baumol and Oates 1975 discuss some of the relevant issues.)

A second important feature of public ownership of resources is the possibility of undertaking public works to improve resource quality. Many of the traditional soil conservation policies of Australian governments fall into this category. Cost-benefit analysis is a method frequently adopted to assess projects of this kind. Two main observations may be made on this approach. First, it is important to remember that accurate quantification, especially of the benefits of such projects, is very difficult, and results must be treated as estimates only. Second, the discussion above suggests that cost-benefit analysis will be of more practical use in ranking competing projects than as a criterion for accepting or rejecting individual projects.

Uncertainty

Uncertainty is one of the most pervasive features of the land-degradation problem. Not only are many of the processes involved in land degradation inherently risky (in the sense that a given land use pattern may have no ill effects in one set of climatic conditions, but be disastrous in another), but scientific and economic knowledge concerning the nature and causes of these processes, their long-term tendency and costs and the effectiveness of possible remedial measures is extremely sketchy. While this uncertainty clearly poses a need for further research, the object of the discussion here is to consider the impact of uncertainty on individual farmer behaviour and on policy options.

In considering the impact of uncertainty on farmer behaviour, I will make the fairly standard assumption that farmers are risk-averse; that is, given the choice between an uncertain income stream and an assured stream with the same mean value, they will prefer the latter. In general, risk aversion may be supposed to lead to more conservative behaviour and this is likely to have positive environmental consequences. For example, risk aversion will engender a preference for mixed farming as against monoculture. Similarly, risk-averse farmers will be less willing to commit resources on the basis of unproven theories such as "rain follows the plough".

However, one aspect of the land-degradation problem interacts with risk aversion in a negative fashion. This is the fact that land degradation occurs over extensive periods. Thus when a farmer is deciding between management strategies, the choice appears to be one between comparatively certain present benefits (eg short-term benefits from overstocking) and uncertain future benefits from policies of land preservation.

Uncertainty is even more significant in constraining the policy choices open to governments. Clearly, governments cannot wait until all uncertainties are resolved before deciding on a course of action. Hence the policy program adopted must be sufficiently flexible to deal with new knowledge as it arises. For example, if the practices which engendered excessive land degradation were clearly identified it might be feasible to prohibit those practices through legislation (this is the main course adopted with respect to road accidents). However, in a situation where the relationship between particular practices and subsequent land degradation is subject to considerable uncertainty, this approach is clearly unworkable. In certain circumstances similar objections apply to proposals involving the creation of rights to pollute, which have been put forward by a number of economists. These issues have been discussed by Quiggin (1984) and Chisholm (1984). In a situation of uncertainty it is important that policy proposals should permit flexible adaptation to new knowledge as well as making good use of existing knowledge.

Another aspect of uncertainty which should be considered briefly is the impact of risk aversion on the decisions of policy makers. As in the case of farmers, this is something of a two-edged sword. Risk-averse policy makers will generally not wish to undertake actions which could lead to environmental catastrophes. On the other hand other risks, such as those associated with pressure from development-oriented groups must be taken into account. Research on the way in which incentives are transmitted through political processes and large organisations is only in its infancy.

Other government policies

The way in which government policies on agriculture can have unintended effects on land degradation has been examined by Blyth and Kirby (1985) and will be discussed in more detail in another paper at this seminar. It is, however, obvious that such policies can have profound effects on farm decision making and that coordination of policies is of vital importance. Such coordination would not prevent all the problems noted by Blyth and Kirby. For example, the expansion of agricultural production is a reasonable objective in itself, but tends to generate increased land degradation in many cases. In this case, it is necessary to review objectives as well as policies.

Concluding comments

This paper has identified discount rates, uncertainty and property rights structures as the major factors determining farm decisons, which in turn largely determine the rate of land degradation. The most important point, however, is not the particular factors which have been identified. Rather it is simply to remember that land degradation is fundamentally the product of human decisions and actions and that policies to remedy it must be oriented towards modifying those decisions and actions.

An economic perspective on government intervention in land degradation

11

Michael Kirby and Michael Blyth

Introduction

According to the public interest approach to regulation, intervention in the market process occurs to protect and promote the economic welfare of the community (alternative approaches to or explanations of regulatory policies are discussed by Posner 1974 and Sieper 1982). Thus, the task of the regulator is to discover instances of market failure—where market practices have resulted in price and output distortions and a misallocation of resources—and to reap the potential gains to society through market intervention.

However, while the concept of market failure is quite well known, even among non-economists, less familiar is the corresponding notion of regulatory failure - that is, where government actions, for a variety of possible reasons, do not result in net welfare gains to the community. Once the notion of regulatory failure is accepted, the appropriate policy response to any economic problem facing society becomes less clear. In particular, there is no guarantee, *ex ante*, that government intervention will improve community welfare.

In the next section, the requirements for an economic rationale for government intervention are outlined. Possible sources of market failure relating to land degradation and some potential pitfalls in government intervention in market decisions, particularly in the context of land degradation, are explored in the following two sections. The major conclusions are contained in the last section.

An economic justification for government intervention

Two conditions are necessary to provide an economic justification of government intervention in private decision making. First, market failure, where private decisions result in an inefficient allocation of community resources (see Bator 1958) must be demonstrated to exist.

However, in practice, there is often difficulty in correctly identifying market failure and sometimes observed market outcomes are erroneously interpreted as cases of market failure. In the context of land degradation, domination of public discussion by purely technical considerations, with little regard to economic aspects, has resulted in misconceptions of the significance of this issue for the community. For example, observations of the mere existence of degraded land do not provide sufficient evidence of market failure; the relevant issue is whether or not the rate of land degradation is optimal (Blyth and Kirby 1985).

Having correctly identified the existence of market failure and, hence, potential benefits from resource reallocation, the second requirement for an economic justification of government intervention in private decision making is to evaluate the costs and benefits of any proposed government intervention in order to determine whether or not there will be a net gain to society. If the benefits do not exceed the costs, the outcome following government intervention will diverge further from the theoretically most efficient allocation of resources than does the unregulated private market outcome.

The possibility of such regulatory failure is often not appreciated, with the result that support for government intervention can be premature and sometimes even misplaced. For example, in the Commonwealth and State Government Collaborative Soil Conservation Study (Aust DEHCD 1978a), which is widely regarded as the most authoritative review of land degradation issues undertaken in Australia, market failure notions are used to justify direct government involvement in land degradation. However, there is no thorough analysis of the costs and benefits of government intervention. Hence, inadequate economic support is presented for the study's conclusion that 'governments have a general responsibility to ensure that an adequate level of soil conservation is maintained and must necessarily be involved in assisting in a variety of ways' (p138).

Possible source of market failure

It is possible to identify several sources of market failure in relation to land degradation in Australia. First, the complexity of the natural world and technological processes means that farmers are unlikely ever to possess complete knowledge of all information relevant to their land management decisions. In particular, it can be argued that private markets may generate insufficient information about the causes and prevention of land degradation because, among other reasons, it is difficult for private researchers to capture fully all the benefits of their research efforts. That is, if information has a public good element the market will supply less than the optimal amount of it (see Newbery and Stiglitz 1981, pp144-8; Samuel, Kingma and Crellin 1983). However, as a result of years of practical experience and substantial government involvement in research and extension, information deficiency may be a less serious problem now than in the past.

Second, problems may occur when there are offsite or external effects associated with particular land management decisions. For example, eroded soil may be washed into watercourses, reducing water quality for downstream users and causing sedimentation problems in water storages. Clearing of native trees and irrigation may result in a rising water table and increased salinity problems on land further away. As offsite effects of land degradation are not borne by the land user and, hence, not always taken fully into account in land management decisions, they are likely to be an important cause of excessive rates of land degradation.

In contrast, however, two of the most popular perceived problem areas are unlikely to be genuine sources of market failure. The first relates to the impact of irreversible land management decisions on present and future generations. There is considerable uncertainty about future technological advances, resource discoveries and availability, consumer tastes and preferences, and product and input prices. As a consequence, private decision makers will be uncertain about the future costs and benefits of their actions, with the result that they, and society at large, may come to regret their decisions.

Concern about uncertainty is particularly relevant when the individual is faced with an irreversible decision; that is, when a chosen course of action forever eliminates some alternative future options. It has been argued that there may be value to the community in the retention of options which might otherwise be foreclosed — the so-called 'option value' (Weisbrod 1964). It is often claimed that this option value is positive (Cicchetti and Freeman 1971). However, subsequent research has found that, in general, the option value of a particular decision can take on any value — positive, negative or zero — and that sufficient information is unlikely to be available to enable empirical estimates of magnitude (Schmalensee 1972; Henry 1974). Thus the concept of option value does not, on balance, provide much fruitful guidance for efficient land use decision making.

Further economic research has considered sequential decision making in the context of irreversibility and, in particular, the implications of gaining extra information with the passage of time (Henry 1974; Arrow and Fisher 1974). Failure to take this information process into account results in a bias toward irreversible decisions. Thus, there is a gain or 'quasi-option value' from postponing irreversible decisions. However, while the concept of quasi-option value provides important guidance to public decision makers who are interested in efficient resource allocation, it would not by itself appear to provide sufficient justification for government intervention in resource management decisions by individuals. As the benefits of the quasi-option value can be appropriated by resource owners, the quasi-option value could be expected to be taken into account by private decision makers interested in maximising the value of use of their resources over time.

For example, choice of cultivation methods may affect soil depth, nutrient levels and the extent of gullying, thus influencing yields and productivity. In such a case, the costs and benefits of individual decisions are borne by that individual. In particular, any benefits from additional information gained from postponing irreversible actions (for example, cultivation methods resulting in extensive gullying) are captured by the farmer and could be expected to be taken into account in the decision-making process.

The second of the popular perceived problem areas concerns the use of market interest rates. When making investment or land management decisions which have effects spread over time, the costs and benefits of the alternatives under consideration must be converted to a comparable basis. The commonly recommended procedure is to convert the future stream of expected net benefits to its discounted present value (Abelson 1979). The net present value, and hence the economic assessment of an investment or management decision, is sensitive to the particular level of discount rate used. Higher discount rates give relatively greater weight to costs and benefits occurring in the near future and would, other things equal, increase the likelihood of judging as unprofitable an investment or land management practice tending to result in high net returns in the more distant future, such as many structural works designed to lessen the rate of land degradation. The choice of discount rate thus plays a crucial role in investment project appraisal and private decision making, both generally and with respect to land degradation.

The question then arises as to whether or not the rates of discount used by individuals in making their decisions accurately reflect society's valuation of future costs and benefits. Some would argue that they do not and, in particular, that private discount rates are too high. If this were true, there would be a persistent tendency for private land management decisions to result in rates of land degradation in excess of those preferred by society.

The view that the social rate of discount is lower than private rates is based on two main arguments. The first is essentially an equity or welfare argument concerning the distribution of resources and incomes among different generations, and the second is an intergenerational market failure argument.

In the case of the equity argument, it is asserted that the present generation has insufficient concern for its own future and for that of future generations (Pigou 1932). Another variant of this equity argument is the assertion that it is 'unjust' that unborn generations are unrepresented in the decision-making processes of the present generation (Neher 1976). According to the proponents of this view, existing members of society are 'guardians or trustees of future generations'.

This intergenerational equity argument can be criticised in many ways. First, the alleged lack of concern for future generations is only asserted to exist. In fact, it is clear that individuals are influenced by considerations of their likely future circumstances, and that mechanisms, such as bequests, exist and are used to express concern for future generations. Furthermore, private market processes enable and encourage the taking of decisions having effects after one's life span, for example, exchange of title to assets before death.

Second, no guidance is provided on the extent of divergence of the so-called social discount rate from private rates. Hence, any policy action based on this equity argument may be little more than an arbitrary tinkering with the discount rate (Fisher and Krutilla 1975).

Third, one implication of this equity argument is that, in order to overcome the problem of individual short-sightedness, society is in need of far-sighted bureaucrats and politicians able to redistribute resources in favour of future generations by using lower discount rates or by taking into account the expected preferences of future generations. However, the argument offers no clear explanation of the mechanism by which such bureaucrats and politicians overcome the myopia allegedly affecting the remaining members of society.

Finally, the concern shown in this equity argument for the plight of future generations seems to be based on a Malthusian notion of resource depletion and a consequent inability of future generations to provide adequately for themselves. However, given the evidence of recent history, this fear appears to be exaggerated. While some resource stocks may be diminishing in a physical sense, technological advances, economies of scale, and product and input substitution have offset this decline so that the community is in little danger of a sudden loss in welfare from this source (Peterson and Fisher 1977). In fact, living standards have been rising over time. Thus, policies designed to decrease present consumption and increase investment activity by lowering discount rates may actually involve a transfer of income from relatively poor present generations to relatively rich future generations.

The second main argument used to support the use of social discount rates is one of intergenerational market failure, developed by Marglin (1963) and Sen (1967). They argue that, although individuals would prefer that everyone invest more for the benefit of future generations, free-riding behaviour results in under-investment and market rates of interest which are too high. Thus, in their analysis, the contribution to the future by the current generation has public good characteristics. Therefore, it may be possible to use government coercion to meet the unsatisfied individual preferences and so make members of the present generation better off by undertaking more long-term investments than each individual finds desirable to make privately. One means of achieving this objective would be the use of lower discount rates when assessing public investments so as to increase the total level of savings in the community.

There are, however, at least two problems with this argument. First, in the absence of empirical studies, it provides little guidance for policy makers and so is likely to involve arbitrary discount rate adjustments. Second, even if the alleged effect does exist and is somehow measured, a question remains as to whether governments can successfully correct the problem. Consider the case in which the government attempts to increase the level of investment in the community by lowering the discount rate applicable to public investment projects. Warr and Wright (1981) argue that private savings are displaced by increased public savings because individuals make their private decisions after taking into account those of the public sector. The 'crowding out' effect may leave the level of total savings in the community unchanged.

Potential pitfalls in government intervention

Demsetz (1969) drew attention to three common pitfalls facing supporters of government intervention and casting doubt on its ability to increase the economic welfare of the community by overcoming market failure problems.

The first involves the observation that the current market allocation is not perfect (often compared with an idealised market outcome as given by economic theory) and the conclusion that government intervention is necessary to correct the observed market inefficiency. Little attempt is made to assess the likely outcome of intervention, which may be even worse than the current situation. However, it is necessary to consider all consequences of intervention, and to assess which real world arrangement is best able to cope with the particular economic problem at hand.

It is generally conceded that lack of knowledge about farm management decisions has been a source of market failure with respect to land degradation. However, informational deficiencies may also hinder the effectiveness of government intervention in market decisions. Chisholm et al (1974, p9) note that government efforts to improve the allocation of resources determined by the market mechanism are constrained by the information that is made available to them about individuals' tastes and preferences. Furthermore, this information is often in a form not immediately suitable for policy purposes. For example, in public discussion, some of the most frequently quoted data are on the extent of land degradation in Australia, in particular the total cost of repairing all degraded land (see Aust DEHCD 1978a). However, these data are of only limited use for policy decisions aimed at maximising economic efficiency, as there would probably be little economic justification in attempting to repair all existing degraded land (Blyth and Kirby 1985).

Lack of information on the part of government can result in the adoption of policies less effective than their textbook ideals. For example, the informational requirements for the setting of an efficient regulatory or standards policy for land degradation are enormous (Chisholm et al 1974, p16). Where there are difficulties in obtaining the required information, uniform standards or mandatory practices, which are usually based largely on technical grounds, are often set. Differences in the costs of reducing land degradation and external effects among individual properties are not taken into account under a system of uniform regulations. Thus, intersource efficiency (where the marginal cost of reducing land degradation is the same for all properties) is unlikely and the cost of achieving a given reduction in land degradation will not be minimised. Chisholm (1985) and McCallum and Blyth (1985) demonstrate the possible welfare losses to society as a result of setting standards based upon incorrect estimates of the benefits of repairing degraded land.

Government policies can also have effects quite removed from their initial areas of impact. The importance of accounting for secondary effects on resource allocation of government policies has been documented by Blyth and Kirby (1985), who found that rates of land degradation can be influenced by general agricultural policies. Thus, government policy itself is an important source of distortion of efficient resource management. For example, Australia's system of land tenure, with its emphasis on leasehold is often inimical to the efficient allocation of the community's land resources and frequently creates incentives for land users to increase rates of land degradation. Uncertainty regarding lease renewal or prior notification of lease termination, combined with lack of full compensation for land value at the end of the lease, discourages conservation-oriented management practices and results in a tendency to faster rates of land degradation (see Young 1985a for further discussion of land tenure and degradation). In addition, government intervention in the pricing of farm inputs and outputs can have secondary consequences for land degradation. In particular, it is likely that the pricing of irrigation water and tax concessions for land clearing (which existed before August 1983), policies which appear to be intended largely to encourage agricultural expansion and regional development, have resulted in increased salinity problems.

The second potential pitfall facing supporters of government intervention is to disregard the costs of regulation. Thus, as it is assumed that market failure can be overcome without cost through government intervention, the identification of market failure is regarded as sufficient justification for government action. However, the existence of market failure indicates substantial market costs in overcoming the difficulties; hence, the important question is whether or not government can do the same with lower costs.

External or offsite effects of land management decisions represent an important source of excessive rates of land degradation. For example, land clearing by particular farms in a catchment area may lead to increased soil erosion, reducing water quality for downstream users and causing sedimentation problems in water storages. When a large number of parties are affected by such externalities, voluntarily negotiated market solutions are unlikely, because of the high transactions costs involved in identifying and negotiating with all the affected parties. The problem is exacerbated by the scope for free-riding behaviour and the frequent difficulty in precisely identifying sources of land degradation. Yet similar costs confront government actions to overcome the externality problems associated with land degradation. Resources must be spent on the administration, information gathering, monitoring, policing and decision making associated with government intervention. For example, demanding information requirements are necessary to implement an efficient system of standards or taxes designed to deal with an externality problem. These requirements are likely to be met only for very simple production systems characterised by, say, few production alternatives and substitution possibilities between inputs, or simple linkages between particular inputs and land degradation (Chisholm 1978). Such circumstances are not common in Australian agriculture. Thus, although the coercive powers of government may offer advantages for coping with free-rider problems (Chisholm et al 1974, p10), the fact that government actions are costly carries the implication that government intervention to cure market failure problems such as externalities may not necessarily provide a more efficient outcome than the unregulated market outcome.

Finally, the third pitfall is to believe that individuals will act differently in different institutional environments. Typically, however, they will continue to act rationally in their own self-interest in a newly created regulatory environment.

It was noted above that subsidised farm inputs, such as those relating to water pricing and land clearing, can encourage increased rates of land degradation. However, even subsidy programs aimed directly at reducing rates of land degradation can have their effectiveness limited by individual responses to the subsidies. First, subsidies lower the private opportunity cost of land degradation by reducing costs of repair. Hence, they provide incentives to adopt practices relatively more conducive to land degradation, thereby offsetting to some extent the direct impact of reduced degradation of the subsidised activities. For example, if farmers find it profitable to increase stocking rates in response to a subsidy on contour bank construction, the reduction in soil loss may be less than expected (Chisholm 1978). Secondly, when only a subset of farm management practices or inputs is subsidised (this is often the case for reasons of administrative ease), the production decisions of land users will be distorted and reflect a bias toward the particular subsidised practices or inputs.

Similarly, the impact of a system of mandatory standards on individual behaviour needs to be carefully assessed. The direct regulatory approach can prove to be inflexible in the face of changing technologies and market circumstances if it allows insufficient discretion in the choice of management practices. For example, it is questionable whether or not a system of mandatory practices would have been sufficiently flexible to permit the increased use of less conservation-oriented land preparation techniques in order to cope with the heavy crop stubble and weed growth associated with the unusually favourable seasonal conditions which occurred in 1983 following the drought.

Supporters of government intervention can also omit to consider the role of those responsible for administering legislation or policy. The responses of these administrators to the incentives faced under the regulatory environment can sometimes lessen the likelihood of outcomes in the public interest. For example, Lindsay (1976) argues that, since it is frequently very difficult to measure a public agency's contribution to the community, its management often faces an incentive to organise production to maximise the perceived value of output. As certain attributes are more readily measured, a tendency exists to distort production by diverting resources towards those attributes. For example, in the case of soil conservation activities, there is an incentive to emphasise structural works compared with less obvious changes in management practices as a means of alleviating land degradation.

Conclusion

The public interest approach to government intervention in land degradation is one of a market correction process, yielding increases in the community's economic welfare. However, attention must be given to the ability of government action to improve the observed market outcome. While a public interest case for intervention is possible, incorrect or incomplete analysis of the problem under consideration may yield a false calculus of the net economic benefits from government intervention.

There are two necessary conditions for an economic justification of government intervention in land management decisions in the context of land degradation. First, it must be demonstrated that market failure exists, resulting in economically excessive rates of land degradation and providing the potential to increase the efficiency of the use of the community's land resources. Second, the benefits from any proposed market correction policy must be demonstrated to exceed its costs. Thus, the existence of market failure is a necessary, but not sufficient, condition and provides only a *prima facie* case for intervention. A thorough analysis of land degradation policy issues must involve consideration of both market failure and the costs and benefits of government intervention.

Consideration of this second necessary condition is frequently ignored. Demsetz (1969) labels this the 'nirvana' approach, under which it is considered that market imperfections can always be cured by government intervention. However, as Chisholm et al (1974, p8) note, the real world government process is very different from the omniscient and benevolent ideal implicit in the nirvana approach and there are many potential pitfalls in government interventions intended to improve resource allocation. To adopt policies likely to improve the community's economic welfare, a comparative institutions approach in which an assessment is made of which real world arrangement is best able to cope with the economic problem under consideration is needed.

12 Abatement of land degradation: regulations versus economic incentives

Anthony Chisholm

Introduction

Land degradation is a complex phenomenon because there is a multitude of possible causes and effects, both onsite and offsite. All human land usage affects the state of the land and its impact is critically influenced by nature's wheel of fortune which brings great variations in rainfall, wind, and temperature patterns. Much land degradation takes place during periods of extreme climatic conditions such as prolonged droughts, severe flooding and so forth. The natural rate of soil erosion in Australia is considered to be relatively low by world standards, but the potential for land usage to accelerate the rate of degradation is high (Olive 1983).

Land degradation problems stem mainly from agriculture and pastoralism which are the major uses of land. However, severe degradation problems are commonly caused by many other activities such as mining and engineering, commercial forestry and woodchipping, real estate development, and some recreational activities (eg off-road vehicles). Resulting damages include onsite loss of soil and nutrients, sedimentation of waterways, dryland salting, irrigation-induced salinisation, desertification, destruction of natural flora and fauna habitats, beach and sand dune drift and erosion, unaesthetic scarring of landscapes and spread of weeds. The costs of land degradation may be fully onsite or fully offsite. However, it will most often result in a mix of both onsite and offsite damages. Thus, it may be seen as a double-edged sword: productivity of the land is reduced and waterways and the air are commonly polluted.

Two forms of excessive onsite costs can be distinguished. First, onsite costs may be excessive in the sense that a land user is adopting sub-optimal management practices that result in forgone profits. That is to say, the present value of assets in land use is not being maximised. In this situation the land user is most likely acting on the basis of imperfect information and the appropriate policy response is to convey relevant information to the owner/operator via extension services. For example,

Dumsday et al (1983) describe a case where dryland salinity could be reduced while increasing farm income by substituting short fallow for long fallow in the cropping rotation. Second, onsite land degradation may be too severe for society as a whole but right for owners/operators given existing government policies and systems of land tenure. Some forms of government intervention, or lack of intervention, may cause a divergence between private and social onsite costs of land degradation. For example, inappropriate government land-tenure systems may cause a degree of 'mining' of the land which is socially excessive, but privately just right. Moreover, some general agricultural policies have side effects which accelerate land degradation (Blyth and Kirby 1985); and some policies specifically aimed at reducing land degradation may be ineffective because they do not properly take account of certain behavioural responses of land users (Chisholm 1978).

Economists generally take the view that before government involvement can be advocated there must be some evidence that an unfettered free-enterprise system is incapable of doing an acceptable job. There may be some forms of market (free-enterprise) failure which government has not adequately corrected and which cause socially excessive onsite land degradation. The major elements of potential market failure in excessive onsite land degradation are imperfect information, imperfect capital markets and intergenerational equity issues, and irreversibilities. Various aspects of these issues are discussed by Chisholm in Appendix A, Kirby and Blyth Chapter 11, Quiggin Chapter 10 and Blyth and McCallum Chapter 4, and are not reviewed in the present chapter.[1] The first-best policy response if any of these forms of market failure do exist is to correct the market failure directly. Information has public good characteristics and there is a clear-cut case for public research and dissemination of information on the causes and effects of land degradation and efficient policies to combat it. If excessive land degradation is being caused by imperfect capital markets, the first-best government policy is to attempt to improve the capital markets.

This author deliberately avoids the important question of the magnitude of onsite and offsite damages currently resulting from land degradation in Australia and the associated socially desirable extent of government intervention. It is hoped that an answer to this question is one which will emerge from the contributions to the workshop as a whole. The present paper proceeds on the presumption (perhaps false) that for much of Australia's land there is a significant *economic* problem which requires government intervention or change of policy. While much of the following discussion of government policy options is in the context of reducing offsite effects, in the main it is also directly applicable to abatement of onsite effects of land degradation. Moreover, there will commonly be a strong correlation between onsite and offsite effects.

The plan of the chapter is first to discuss the nature and economic implications of offsite damages and then to review the broad characteristics of economic incentives and regulations and the economic conditions that require to be satisfied for an efficient solution to land degradation problems. The main part of the paper provides an analysis and discussion of the major types of economic incentives and regulations.

Offsite damages

Offsite damages occur when a land user's activities have detrimental spill-over (external) effects on other parties.[2] These external effects represent a failure of decentralised (market) decision-making processes insofar as land users do not take into account the incidental damages their activities impose on other parties, which social costs, if they were taken into account would result in different and socially preferable land management decisions. The costs to society of certain land use activities exceed the costs to private owner/operators and the aim of government should be to internalise external effects so that land users act as if they were bearing the full social costs of their actions.

In some situations, external effects will be confined to a small number of affected parties and there will be a clear incentive for parties to get together and negotiate a cooperative course of action. Courses of action could include the internalisation of external effects through one land owner purchasing the land of others, or the management of an area of land owned by a number of farmers as a 'common property' resource. However, even when numbers are small there are reasons why voluntary negotiations may break down. First, if property rights are not well defined the parties will have difficulty reaching agreement on cost-sharing arrangements. The role of government and the courts should be to define clearly property rights and liability rules. Second, even when the rights of damaging and damaged parties are clearly defined a damaging party has an incentive to overstate the costs of reducing external effects while the damaged party has an incentive to overstate the costs imposed by the damages. Such attempts by parties to out-smart one another by devious strategies are not conducive to successful negotiations. Opportunities for successful negotiations may be enhanced if soil conservation agencies provide technical advice and act as an arbiter in disputes that may arise with respect to such matters as cost-sharing arrangements.

Most of the major external effects associated with land degradation, however, involve large numbers of individuals. Negotiated solutions will not be reached because of high transactions costs and because each individual has a strong incentive to be a 'free-rider' in any negotiations. Consequently, to establish a major improvement in the allocation of resources to reduce damages from land degradation it is usually presumed that government will need to intervene directly through the use of its fiscal or regulatory powers, or both.

Economic incentives versus regulations

The primary difference between fiscal incentives and regulations is that the former operate on relative *prices* while the latter involve the direct manipulation of relative *quantities*. Economists have generally been critical of the use of command-and-control regulations and have advocated the use of various price-based incentives, although at the purely conceptual level the difference between a price-based incentive and a direct regulation is not as great as is commonly believed. A direct regulation is essentially an *implicit tax*, usually at a zero rate up to some point at which a large discrete charge (eg a fine) occurs if the regulation is violated.

The essential reason why economists advocate the use of price-based incentives is that, compared with regulations, the incentives generally make efficient use of scarce information. To understand this point it is helpful to summarise the main characteristics of an *ideal* economic solution to the land degradation problem. An ideal solution would be one which attained a socially optimal level of land conservation/ degradation both now and over time, at least social cost. From an economist's perspective, the efficient level of land degradation in a particular region may be greater than, or less than, the 'natural' rate. The efficient level of land conservation/degradation is where the sum of the total damages (offsite and onsite) plus the costs of abatement are at a minimum.

A number of conditions will be satisfied to a degree when an efficient solution to land degradation is found. First, each land user's costs of land conservation will be equated, at the margin, for all methods of conservation. A balance will be struck between investment in structural conservation works, adjusting cultivation techniques, altering stocking rates and so forth. Second, the costs of land conservation, at the margin, should be fairly equal between land users within a reasonably homogeneous region. That is to say, after land users have responded to conservation incentives there should not be a situation where, for any given small change in the regional level of land conservation/ degradation, some land users would incur high costs and others low costs. Third, the policy instrument should ensure that a continuous incentive is provided through time to land users for technological change and adaptation towards more efficient ways of controlling land degradation. It is not commonly recognized that the rate of innovation and technological change is endogenous to the incentive structure provided by the abatement policy. Much policy debate has taken a too static view of the world. In this respect, the policy should also be sufficiently flexible to allow a rapid and efficient response to critical natural events like droughts. Fourth, administrative costs, comprising mainly information costs and monitoring and enforcement costs, should not be excessive. In brief, an ideal incentive or regulation should meet the goals of efficiency, equity, and simplicity.

The above set of ideal conditions provides a benchmark (and checklist) against which the effectiveness of alternative policy instruments can be judged. In most real-world situations the best practical policy will be significantly influenced by factors such as measurement and monitoring costs and it will clearly fall short of fully attaining the first three ideal conditions given above. The major advantage of an appropriate price-based incentive scheme is that, unlike most regulations, fairly efficient outcomes can be achieved in some circumstances without an administering agency requiring any knowledge of a particular land user's least-cost method of abating land degradation.

Finally, it should be recognised that the reason why certain markets to handle offsite effects from land degradation do not exist is because transaction costs are in excess of potential gains to the respective parties from negotiating and trading. The government, particularly through its coercive taxation powers, has some advantages over the market because, if necessary, it can force everyone to contribute to the cost of land conservation. It thus has a means of coping with the 'free-rider' (non-excludability) problem. However, the very factors causing transactions costs to be too high for the emergence of certain markets will also often prevent a well-intentioned government from achieving an outcome in which the benefits exceed the costs of its actions. There is reason therefore for treating with more caution than usual the presumption that government will necessarily improve on the market's results. In some situations the best policy will be to do nothing. A detailed discussion of these issues is given by Chisholm et al (1974).

Charges, subsidies and transferable rights

There is a wide spectrum of possible economic incentives for controlling land degradation that have similar efficiency characteristics but different income distributional effects. To simplify the exposition and highlight the conceptual issues, the major types of incentives are first outlined on the presumption that offsite damages caused by each land user can be directly measured and monitored. This assumption is subsequently relaxed and the implications for policy of the non-point nature of most forms of land degradation are considered.

A pure charge (tax) system would impose a charge on all units of, say, soil loss at a rate equal to the marginal social cost of the resulting offsite damages. A pure subsidy scheme would pay a subsidy on each unit of soil loss abatement at a rate equal to the marginal social benefit of reduced offsite damages.

In principle, a tax and a subsidy, set at the same marginal rate are both capable of attaining the optimal level of soil loss abatement at least social cost. From an efficiency viewpoint, it may be argued that the 'stick' and the 'carrot' are equally efficient instruments for controlling land degradation. Moreover, in the context of controlling point-source pollution, it is commonly claimed that these forms of economic incentives minimise the information required by the protection agency. The agency, it is argued, requires information only on the marginal benefits of abatement. The agency need have little knowledge of the optimal level of abatement; the physical, biological and chemical processes causing pollution damages, or of the methods and costs of alternative abatement procedures. The voluntary responses by firms to an appropriate system of pollution charges (or subsidies for pollution abatement) would lead to an efficient overall level and method of abatement.

Conservation standards

In some situations a reasonable valuation can be made of offsite damages and the contribution of individual land users may be able to be identified and directly monitored: for example, the siltation of a reservoir in a catchment area with a single, or a small number, of land users. It is usually claimed, however, that it is either impossible or too costly to value explicitly offsite damages and the social benefits of their abatement. Instead, regional conservation standards may be set on the basis of political/bureaucratic judgment in much the same way as a government determines priorities and goals with respect to education, health care, and defence. The specified standard may, of course, deviate substantially from the *true* social optimum point at which marginal social benefits equal marginal social costs. The aim is to achieve the standard at least social cost. It is possible to devise a charge or subsidy scheme that has this desirable attribute. The system of charges (subsidies) is efficient and equitable in the sense that each land user is given an incentive to select the least-cost mix of strategies to reduce soil loss on their property and, after responses have been made, each land user within a region faces the same costs of abatement at the margin.

The most likely regulatory procedure, in practice, would be for the agency to determine first the recent historic level of soil loss for each farm and then command a uniform percentage abatement of soil loss (say 40 per cent) by each property owner. Where there is considerable diversity of farms and of the resource constraints faced by individual farmers, the above procedure is inequitable and inefficient. After adjustment the marginal costs of soil-loss abatement will vary substantially between farms. Studies of industrial pollution in the United States (Kneese et al 1971) show that cost-savings of the order of 50 per cent could be made by using economic incentives rather than direct regulations to attain environmental standards.

The above form of regulation is essentially a system of 'fixed' soil-loss rights that have been allocated free of charge to each land user. With a system of fixed rights, a land user has no incentive to reduce soil loss below the level of their quota even when the costs would be small. If, following their initial allocation, such rights were permitted to be transferred, land users facing high abatement costs could purchase rights from those with low abatement costs. Buying and selling of rights would ensure that the marginal cost of abatement for each land user was approximately equal to the market price of a right. The direct regulation has been converted to an economic incentive by simply permitting the transfer of abatement rights.

Charges, abatement subsidies, and transferable rights are the three major forms of economic incentives. Each type of incentive implies a particular set of property rules and associated income distributional effects. These effects are illustrated in Figure 20, where the sum of total abatement costs plus total damages (OSR) are at a mimimum at point P^*. Charges represent the most extreme form of the 'polluter-pay' principle. Land users incur all abatement costs (P^*SR) and are charged for each remaining unit of soil loss causing offsite damages ($OTSP^*$). This outcome would also occur if a quantity (OP^*) of pollution rights were auctioned. On the other hand, a full subsidy scheme effectively grants to land users the right to pollute. Government must pay or compensate each land user for each right to pollute that is not used (ie for each unit of soil loss abatement). The government effectively pays the full abatement costs and land users are not charged for any remaining units of soil loss. The implicit set of property rules with a system of transferable rights, allocated free of charge to existing land users, lies between the charges and subsidy schemes. Land users incur full abatement costs, but as a group they are not charged on those units of soil loss that continue to cause offsite damages.

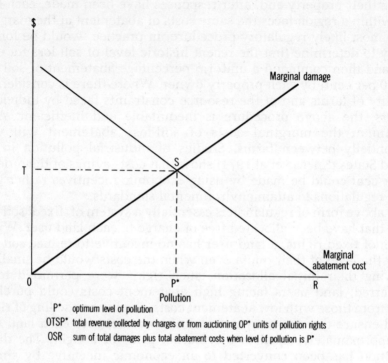

P*   optimum level of pollution
OTSP* total revenue collected by charges or from auctioning OP* units of pollution rights
OSR   sum of total damages plus total abatement costs when level of pollution is P*

Figure 20 Distributional effects of alternative pollution control policies.

In principle, these economic characteristics and the choice between the policy instruments hinge on how government views the income distributional effects of each incentive. Clearly, land users would prefer subsidies to transferable rights and they would be most opposed to charges. Administrators are also likely to be attracted to subsidies for this type of incentive makes for a more harmonious working relationship between land users and administrators. However, it will be pointed out later in the chapter that problems can arise with the use of certain forms of subsidies.

Non-point runoff

So far the exposition has been simplified by assuming that offsite damages caused by each land user can be directly measured and monitored. But most forms of offsite damages caused by land users arise from runoff of sediment, nutrients and chemicals. The runoff is typically of a non-point nature so that the use of monitoring devices similar to those used for measuring pollution from point sources (eg stack emissions and piped effluents) common to industrial production is usually not possible. The only feasible target for purposes of measurement and monitoring are the factors and activities influencing non-point runoff. This adds substantially to the information required by a regulatory agency since it now needs detailed information on the physical and biological factors and land users' activities that determine soil loss and other forms of degradation.

The Universal Soil Loss Equation (USLE) is a simple relationship expressing average annual soil loss as a function of six major erosion factors: a rainfall factor expressing the erosive potential of the rainfall in the locality; a factor reflecting the inherent erodibility of the particular soil type; a topographic factor that accounts for the combined effects of length and steepness of field slope; a factor that introduces the effects of land use and management variables; and a factor for explicit soil conservation practices such as contouring (Wischmeier 1976).[3]

The essential feature of USLE is that, in principle, it allows efficient policies to be based on the monitoring of those factors and activities which determine soil loss rather than the soil loss itself. Instead of monitoring soil loss directly, the individual determinants of soil loss would be monitored. To measure soil losses accurately from individual farms, using USLE, a regulatory agency would require detailed knowledge of the physical and biological characteristics of each farm and of the quantitative impact on soil loss of all combinations of farm inputs, outputs, and management practices. Economic incentives would be applied to the measure of soil loss estimated by USLE. While the government agency requires knowledge of the physical determinants of soil loss it is not necessary for the agency to have detailed information on the firm's cost and profit relationships.

In responding to an incentive based on the measurement of soil loss by USLE, farmers would first need to inform themselves on the technical characteristics of the equation and the way it applied to their farms. In determining the least-cost mix of soil-loss abatement strategies a farmer will first consider the extent of each activity's physical contribution to soil loss or its abatement. To aid this process, it is clearly desirable that those responsible for applying USLE and administering the scheme at an individual farm level be good extension workers.[4] Farmers, on the basis of technical information on the determinants of soil loss embedded in USLE, together with their detailed knowledge of the farm production relationships, the substitutability between factors of production, and input and output prices would then determine the least-cost (profit-maximising) response to the economic incentive.

To recognise the full implications of this point it is most important to keep in mind that soil is a primary input in all land use enterprises. Consequently, all forms of land use activities have a greater or lesser impact on the state of the soil. In agriculture, for example, a farmer may respond to a tax (subsidy) on soil loss (abatement) by investing in structural and direct methods of soil conservation like contour banking, planting trees, and fencing off erosion-prone gullies. In addition, changes may be made to a wide array of management practices. Farmers may change their product mix, reduce stocking rates, modify crop rotations and grazing systems, change methods of cultivation, adjust rates of fertiliser or irrigation water usage, alter their strategies toward drought, and adopt more stringent pest and weed control procedures.

In circumstances where there is considerable diversity among farms and among the resource constraints facing each farmer, the least-cost soil loss abatement procedures will vary considerably from farm to farm. Given adequate access to good extension advice on soil conservation techniques and other management practices, the individual farmer is in the best position to determine the least-cost mix of strategies to reduce soil loss on his property. It would be prohibitively costly for a centralised agency to acquire information on physical and biological processes of land degradation and have complete knowledge of cost and profit relationships for each individual farm. Yet this is precisely the set of information that a regulatory agency would require in order to determine a truly efficient mandatory soil-loss abatement program for each farm.[5]

Administration and enforcement of incentives would require close interaction and cooperation between farmer and government agency. It is important that agents be perceived by the farming community primarily as sympathetic extension workers rather than bureaucratic regulators. We now turn to consider some problems associated with the practical application of USLE.

Influence of climatic fluctuations and uncertainty

The USLE estimates long-run average *annual* soil losses for specific combinations of physical conditions and management practices. Climatic fluctuations will cause the degree of land degradation and offsite damages associated with a particular set of land inputs and management practices to fluctuate through time. Land degradation and offsite damages may suddenly and markedly increase as a result of heavy storms and flooding. On the other hand, large changes in the level of onsite degradation and offsite damages may be a gradual cumulative process over time, as with the development of a major drought. Fluctuations in offsite damages may be the result of variability of onsite degradation, or changes in the capacity of offsite environments to absorb the effects of various forms of spillovers arising from land degradation occurring on other sites; for example, the capacity of a waterway to absorb saline runoff is dependent on the level of stream flow.[6]

Given that there is some evidence that much land degradation occurs in atypical years, USLE should ideally provide quantitative estimates of the determinants of soil loss in atypical years as well as for the average year. Drought policies, for instance, are likely to exert a major influence on land conservation/degradation, yet USLE long-run average *estimates* of annual soil loss provide information of only limited use for formulating efficient drought policies to combat land degradation.

Land degradation caused by severe floods or bush fires which usually occur suddenly with very little, if any, warning requires a command-and-control response to an emergency situation. Certain forms of direct action and regulation are appropriately triggered at the times of such crises.

At other times, certain states of nature may occur that require incentives/regulations to be sufficiently flexible so as to bring about efficient responses from land users to a particular state of nature. Certain property rights or entitlements, for example, should be specified as being dependent upon states of nature. A system of transferable discharge permits, for example, is used for the management of water quality in the state of Wisconsin. The entitlement to discharge effluents into a river is dependent upon the levels of water flow and water temperature. When water flow is low and water temperature high a discharge permit may entitle the holder to only 50 per cent of the normal entitlement. The details of an entitlement are clearly specified in advance so that the permit transfer market can operate efficiently.

Another important issue relates to the problem faced by a regulatory agency when it is forced to act with imperfect information. The influence of uncertainty on choice of policy instrument has been discussed elsewhere at some length by the author (Chisholm 1985).

Finally, it seems likely that for some land, information is not currently available to estimate soil loss even roughly using some variant of USLE. Edwards and Charman (1980) claim that USLE imported from the USA is presently inappropriate under Australian conditions. It is possible to identify a few inputs or activities that are causally linked to soil loss, but with existing information it is not possible to quantify the known links or even be confident that all the important determinants of soil loss have been identified. Moreover, it would appear that even when we have the necessary technical knowledge to apply the USLE the costs of applying the equation in many areas may be high. In these circumstances, the only practical form of intervention available is to apply regulations or incentives to factors or activities which can be identified as being strongly linked with soil loss. However, the fact that *all* farm inputs and management practices potentially have some impact

on soil loss should caution us against possible problems caused by tying
regulations or incentives to a single or a few inputs/activities.

Incentives and regulations applied to some specific activities

For some forms of agricultural runoff there is clearly a strong link
between a particular activity, or the use of a particular input, and offsite
damages: for example, offsite damages caused by the use of fertilisers
or pesticides. The obvious targets for the application of incentives or
regulations are the inputs of fertilisers or pesticides. The control of
pollution damages from pesticide use is a more complicated problem
than the control of fertiliser runoff. The sheer number of pesticides with
differing properties poses a formidable information problem in
choosing, for each situation, which pesticide is most efficient from
society's viewpoint. Furthermore, pollution damages caused by
pesticides are typically sensitive to the method of application and
climatic conditions at the time of application. It is these characteristics
of pesticide use which have led to proposals for a control system of
incentives and regulations combined with a licensed operator scheme
(Longworth and Rudd 1975). Runoff from fertilisers and pesticide use
is a major problem in countries such as the United States where these
inputs are very intensively applied over large areas of land. The
extensive nature of most Australian agriculture tends to limit major
problems of fertiliser and pesticide runoff to a few intensively farmed
areas.

Serious land degradation problems occur in Australia as a result of
salinity problems associated with irrigation, dieback of native trees on
farms, and dryland salting and stream salinity caused by clearing of
native vegetation. While the causes of tree dieback are complex and
non-specific, there are clearly identifiable linkages between irrigation
and salinity and between the clearing of native vegetation and dryland
salting and stream salinity.

The case of dryland salting provides some useful insights. Economic
aspects of dryland salinity control have recently been analysed by
Edwards and Lumley (1985) and Dumsday et al (1985). Consider a rural
region where tree clearing on one property causes dryland salting on
other properties. Each farmer is both a potential generator and recipient
of externalities, since farmers may cause salting to occur on each other's
land. The appropriate target for the incentive or regulation is clearly the
tree-clearing activity, though the relationship between tree clearing at
one location and the pattern of induced dryland salting at other locations
is complex and not well understood. We initially focus on dryland

salting and assume that tree clearing causes no stream salinity.

The imposition of a charge on offsite salting damages caused by tree clearing would, in principle, bring about an efficient level of clearing on each property. If reafforestation is an effective means of abating dryland salting, the tax base is a farmer's current area of cleared land rather than the activity of clearing land (Hodge, 1982).[7] It is likely that although there is a general gain to society from more efficient land use, farmers as a group will be worse off. It is true that both society and farmers as a group benefit from the reduction in tree-clearing activity; but the revenue collected from the charges that benefits society at large will often exceed the benefits to the group for which the external effects have been internalised.

The likely political unpalatibility of this outcome could be overcome if a way could be found to return the tax revenue to the group of farmers in such a manner that it did not in any way alter the farmers' responses to the incentive. One procedure would be to distribute the proceeds from tree-clearing charges to farmers according to the area of land owned by each farmer. This redistributional procedure would leave intact the incentive provided by the charges to reduce tree clearance.

This approach has been suggested in the interesting context of a common property approach (Quiggin 1986b). The common property is the total area of land subject to dryland salting owned by a group of farmers. Maximisation of the value of this asset suggests that the group of farmers themselves should levy charges on tree clearing on the basis of reduction in land values to other users. Revenues would be returned to the farmers in the group on the basis of the area of owned land. The effective operation of a common property approach requires the voluntary cooperation of all farmers in a region and this is likely to be achieved only when the number of farmers is fairly small.

The existing policy of some states of prohibiting tree clearance as a means of combating dryland salting represents the most stringent form of regulation. It may be justified on the grounds that in some regions the social costs of clearing any trees is known to exceed the social benefits. It is more likely justified as a short-term emergency measure to provide time for more information to be gathered and analysis undertaken. It seems unlikely that outright prohibition of tree clearance is a long-run optimal policy.

The prohibition on tree clearance effectively freezes rights in 'cleared land' at the existing level of cleared land on each property. One procedure for relaxing the prohibition would be to permit transfer of cleared-land rights between land users. Farmers who incurred high costs, in terms of forgone profits, from maintaining their uncleared land area could then purchase additional cleared-land rights from farmers with lower forgone profits. Farmers selling, say ten hectares of cleared-land rights to another property owner would be required to increase the area of trees on their property by ten hectares. If some tree-covered areas are particularly sensitive, ie clearing them would generate large offsite damages, the transferable rights market could operate subject to maintenance of the sensitive areas in an uncleared state.[8]

When the transferable rights market was in equilibrium the cost at the margin of limiting tree clearance would be equated for each land user with the market price for a cleared-land right. The market determined price for a cleared-land right provides valuable information to government on the costs of maintaining the existing area of uncleared land. By comparing the transfer price of cleared-land rights with the perceived benefits of maintaining the existing level of uncleared land in a region, a government agency can make reasonably rational decisions whether to increase, or decrease, the aggregate number of cleared-land rights.

The existing pricing mechanisms for irrigation water provide one of the most notable examples of a system which causes inefficient resource allocation. Generally, the price paid by farmers for irrigation water is substantially below its full costs of supply, even when no account is taken of the added offsite salinisation costs.

The total amount of irrigation water supplied to farmers in a region should reflect its full social costs of supply, including costs of salinisation. Entitlements to the available water could be auctioned, or distributed free to present irrigators on the basis of their historical patterns of water use. Market transfer of water entitlements would be subsequently permitted. Water rights would be legal property instruments, vested in the individual and negotiable independently of the land.[9]

Entitlements conferred on the holder of a transferable right should be clearly specified. This point can be well illustrated in the context of transferable entitlements for irrigation water (Randall 1981). Rights should be specified so that the duties and obligations of the water authorities are clear and the holder of the right has secure expectations. Any water charges levied by the authorities would need to be known in advance and be reasonably predictable well into the future. Examples of other issues that would need to be resolved are the time span of the entitlement, and possible provision for rental rights, and the method of handling the stochastic nature of water availability.

Subsidies and compensation

In terms of pure theory, charges and subsidies have a different distributional impact, but they can be equally efficient instruments for attaining abatement of land degradation. In practice, however, there are significant differences between charges and subsidies and some major pitfalls in the application of the latter. Discussion of these issues is particularly important because subsidies for soil conservation are currently the major economic incentive in actual use and political pressures seem likely to ensure their continued use.

A subsidy on say soil loss abatement requires information on the existing level of soil loss for each land user. There is a potential problem that land users, anticipating the introduction of a subsidy scheme, may deliberately undertake activities to increase soil loss to attain a more favourable benchmark for the subsidy. Charges, unlike subsidies, do not require the initial establishment of a benchmark for each land user providing the natural rate of erosion is close to zero. More importantly, charges provide a correct incentive structure for entry and exit of land users between particular types of land use.

Following the introduction of a charges scheme it would be expected that some land previously used for cropping would become grazing land while some marginal lands would cease to be used for agriculture or pastoralism. Subsidies, on the other hand, may preserve, or even increase, areas of marginal lands in uses which are profitable to land users, but uneconomic from the viewpoint of society. To avoid this problem it would be necessary to continue payment of a soil loss abatement subsidy on marginal lands which subsequently ceased to be farmed. The subsidy base would be the difference between the initial benchmark level of soil loss for the land and the 'natural' rate of soil loss. If abatement subsidies did not apply to marginal lands that were permitted to revert to their natural state there would be an inappropriate incentive for property owners to continue farming some marginal lands solely to receive the subsidy.

In some circumstances, especially when the costs of monitoring land degradation for particular types of land use are high, zoning may be the most appropriate policy for determining boundaries between *broad* categories of land use. For example, land which over a wide range of economic conditions is judged to be suitable for cropping, pastoralism, and non-agricultural usage only could be classified. Zoning of land use can be an effective strategy, but in the author's view it is a fairly blunt policy instrument and its use should be restricted to classifying broad categories of land use. It is also most important that any zoning policies are based on both bio-physical and economic criteria. Some land in the semi-arid areas, for example in the Kimberley region (Dillon 1985), may be zoned as unsuitable for pastoral usage because in both an economic and bio-physical sense it is 'marginal' land for pastoral use.[10]

In the pastoral zone, particularly the semi-arid areas, land degradation appears to be primarily an onsite problem. The level of land degradation could be measured by jointly monitoring vegetative cover, depth of topsoil and stocking rates. Regional land quality/degradation benchmarks could be set in terms of these parameters. Property owners could then be subsidised for rates of land degradation below the regional benchmark and taxed if rates of degradation on their land exceeded the regional benchmark. If subsidy/charges were levied on an annual basis, allowance would need to be made for the stochastic influences of weather, particularly severe droughts. For government leasehold land, subsidies (charges) could be simply deducted from (added to) the lease payments. At the time of termination of a lease, compensatory payments should be made by government, or by a lessee, depending on whether the land is in a better or worse state than the regional benchmark. The calculation of compensation at the time of termination of a lease would, of course, need to take account of previous subsidies/charges.

The regulations applying now to much government leasehold land specify maximum stocking rates and impose very severe penalties on land users exceeding set levels. However, penalties are so severe that regulators only very rarely enforce them. As a consequence, leasehold land is commonly stocked at rates well in excess of the specified maximum. Moreover, there is no incentive for any land user to stock a property at a level below the maximum rate. The regulations have largely failed (Young 1983 and Chapter 8). The subsidy/charge system described above has a much more finely tuned structure than the existing regulations and provides a continuous incentive to a land user for abatement of land degradation.

Subsidisation of selected conservation measures

There are important problems associated with the present practice of subsidising only a few selected soil conservation measures. The subsidisation of particular forms of soil conservation will provide an incentive to a land user to undertake greater investment in soil conservation only when the investment increases a landowner's profits. In some situations the costs of land degradation, and thus the benefits from introducing conservation measures, will be completely external to the land user causing the damages. The land degradation will affect neither revenues nor private costs of production. For example, in the case of beach and sand dune degradation caused by the operations of a sand-mining company, so long as the subsidy is less than 100 per cent of the cost of conservation, there is no incentive for the company to undertake it, except perhaps as a public relations gesture. Even where there are internal benefits from soil conservation, so long as the bulk of the benefits accrue to others, the rate of subsidisation required to induce the optimal level of degradation abatement will need to be high.

A major vehicle for subsidising soil conservation activities is the Australian Income Tax Assessment Act (Section 75D) which allows 100 per cent deduction (immediate write-off) of capital expenditures on structural works for soil conservation, but which does not permit similar full deductibility of certain capital expenditures associated with good land management conservation practices. For example, construction of levee banks to combat soil erosion and the construction of surface or subsurface drainage works to control drainage and salinity problems are eligible for 100 per cent immediate write-off for income tax purposes, but expenditures on minimal tillage cultivation equipment and direct-drilling equipment are not.

The potential problems associated with subsidising a few selected soil conservation measures arise because *all* land inputs and management practices have some impact on land conservation/ degradation. Consequently, if government subsidises only structural forms of soil conservation the likely outcome will be over-investment in structural forms of soil conservation relative to adjustments in land management practices that reduce land degradation. A similar criticism can be levelled against large mitigation projects, which are often fully publicly financed, when the investment takes place without sufficient attention given to alternative methods of abatement, or to the behavioural responses of the land users. Kelly (1985), for example, claims that if money had been spent on research techniques for rationing irrigation water to plants, rather than on investment in expensive engineering works designed to prevent saline water seeping back into the Murray River, society would be better off. Existing policies provide quite inadequate incentives to use irrigation water sparingly in the Murray River basin.

This author has argued previously that '... many forms of subsidy, if expected to be permanent, provide incentives for farmers to alter farm-management procedures towards those which have the potential to cause higher levels of soil degradation, but which are now more profitable because the subsidy has lowered the *private* costs of repairing, or containing, the damage'. In other words: 'Subsidies will induce more investment in soil conservation, but at the same time, they are likely to increase the *need* for more investment in soil conservation' (Chisholm 1978, pp6-7).

The subsidisation of structural forms of soil conservation, or of water and fodder reserves to combat drought, for example, may be largely ineffective in reducing land degradation because farmers respond by increasing stock numbers (Chisholm 1978; Freebairn 1983).

The subsidisation of investment in structural forms of soil conservation that increase the onsite productive capacity of land are, of course, very attractive to land users. This is particularly true when the subsidy is unconditional in the sense that the land user is free to adjust the levels of all other activities to maximise private profits. To take full advantage of the increase in the physical productive capacity of the land a land user will generally increase other inputs and thus use the land more intensively. Preventive measures of land degradation that reduce offsite damages, but do not substantially increase the onsite productive capacity of the land, are a much less attractive means of conservation. Moreover, in a number of states responsibility for soil conservation is vested in the Department of Agriculture which traditionally has had a production orientation. From the viewpoint of these departments, subsidisation of highly visible structural forms of conservation have the advantage of giving the appearance of 'getting the conservation job done' while at the same time usually increasing the onsite productive capacity of the land.

The provision of drought subsidies for water and fodder reserves gives farmers an incentive to carry more livestock on their properties since the private costs of carrying a livestock unit through a drought are reduced. Farmers will thus carry more livestock into a drought and retain more stock on their properties for the duration of the drought. There may well be a resultant increase in land degradation. Protection of the land/soil component, which is the least renewable component of the animal-plant-soil system (Costin 1983), is not a clearly enunciated and specific objective of Australia's drought policy. There is insufficient recognition that in some drought situations the socially optimal strategy would be to move livestock out of a drought stricken region.

It is beyond the scope of this paper to discuss optimal institutional structures for controlling land degradation (see Junor and Watkins Chapter 14). However, where there is a wide array of activities that interact and influence soil degradation, unless all activities are appropriated subsidised or regulated, not only will the mix of soil conservation measures not be cost-minimising, but there is also a potential danger that no general reduction in soil degradation will be achieved.

It should be recognised that, if government policy includes the payment of subsidies for soil conservation, the fact that there are few, if any, ways of collecting revenue to finance the subsidies, without introducing further distortions in resource allocation, needs to be taken into account. The major source of government revenue in Australia is income tax. Studies of the distortionary costs of the income tax in Australia suggest that, at the margin, it costs about 40 cents to raise an additional dollar of income tax revenue. Policy makers should be aware of these distortionary costs when they are deciding the extent to which subsidies should be used as an instrument for controlling land degradation.

Income tax

The Income Tax Assessment Act (Section 75D) is currently a major vehicle for subsidising certain structural soil conservation activities. An anomaly existed in the Act until 1983 in that farmers were allowed under Section 75A of the Act to deduct from their taxable income the full cost of land clearing in the year in which it was incurred. This taxation concession substantially reduced the private cost of clearing and made it profitable to clear greater areas. Removal of vegetation is now recognised as a potential major cause of land degradation. Within the same Section of the Income Tax Assessment Act there was thus, for many years, two separate taxation concessions applying to land use which were directly opposed to each other with respect to their effects on soil conservation (Blyth and Kirby 1985).

The income tax system has a number of attractive practical features as a vehicle for providing incentives for conservation. It has the virtue of allowing Commonwealth government involvement in a fairly simple and direct manner through a long-established fiscal device.

For example, Section 75A of the Income Tax Assessment Act could be resurrected, but in a quite different form. Taxation concessions would be given for *not* clearing vegetation cover, rather than for clearing land. The scheme could be applied, for instance, to the problem of dryland salting and stream salinity caused by the clearing of trees and native vegetation. The existing proportion of cleared to uncleared land on each farm would be a possible benchmark for the income tax concession. Land users would receive a tax concession on areas planted to new trees or on areas on which native vegetation was permitted to regenerate naturally by say, fencing off livestock. An income tax surcharge would be payable on areas of established trees which were cleared.

An income tax concession-surcharge system could also, in principle, be applied to areas such as the arid and semi-arid zones. If it is possible to establish a socially desirable benchmark with respect to such factors as vegetative cover, depth of topsoil, and stocking rates, which is fairly uniform between properties this could be used as a base for the concession-surcharge. An income tax concession would be given to a land user when the state of the land was better than the specified benchmark according to the above indicators of land conservation/ degradation and an income tax surcharge would apply to land which was more degraded than the established benchmark. Recent advances in remote sensing (satellite imagery) techniques may make monitoring of vegetative cover in arid and semi-arid regions a reasonably low cost enterprise.

From the viewpoint of both efficiency and equity it is desirable that income tax concessions to stimulate conservation are given in the form of tax rebates rather than as deductions from taxable income.[11] Providing that a land user is liable for income tax, tax rebates provide the same dollar incentive regardless of a taxpayer's marginal tax bracket whereas the value of a deduction from taxable income varies with the tax bracket of an individual.

A further point which is relevant to the issue of charges versus subsidies relates to the problem of uncertainty and the likely significant errors involved in the measurement of offsite damages and their attribution to particular sources. When measurement of offsite damages caused by a particular land user is subject to considerable uncertainty (error) the legal basis for charges (taxes) may be very tenuous. Equity issues are raised and an appeals process would almost certainly be required. In circumstances where there is substantial uncertainty and potential errors of measurement of offsite damages it would appear that legal considerations favour the use of subsidies rather than charges.

Compensation and expectations

A number of economists (eg Wills 1985) have argued that compensatory payments should be made to land users in some circumstances where government action reduces their property rights and thus imposes costs. For example, the prohibition on clearing native trees and vegetation in some regions (eg Western Australia) is accompanied by compensatory payments to farmers. A critical factor which has not been adequately discussed in the literature is the role of expectations.

In a well-functioning land market, land users formulate expectations with respect to future commodity prices, input costs, and government policies. Land users will seek to maximise the present value of their net worth by putting assets to their highest value use. The highest value use of assets is determined by discounting the estimated net income streams produced through time of alternative use patterns of assets. The particular pattern of asset use which maximises net present value may then be selected.

Expectations of future commodity and input prices and government policies are continuously revised so that any major change in expectations is ultimately reflected in a change in the market value of land. If a government is pursuing consistent and reasonably predictable policies market land values will reflect these policies. In selecting an efficient set of economic incentives/regulations to combat land degradation a government should seek to avoid introducing policies which impose significant and unanticipated losses, or windfall gains, on those owning land at the time of the policy announcement. The set of property rights implied by the chosen environmental policies should as far as possible be specified so that land users have secure expectations.

When environmental policies are in place, an equity case for separate compensation can be made only for situations in which fairly major and quite unexpected changes in government policy occur. The identification of these situations is, of course, ultimately a matter of political judgment. The author's view is that, through unanticipated changes in policy, governments impose windfall losses/gains, and associated changes in property rights, on people in all spheres of economic activity (eg investors in government bonds) and that a government should only make compensatory payments in special circumstances.

Public agencies and government

The fact that government general policies relating to land and agriculture may have secondary effects which exacerbate the land degradation problem has already been briefly discussed. However, in the area of pollution generally, public sector agencies themselves are a major source of environmental damage. Moreover, it is claimed that in some countries public agencies are among the worst violators of regulations designed to protect the environment; for instance, this claim is made for the United States by Baumol and Oates (1979). It appears that governments do not have the same legal powers they apply to private firm violators and that they are unwilling to use their legal powers on a public agency that violates an environmental regulation. The above researchers recommend that governments should consider applying pollution taxes to the polluting activities of public sector agencies. The efficiency of the response of public sector agencies to pollution taxes is, however, less certain than the response of private firms because public agencies are non-profit maximising.

In the context of land degradation a distinction needs to be made between the regulation of activities of private agents on publicly owned land and the regulation of activities of public agencies on publicly owned land. In the former category, a government could give a directive to a sand-mining company, for instance, that a mining lease will be issued only on condition that conservation measures, designed to return the site as nearly as possible to its original state, are undertaken after mining is completed. Similarly, standards may set limits on soil erosion and sedimentation of waterways in state forest areas being used for woodchipping. To meet these standards a woodchipping industry may adjust harvesting techniques, replant cut-over areas or implement certain structural forms of soil conservation. In both of these cases the firm, after consultation with the relevant public agencies, should be left to choose the least-cost method of meeting the standard.

Controlling land degrading activities of public agencies or governments is, of course, a complex problem. Conflicts between public agencies within a state are involved as are conflicts between a state government and the Commonwealth government. The Commonwealth government was able to use its power on the granting of export licences to stop sand mining at Fraser Island, but was unwilling or powerless to prevent land degradation associated with the construction of the Daintree road in Northern Queensland.

By and large, the general principles pertaining to public agencies and land degradation are reasonably clear though their implementation is fraught with difficulties. First, the number, size and structure of public institutions dealing with land degradation in Australia should reflect the fact that there is a wide array of forms and causes of land degradation. All land use activities have some influence on the state of the land and degradation of land may cause offsite degradation (pollution) of other land, waterways and the air. There are important and complex interrelationships and a multi-disciplinary approach to the problem is essential. Water, for instance, is the primary transport medium for sediment and nutrient runoff from the land. Thus water pollution is a direct result of land degradation. Cognisance of these factors strongly suggests that there is undue fragmentation of public agencies for land and water in Australia.

The size of catchments (total catchment management), for example, ranges from small reservoirs to large river systems that cross several state boundaries. Conservation jurisdictions should, as far as possible, reflect the geographic area or region over which the major physical and biological interrelationships and relevant external effects occur. For large ecosystems, like the Murray River basin, where land and water use practices at one site may cause offsite effects many hundreds of miles away, there is a requirement for a very high level of inter-agency and inter-government cooperation in acquiring and sharing information and formulating policies.

Finally, many large public and private projects in the area of forestry, mining, road and powerline construction, national and state parks, and coastal land development and management require careful planning and cooperation beween a number of agencies and parties. Land degradation and other environmental effects pose a potential problem for all the above types of projects. Use has been made of environmental impact statements for assessing the environmental effects of large-scale private and public projects (see Formby 1977 for a critical evaluation of the use of environmental impact statements in Australia). The central purpose of the statements should be to help government to decide whether or not a particular project should be allowed to proceed, after taking due account of environmental damages. Ideally, this would involve using benefit-cost analysis to put a money value estimate on the environmental damage associated with each of the main alternative ways of undertaking a particular project. It would also be necessary to determine the optimal level and form of environmental damage abatement expenditures for each development strategy. With this information, the particular strategy which maximises the net present value of the project—when full account is taken of abatement expenditures and the environmental damages that remain—can be chosen. Private enterprise (or government) would be permitted to proceed only with those projects having a positive net present value. In practice, of course, it is commonly not possible to put a monetary value on some environmental damages. A detailed physical description of such intangibles is the best that can be done. Such intangibles should be taken into account by policy makers when they use their judgement to set an appropriate environmental standard for a particular project.

Criticisms have been levelled against public agencies for giving undue prominence to purely technical factors when deciding on methods of construction and placement of roads, powerlines etc and the building of large mitigation works for conservation purposes. Benefit-cost analysis has many weaknesses and limitations, but there does not appear to be any other reasonably coherent and rational framework for guiding policy makers (see Appendix A).

Conclusion

Policy prescriptions for the control of land degradation are difficult to formulate due to the non-point nature of most forms of onsite and offsite damages. An endeavour has been made in this chapter to review the major policy instruments and point out some areas of potential application. Various weaknesses in existing forms of government intervention have been identified and discussed.

The application of incentives and regulations to control non-point forms of land degradation requires that damages be measured *indirectly* via the identification and monitoring of those factors and activities that are responsible. In some situations there will be a single, or small number, of clearly identifiable activities causing degradation that are an obvious target for a regulation or economic incentive: eg tree clearing or excessively high stocking rates. It has been argued that in these cases the particular forms of existing regulations are commonly inefficient or unenforceable.

Perhaps the most significant failure of existing policies is that economic incentives and regulations typically apply to only a few selected inputs or activities (eg investment in structural soil conservation measures) in circumstances where many activities have an impact on soil conservation/degradation. Under such policies, land users will not only not select the socially least-cost methods to combat land degradation, but they may also respond by altering levels of other activities (eg increasing stocking rates) with the result that there is little if any abatement of degradation.

Effective policies to control land degradation must take account of land users' behavioural responses. At the very least, there is a need for subsidy payments for structural soil conservation measures, for example, to be made conditional upon a land user having a 'reasonable' behavioural response with respect to cropping practices, stocking rates etc.

A further step towards the design and implementation of policies for soil conservation would be the widespread development and application to Australian conditions of models like USLE and SOILEC. The potential for large scale application of such a system, at the regional and individual farm level, will depend, of course, on the costs and benefits of its development and administration. Close links and good working relationships between researchers, extension workers and land users would be most important and these links are already fairly well established in the Australian rural sector.

The ultimate stage of the above process would be for extension workers and farmers jointly to apply these models to help determine least-cost (most profitable) farm plans and management practices which meet 'safe minimum standards' of conservation. In this process it is most important that the dynamic nature of land use be fully recognised and that successful land management typically requires rapid responses to changing seasonal and economic conditions. Commonly, there will be many technically feasible farm plans and the least-cost plan will depend upon expected product prices and input costs. It is the individual land users who bear the responsibility of the financial outcome of implementing a particular farm plan and it is they who should be permitted to exercise their judgment freely in making the final choice of the least-cost strategy to control degradation.

Endnotes

1. For further treatment of these issues see Dumsday and Edwards (1984), Musgrave (1984a), and Blyth and Kirby (1985).

2. Some forms of onsite land degradation may have beneficial offsite effects: eg the depositing of topsoil on alluvial river flats. Indeed, it is conceivable that in some instances the beneficial offsite effects may outweigh the onsite degradation costs.

3. It should be noted that the soil loss predicted by USLE is that soil moved off identified slope segments. Soil loss should be distinguished from field sediment yield. A field's sediment yield is the sum of the soil losses on the slope segments minus depositions in depressions within the field, along field boundaries etc. USLE does not account for this deposition and thus overestimates the soil transported offsite. Furthermore, erosion is not synonymous with soil loss. The soil loss from a contoured and stripcropped slope, for instance, is considerably less than the amount of soil eroded between the sod strips.

4. Extension workers should also be familiar with models like SOILEC.

5. See De Boer and Gaffney (1976) for a critical review of some aspects of the mandatory land use regulations applied at that time on the Darling Downs which they claim to have been based primarily on technical considerations. Research results from their linear programming study suggested that the major impact of the mandatory program on farm costs was caused by restrictions on forage cropping. The regulations led to increased summer feed supplies (pasture) which is normally already in surplus supply and it seemed most unlikely that soil loss abatement was being attained at least cost.

6. For a good discussion of runoff pollution of waterways see Olive and Walker (1982).

7. Hodge proposed a system of transferable rights for cleared land. Greig and Devonshire (1981) had earlier proposed the use of tree clearing taxes on land users causing offsite saline damages.

8. A similar system of transferable cropping rights could be applied on marginal cropping lands.

9. A system of transferable water rights for irrigation has been operating in recent years in South Australia and appears to be working well (Tuckwell 1984).

10. It is worth noting in this context that agricultural policies in overseas countries, particularly in the European Economic Community and the United States, exert a considerable influence on land use in Australia via direct and indirect effects on demand for Australia's agricultural exports. The impacts of agricultural policies on world trade and the pattern of production and consumption for five major agricultural commodity groups have been modelled by Chisholm and Tyers (1985). Insofar as foreign agricultural policies influence the pattern of land use in Australia they will also affect land conservation/degradation. Foreign policies which result in high export prices, say for Australian wheat, may cause wheat production to expand into marginal areas with a consequent high level of land degradation. However, on the basis of this observation it would be wrong simply to assert that high export prices are a bad thing for land degradation and low prices are a good thing. Clearly, if export prices are too low farmers will be under pressure to reduce their level of investment in various forms of soil conservation.

11. Since writing this chapter the author has become aware of a recently published research study by Haynes and Sutton (1985) which provides a comprehensive treatment of this topic.

Commentary

Geoff Edwards

The four papers (including Young, Chapter 8) on which I have been asked to comment all exhibit characteristics of the economist's approach. However, they cover a wide field. My approach is to focus on some issues which are central in the economics literature on the use of resources for the social good. Some implications of received economic knowledge for efforts of governments to deal with land degradation are noted, drawing on the content of the four papers. One or two queries are raised about the arguments in those papers. Because the paper by Chisholm provides the most detailed treatment of most of the main economic questions which arise in the context of land degradation, my comments contain more references to his paper than to the papers by Young, Quiggin, and Kirby and Blyth. The issues considered in my comments are:

(a) Determining the problem and the solution: two approaches in the social sciences;
(b) Property rights; and
(c) What can governments do?

Before dealing with these issues I make the observation that, perhaps understandably, the authors of the four papers have concentrated on the more recent literature. A little delving in older publications reveals that a good deal of what we know about the economics of soil conservation is not new knowledge. In particular I commend to the reader Campbell's (1948) article on subsidies for soil conservation.

Determining the problem and the solution: two approaches in the social sciences

Two distinct ways of thinking about resource use issues with which there is concern can be identified in the social sciences literature. For convenience these may be called the 'market imperfections approach' and the 'cost-effectiveness approach'.

The market imperfections approach

A fundamental proposition accepted by the supporters of this approach is that competitive markets, when they function well, will generate an efficient use of resources. A necessary condition for establishing that the market is not allocating resources efficiently is the identification of a market imperfection which prevents the market performing in accordance with the fundamental proposition. There are many potential sources of market failure. Among the more commonly mentioned are inadequate competition; deficiencies in the capital market; information gaps; and externalities, including public goods. Demonstrating that a genuine impediment to the efficient working of the market exists is not a *sufficient* condition to justify government intervention in the market. A sufficient condition is that the expected benefits to society from intervention exceed the expected costs. Benefit-cost analysis is thus regarded as an essential methodology for assessing the case for government intervention to improve resource use.

The market imperfections approach appears to fit most easily with the view that governments give equal weight to changes in the welfare, as measured by dollars, of each member of a society. This is sometimes called the 'public interest' interpretation of government. It supposes that government intervention in markets is motivated solely by a desire to achieve efficient use of society's resources — that is, to raise the aggregate level of real income. Any redistribution of income that is desired will be obtained, on the public interest interpretation, by general taxation and income transfer policies rather than by intervention in particular markets (Sieper 1982).

The market imperfections approach, with the use of benefit-cost analysis, is the most generally supported approach to the study of resource use concerns in economics. It is the approach taken by Kirby and Blyth. They point out that '. . . observations of the mere existence of degraded land do not provide sufficient evidence of market failure; the relevant issue is whether or not the rate of land degradation is optimal' (p 222). Kirby and Blyth consider that the main candidates for the title 'impediments to efficiency in rural land use' are government policies (including land tenure arrangements, taxation and subsidies for irrigation water) and external diseconomies.

Note that the first of these impediments represents failure of government rather than failure of the market. Government failure was also seen by Young as endemic in the operation of leasehold tenure in arid Australia. None of the 27 types of lease extant in arid Australia met more than two of the nine conditions identified by Young as necessary for effective control of land use in arid Australia.

Given that benefit-cost analysis is most important when it is forward looking and carried out before a government acts, it is important to recognise that it is *expected* benefits and costs that are relevant. There will usually be a substantial amount of uncertainty about the estimates, which is the reason why sensitivity testing to variations in important parameters is widely advocated and practised. It is worth noting that retrospective benefit-cost analysis of past projects and policies may also be useful, largely because it assists in identifying unreasonable assumptions and inconsistencies between evaluations of different projects/policies, that reduce the accuracy of benefit-cost assessments.

The cost-effectiveness approach

With this approach the objective, or the level of an activity that is to be achieved, is determined outside the market system, with the role of economic analysis being confined to comparing the costs of different ways of achieving the goal. One non-market approach to determining how much land degradation should be accepted—or how many resources should be allocated to reducing damage—would be to rely on the judgment of experts. Burton (1983) is prepared to say it is 'self-evident' that more resources are needed in the soil conservation area. He saw an urgent need for about five times as many trained professionals working on soil conservation as there were in 1983.

A less extreme view is taken by Chisholm in his chapter. Chisholm '. . . proceeds on the presumption (perhaps false) that for much of Australia's land there is a significant *economic* problem which requires government intervention or change of policy' (p232, original emphasis).

The cost-effectiveness approach is adopted explicitly by Quiggin. He sees it as impossible to determine the optimal rate of land degradation, or the optimal quantity of resources to allocate to reducing damage to land, by means other than political ones. Baumol (1972) has also taken the view that it is reasonable to view levels of environmental quality as political decisions, with the efforts of scientists and social scientists being concentrated on achieving the targets at the lowest possible cost.

Because resources allocated to protecting or repairing the land resource are at the expense of resources for education, transport, defence, quarantine and other uses, their allocation inevitably involves political considerations. With decisions determined largely by political considerations, many may see the public interest view of government behaviour as less compelling than the private interest view which emphasises the political benefit-cost calculus performed by those in governments and bureaucracies. Clearly, if the private interest paradigm is the more relevant, it can be expected that the effects of government decisions on the distribution of income and wealth— ignored in the market imperfections approach—will be relevant in making those decisions. Since this applies to decisions on choice of policies as well as to decisions on how much abatement of land degradation is desired, it is not clear that governments, or their agencies, will be much interested in what social scientists and scientists have to say about cost-effectiveness. It is true, however, that the better the case these scientists can state for expecting a particular course of action to generate benefits to society in excess of the costs the greater the attention governments and bureaucrats will have to give it.

Let me attempt a summary. For economic efficiency, which encompasses efficiency in allocating resources to the prevention and restoration of degraded land, the market imperfections approach is in principle preferable to the cost-effectiveness approach. The greatest deficiency in the cost-effectiveness approach is that the politically determined level of land degradation may be far removed (in either direction) from the level that gives a balance between marginal social costs and marginal social benefits. But this approach makes fewer informational demands on members of our disciplines. If, as is widely agreed, the most important class of market failure is externalities, the major additional information needed from our disciplines to apply the market imperfections approach is the extent to which the external effects should be internalised. The more successful scientists, economists and others are in evaluating the social damage from external diseconomies of land degradation, and the social costs of reducing that damage, the stronger will be the case for — and the likelihood of — placing greater reliance on the market imperfections approach and less on the cost-effectiveness approach.

Property rights

Property rights are '... one's effective rights to do things and his effective claims to rewards (positive or negative) as a result of his actions' (McKean 1972 p177). Property rights may have the legal sanction of the state or they may rest on less formal sources of power, such as custom. In either case, property rights determine the expectations a person can reasonably hold in his dealings with others. From the point of view of the individual decision maker, what are private costs and what are external costs depends on the specification and allocation of property rights. If a farmer has the right to clear trees on his property he can be expected to disregard any external costs which this imposes on others through the salinisation it induces. If the right to be free of such salinisation effects is vested in those who would be harmed by the removal of the trees — and if the rights are enforced — the farmer will be compelled to include in his own costs, the costs of preventing damage to others, or of compensating them for damage costs.

Once property rights are specified and allocated, there are various ways of protecting them. Three ways that are important are outlined here (see, for example, Bromley 1982a).

First, A's property rights may be protected by stipulating that no one can infringe them without first obtaining A's consent. This is the protection involved with traditional property rights. With such rights the party wanting to infringe them has to take the initiative in negotiations with A and will usually have to bear most of the transaction costs involved (Bromley 1982a).

Second, it may be possible for other parties to act contrary to A's interest without obtaining his consent, but knowing that they will have to compensate A for the damage experienced. The compensation would be determined by an independent body. In this case A's rights are protected by a liability rule.

Property rules and liability rules are both of limited use for protecting property rights when dealing with the major classes of land degradation involving non-point externalities, a point made clearly by Chisholm. One reason for this is that the property rights are often not well defined, so that it is not clear who must initiate the bargaining and meet the bulk of the transaction costs. Even if the relevant rights are clearly specified and assigned other formidable problems exist. Partly because of imperfect knowledge of the physical relationships involved, it is impossible or at least very costly to determine the extent to which particular decision makers are contributing to externalities experienced by others. The need for cooperation between perhaps many individuals, with the associated free-rider difficulties, is a further problem. These weighty problems seem not to be acknowledged in the discussion of liability rules by Bradsen and Fowler.

The third way of protecting A's property rights is through an inalienability rule. This prevents others from damaging A's property at any price. Chisholm makes the point that prohibition of tree-clearing, an example of inalienability, may be warranted '... as a short-term emergency measure to provide time for more information to be gathered and analysis undertaken (p 235) but he considers prohibition is unlikely to be optimal as a long-term policy. It appears reasonable to expect that Chisholm's view, with which few economists would disagree, applies also to prohibitions of other activities that may contribute to land degradation — including, for example, cultivation of steep slopes and fallowing. Inalienability is often seen as an appropriate way to reduce the incidence of highly dangerous substances in the environment. Zoning is a less extreme approach to this problem than a general prohibition. Braden (1982) suggests that inalienability is suitable where a right has such a high value to its collective owners that little exchange would occur if it were allowed. The problem lies in deciding where this situation exists. Even if a goverment or government agency correctly gauges at a particular time that protecting certain rights by an inalienability rule would result in decisions very similar to those resulting with a property rule or a liability rule, the prohibition could be made inefficient by subsequent developments in knowledge, technology or preferences. Regulations, of which prohibitions are an extreme form, often endure long after the balance between their costs and benefits shifts dramatically from that existing when they were introduced.

The need for property rights to evolve

It will be clear from the above discussion that the specification of property rights is not something that can be reasonably viewed as static. In fact, it is to be expected that the efficiency with which a society meets the aspirations of its citizens in the long run will depend to a large extent on the adaptations made to property rights in response to technological developments, newly discovered relationships and changes in community values (Demsetz 1967). As Braden (1982) notes '. . . rights in agricultural land are no exception to this process' (p26).

Approaches to the allocation of new property rights

Two approaches to the allocation of property rights can be distinguished in the writings of economists and others on this subject. One approach emphasises equity or fairness. I include Buchanan's (1975) 'contractarian' approach and Rawls' (1971) 'veil of ignorance' approach to determining rights in this category. The other approach puts the emphasis on efficiency. That is, the objective is to allocate property rights to the party that values them most highly. Posner is the person most strongly associated with this approach (eg Posner 1972).

We know from Coase (1960) that when property rights are transferable and transaction costs are low it can be expected that resource allocation will ultimately be independent of the allocation of property rights. But, as Chisholm notes, low transaction costs are generally not a characteristic of the sorts of exchange that come under consideration in the context of land degradation. This means that, if efficiency is the criterion, rights need to be allocated initially to the party to whom they are most valuable. So, if the right to clear trees is more valuable to the farmer on whose land they grow than the right to stop their removal is to those opposing removal, it is efficient to assign the right to the farmer. Allocation of property rights according to efficiency will not always be consistent with the widely supported 'polluter-pays principle'. Chisholm implies that efficiency considerations may be relevant in assigning property rights when he accepts that efficiency gains from reductions in income taxes may be seen by policy makers as an argument in favour of taxing land degradation rather than subsidising its abatement.

There are, however, some strong arguments against assigning property rights on the basis of efficiency (eg Randall 1983). Rights would be insecure, because they could be changed with changes in technology or relative prices or both. Security of rights is important for economic decisions, especially decisions in such areas as saving and investment. Security of rights is also important for long-term social stability. The absence of this security could be expected to increase resources devoted to bringing about changes in the rules of the game. In Randall's words, '. . . rights which shift with the benefit-cost numbers would tend to discourage voluntary exchange, while encouraging efforts to generate and gain recognition for the kind of benefit-cost data which would ensure reassignment of rights in a Coase-Posner world' (Randall 1983, p144).

Another problem in allocating property rights according to the efficiency criterion is that income effects may be significant. If this is so the allocation of property rights influences the pattern of resource use that is most efficient, even if transaction costs are zero. This causal relationship between property rights assignment and efficient allocation of resources is considered by some to render benefit-cost analysis hazardous or even inappropriate for evaluating projects or policies that involve rearrangement of property rights (eg restrictions on rights of road users, regulation of chemical additives in animal feeds, controls to reduce erosion or salinity) and hence the distribution of income flows (eg Lang 1980, Bromley 1982b).

What can governments do?

At least three important areas for possible government action are raised in the papers under discussion. These are government review of the consequences of its own agricultural and other policies for land degradation; research and extension; and policies directed specifically to reducing land degradation.

Review of government policies

In the four papers under discussion several instances have been given of agricultural policies which have added to soil erosion or salinity or to both. According to Young, closer settlement policies which restrict the size of pastoral holdings in Queensland and New South Wales have resulted in a deterioration in the condition of the land resource, and the use of maximum stocking rates in South Australia may have increased the stocking rate by up to ten per cent. The adverse effects on soil erosion and salinity of tax incentives for land clearing were referred to by Kirby and Blyth and by Chisholm. In all four papers reference was made to adverse effects of drought assistance policies. Policies of assistance based on livestock carried, or fodder used, both discourage early destocking in a dry period and encourage higher stocking rates at all times. Other critics have pointed out that drought assistance may make future droughts worse by weakening incentives for farmers to act to protect themselves against such events (eg Freebairn 1983, Aust IAC 1983). Subsidised irrigation water, which induces excessive use and increases salinisation, is another policy which is in conflict with maintenance of the environment.

Of course, the fact that a policy has some unfavourable effects does not necessarily mean it is inconsistent with efficient use of resources. Decisions at the family, firm or national level usually involve some bad effects (costs). It is important that such costs be taken properly into account by governments in assessing policies. Sometimes agricultural policies or general economic policies will have *favourable* effects on the soil resource; such benefits should be included in evaluations of the policies. A question that warrants consideration is whether adoption of the policy favoured by farmers and economists of reducing protection for manufacturing would reduce or increase land degradation. Various effects are involved. An increase in the profitability of agriculture due to removal of assistance to manufacturing (or to higher world prices for farm commodities) appears likely to reduce onsite soil deterioration occurring on existing farm land and of which the farmers concerned have full knowledge (the increase in the value of the soil provides an incentive to reduce this class of land degradation). The effect on onsite damage about which landholders have imperfect information appears more uncertain. Offsite damage could be expected to change in the same direction as onsite damage in many but not all situations. The induced expansion of farming into marginal lands caused by higher profitablity would tend to increase both onsite and offsite land degradation. The overall effect of increased profitability on land deterioration seems to be unclear and deserving of further consideration.

Within the existing institutional arrangements the imperfections that are most evident are those introduced by governments rather than those inherent in the market system. It may be easy to reach agreement on this. That does not necessarily mean that the policies can be readily changed, especially if governments are motivated by private interest considerations.

Research and extension

There appear to be situations in which farmers could simultaneously increase their profits and reduce the external costs they generate by changing their production mix or their management practices (Dumsday et al 1985). According to the market imperfections approach, extension of information on the more profitable production systems may be appropriate—depending on the relative size of benefits and costs—in these situations. However, it needs to be recognised that taking advantage of profit-increasing opportunities currently forgone because farmers are unaware of them would sometimes result in larger external diseconomies of land degradation. That is, ignorance of profitable changes in production or management practices is not confined to those that would have favourable effects on other parties.

In view of the small amount that is known about the contributions of different landholders to externalities of land degradation and about the value of the damage arising from externalities, further research in these areas is important for better decision making on policies.

Policies: taxes, subsidies and regulations

A vast literature has developed on environmental economics in the last 20 years. The advantages and disadvantages of various market approaches (taxes and subsidies), regulatory approaches and combinations of the two (eg marketable pollution quotas, performance standards with penalties and rewards for performing worse or better than the specified standard) have been analysed closely. However, the work of economists has focused almost entirely on point pollution. The land degradation problems of major concern are usually problems of non-point pollution. That is, it is either impossible or excessively costly to determine the contributions of individiduals to external diseconomies such as sedimentation and salinisation. The important book on environmental economics by Baumol and Oates (1975) contains only half a page on non-point pollution.

Because it is normally infeasibile to measure the output of environmental bads (eg soil loss, salting) from individual farms, policies which are most efficient ('first best') for internalising environmental externalities generally cannot be used. That is, policies of taxing output of the pollutant, subsidising reductions in output of the pollutant, or use of pollution quotas have little if any relevance to problems of land degradation given present knowledge. Nor can regulatory measures, which are generally less preferred by economists, operate on the output variable of ultimate interest.

Feasible market, regulatory and mixed approaches to reducing land degradation must operate on variables that are *related to* the variable of primary interest — eg production activities, inputs, management practices. All the issues that arise in considering the choice between various policies operating on output of pollutants in the case of point pollution exist also in assessing the feasible ways of reducing non-point pollution. In particular, the choice between penalties for damaging the environment and rewards for behaving in ways that reduce damage can be seen as depending on society's views on who has what property rights. Chisholm's point about subsidies suffering the disadvantage that revenue must be raised by means of taxes, which themselves result in economic costs of over 30 cents for each dollar of revenue raised, applies to subsidies on use of inputs, production of commodities or use of practices that reduce land degradation, as well as to subsidies paid directly on damage abatement. Use of penalties not only reduces the need to use distorting taxes for revenue puposes, it in fact allows such taxes to be reduced, thereby generating efficiency gains. (This supposes that the revenue from the penalties goes into government revenue and is not used to compensate those suffering from land degradation or hypothecated for other specific purposes such as improvements to damaged land). Chisholm's suggestion that tax rebates are a more efficient way of subsidising soil conservation than tax deductions is questionable. The use of rebates would introduce a distortion between spending on deductible items and rebateable ones. Farmers with a marginal tax rate higher than the rebate would find it profitable to spend less on soil conservation and more on deductible items than if all expenditures were tax deductible. Farmers whose marginal tax rate was lower than the rebate would gain by reallocating spending from deductible outlays to soil conservation.

Policies that operate on inputs, activities or practices have an important disadvantage not shared by policies that work on output of pollution or environmental damage. This is that there are *many* things which a farmer can do that will reduce the external diseconomies which his activities generate. Read (1984) tabulates some of the possibilities for the control of salinity problems in Victoria. Unless policies—be they taxes, subsidies or regulations—provide incentives for farmers to change their behaviour in *each* of these ways to an extent which reflects their contribution to reducing external damage, any given reduction in damage will not be achieved at the lowest possible cost. Feasible policies will invariably be directed at one or at most a few factors thought to be important contributors to damage — eg tree removal, cultivation of steep slopes. This problem is raised by Kirby and Blyth and discussed at greater length by Chisholm in this volume and elsewhere (Chisholm 1978).

Undoubtedly, feasible policies will be very imperfect. This represents a marked difference between policy for control of land degradation and, say, industry assistance policy. In the latter case a policy can be defined (something close to free trade) that would be best for Australia from the view of efficiency. It is not yet clear whether governments can use taxes, subsidies or regulations that reduce land degradation — beyond changing government policies that generate costs in land damage that are excessive in relation to the benefits — in ways that increase efficiency. Greig and Devonshire (1981) estimated that a tax of about $88 per hectare on land clearing in the Loddon catchment in Victoria would internalise stream salinity externalities. Further case studies, incorporating simulation studies of the sensitivity of net benefits from government intervention to assumptions about substitution relationships between activities, inputs and management practices would help in defining the scope for economically beneficial government intervention to conserve the land resource.

Furthermore, the structure of taxes, subsidies or regulations that was optimal—or even 'reasonable'—would vary both between regions and between farmers within a region because of differences in soil types, topography, management and other factors. The problem is thus less conducive to treatment by means of a uniform nation-wide policy than is the problem of achieving an efficient industry structure, or efficient use of Australia's oil resources. The relationship between the benefits and the costs of feasible government intervention is also certain to vary substantially between locations.

Finally, a few words are in order on another approach that is seen by some as the best way to tackle problems of soil loss and deterioration: viz, education. Four points are made. First education is likely to be effective only when there is a gain to the individual concerned. This gain may be in terms of higher current income or reduced deterioration of the individual's own land, or both. It is unrealistic to expect that education on ways of reducing offsite damage will generally result in changes in land use unless there are benefits to the individual also. Second, the very gaps in knowledge which prevent the application of other approaches to land degradation will often preclude also an efficiency-increasing education program. Third, as already noted, there appears to be no basis for expecting that the areas where education would give the greatest gain to the individual landowner would reduce external costs of land degradation rather than increase them. Fourth, the question of property rights—or who should pay—is as relevant in considering the education approach as it is in thinking about using taxes, subsidies or regulations.

Concluding comment

Thinking carefully about the issues that are relevant in deciding how best to conserve the soil resource is more difficult than analysing how to deal with pollution from a factory. The non-point nature of most land degradation problems means that it is not feasible to penalise the person imposing external costs on others in direct proportion to the size of those costs. Resort must be had to policies which operate on variables—such as production activities, inputs and management practices—that are related, albeit imperfectly, to the variable of prime concern. This will inevitably involve inefficiencies and rough justice. Whether it is better than doing nothing is unclear: the answer may vary from situation to situation. No objective answer can be given to the question 'should sticks or carrots be used to achieve a reduction in external diseconomies of sedimentation, salinity etc?' However, the widely accepted value judgment that 'the polluter should pay' implies that penalties rather than subsidies are appropriate. The polluter-pays value judgment implies that the property rights of a landholder do not extend to behaviour which causes physical damage to the property of others.

V

Pressure groups,
public agencies and
policy formulation

13 Pressure groups and policy formulation

Pressure groups
John Ballard

On one dimension of land degradation the papers in this volume are in agreement: the issues raised are exceedingly complex. This reflects infinite local variation in the character of land degradation, which often requires multi-disciplinary analysis and the costs of which are difficult to assess and assign. This fact alone ensures that a wide range of public and private agencies and interests are involved, making the aggregation of interests and the derivation of consensus on the nature of problems and solutions exceptionally difficult to achieve.

Much of the discussion of policy on land degradation concerns the relative virtue or necessity, or both, of relying either on regulation or on voluntary cooperation, with the latter either encouraged or not encouraged by incentives. Very few commentators raise the possibililty of delineating the issues on which one approach or the other might be most appropriate, yet it is clear that actual practice is a complex amalgam of both approaches. It is also clear that, barring crisis, policy change will continue to be incremental and that neither regulation nor voluntary action will be adopted as a global prescription.

The problem considered here is that of analysing the role of non-government organisations, specifically pressure groups, and their scope for influence, given the disaggregated structure of policy making. I argue the need for considering greater integration of policy making, first to identify more clearly the wider costs of land degradation, second to maximise areas of agreement on appropriate action on specific problems, and third to assist implementation of policy, whether through regulation, economic incentives, or education. Although it is evident that much of the action taken must be locally varied, or even site-specific, it is also evident that coherence in understanding the causes, costs and potential remedies of land degradation can best be achieved through an integrated forum with continuing linkage to all levels of decison making. Since the sources of expertise and the capacities for educating and mobilising public and farmer opinion often lie outside government agencies, it is vital that non-government organisations be incorporated into such a forum.

Pressure groups and land degradation

There appears to have been no substantial research on the activities of
non-government organisations concerning issues of land degradation.
This may be due in part to the recency of recognition of degradation as
a general category to be dealt with in more than a local and piecemeal
fashion. Land degradation is largely the result of activities related to
land development and management and these activities have, since the
earliest period of Australian history, held a central political legitimacy
not easily challenged. The costs of degradation were not perceived as
being borne by the groups, primarily urban, which were mobilised in
opposition to rural development interests, so land degradation did not
reach the party political agenda. In the rare instances in which costs
were perceived in aggregate fashion, eg in the case of South Australian
water supplies suffering from up-river Murray irrigation projects, these
were framed not in terms of degradation but of rival interests in
development (Kellow 1985).

Rural development interests have been well institutionalised for over
a century at state level through commodity-based producer associations,
political parties and government departments catering primarily for
these interests. Differences among rival commodity groups, however,
prevented the amalgamation of associations at both state and
Commonwealth levels until the 1970s. Only after the establishment of
the Industries Assistance Commission (IAC) created a need for technical
economic submissions on the wider costs of tariffs and subsidies did the
process of integrating rural interests into peak associations reach fruition
(Trebeck 1982). The formation of the National Farmers' Federation
(NFF) in 1979 provided rural development interests with a structure
linking state commodity bodies to a highly professional Canberra office
capable of negotiating on the whole range of rural interests. The success
of the NFF in mobilising the rural sector in defence of its wider economic
interests has given it a broad legitimacy and credibility among its
constituents and a capacity to educate on the wider costs of land
degradation.

The history of the mobilisation of environmental interests is very
different, yet the structures that have evolved resemble those of rural
development interests. The emergence in Australia of an awareness of
public interest in the maintenance and restoration of environmental
resources, distinct from rival interests in the exploitation and
'development' of those resources, has not been adequately traced. It
seems clear, however, that there was only limited assertion of this
awareness before the 1960s. The Australian Conservation Foundation
(ACF), established in 1965 as a federation of state and local bodies and
of individual members, had an initial focus on wildlife preservation and

national parks. The first major environmental confrontations over Lake Pedder (Davis 1980), the Great Barrier Reef (Wright 1977), Clutha (Hagan 1972), the Green Bans (Roddewig 1978) and Fraser Island (Roddewig 1978: p 118-139) involved locally concerned groups, with the ACF providing a broad umbrella of support and expertise. All these disputes involved proposals for abrupt disruption of the natural environment through development projects and none was concerned with agricultural or pastoral development, where land degradation is typically a dispersed and long-term process.

The creation of government agencies and commissions specifically responsible for environmental protection and planning was a phenomenon of the 1970s and particularly of the Whitlam government. They were part of a more general concern with broad sectoral planning and the examination of wider costs, which produced institutions as varied as the IAC, the Department of Urban and Regional Development and the Royal Commission on Australian Government Administration. The ACF in the mid-1970s also entered into the process of broad planning and began to insert land degradation from agricultural and pastoral practices onto an environmental agenda largely preoccupied with coastal nature preservation.

The joint involvement of farmer and environmental interests in broad issues of land degradation arises thus from increasing awareness of wider costs by government agencies and by peak organisations representing both interests. Whereas any form of land development, eg of irrigation or arid lands development, was previously seen as cost-free, the costs of degradation are now increasingly publicised. In Kellow's terms, borrowing from Lowi's typology of policy arenas, what was previously a privatised distribution of benefits has been socialised into a contested public issue of redistribution (Kellow 1981). The same process can be seen in the shift from privatised distribution of tariff benefits in the 1960s to contested redistribution, with full exposure of the costs of protection, under the aegis of the IAC. The growth of farmer awareness of protection costs has been contemporary with the growth of public (including farmer) awareness of the costs of land degradation.

A suggested framework for policy formulation

Land degradation is now established on several agendas, but these are diffused, located at Commonwealth, state and local levels and framed in terms of policies with a wide variety of other labels. Attempts have been made, notably by the ACF in its National Arid Lands Conference (Messer and Mosley 1983), by the Australian Institute of Political Science (AIPS) in its conference on Governing the Murray/Darling Basin

(Mant 1985), and by the conference on which the present collection of studies is based, to bring together various disciplinary and policy approaches and to establish a unified discourse on the issues of land degradation. Despite participation in these conferences by officials involved in the framing and implementation of policy, there is no assurance that the areas of consensus revealed will be translated into action. The new Soil Conservation Council provides the first official forum for Commonwealth-state consultation, but if it resembles other ministerial and official councils it will coordinate only government action at the highest levels. It will also lack involvement by non-official organisations, continuity of staffing and meetings, and flexibility in considering the widest range of factors and costs.

The task force approach to the issue of salinity taken by Victoria (*Australian Financial Review*, 6 November 1985) and initiatives for the establishment of a new Murray-Darling Basin Authority offer the most promising and flexible machinery for coping with land degradation on a regional basis where perceptions of crisis have broken through the barriers of jurisdiction. It is unlikely, however, that these approaches can be replicated where problems are not defined both regionally and in terms of a major economic and political crisis.

The chapter by Bradsen and Fowler with commentary by Barker in this volume indicate the difficulties in establishing broad administrative agencies covering the full realm of land issues and each suggests the greater practicability of consultative and advisory mechanisms. The problem is that of establishing and institutionalising a forum in which the wider costs of land degradation can be identified and assessed and optimal solutions negotiated. Such a forum would need continuity of staffing, the capacity to commission research and evaluation, and continuing links with all levels of government and with organisations representing affected interests.

The IAC provides elements of a model for this kind of forum. Its impact in raising the level of debate on protection, in establishing assessment of economic costs as the basis for common discourse, and in mobilising technical competence among all interested parties (including state governments) makes it an attractive model (see Warhurst 1982). If jointly established by the Commonwealth and state governments with funding comparable to that of the IAC a Land Conservation Commission could receive references on issues from both levels of government. It would also need the power and capacity to initiate inquiries of its own and to carry out or commission research into problem areas or the evaluation of promising programs and local initiatives.

Given the complexity of land degradation issues, an institution of this kind could not be expected to produce standardised solutions. What it should achieve is a sharing of relevant scientific, economic and administrative experience and the negotiation of understanding among the relevant interest groups of costs and the variety of policies and instruments most effective in coping with them. Incorporating the NFF and ACF and other relevant groups into the structure of inquiry and negotiation on a continuing basis may not avert disagreement or even confrontation, but should at least raise the level of discourse. These organisations are already repositories of substantial expertise and experience and, to the extent that they maintain capacity for educating their constituents, they can be the most appropriate channels for raising awareness among the relevant publics about the costs of practices leading to land degradation. This would appear to offer the best possibility of placing the broadest interpretation of conservation issues permanently on the public agenda.

Non-government organisations: a synopsis by the National Farmers' Federation

Andrew Robb

Pressure groups have a legitimate role to play in policy development. Groups such as NFF represent the interests of a specific sector of the community and the economy, and in land degradation policy are directly involved in the problem and its solutions. Much of the responsibility for policy representation on land degradation matters has been accepted by state farmer organisations because the major responsibility for land management rests constitutionally with state governments. Recent federal initiatives such as heritage legislation and the development of the National Soil Conservation Program have brought the federal government into the land degradation issue and accordingly it has become a higher priority for NFF.

As a policy development problem land degradation is more difficult to come to terms with than many other political issues. It is a problem which:

• has effects beyond the normal period of political consideration and even beyond the time span of a generation;
• is difficult to analyse in an economic context because of the very long-term effects;
• involves social and amenity values which are difficult to assess intuitively or economically; and
• still has many technical uncertainties surrounding it.

The farmer organisations, like many other pressure groups and sections of the community, are attempting to come to terms with these difficulties and to develop policies which are appropriate to the community and government concern about land degradation. In the policy context it is important to realise that farmers are receptive to the objectives of land conservation and with a few exceptions are seeking to improve their management of the land and are critical of those farmers who wilfully degrade their land.

There is a general opposition among farmers to rigid and draconian regulation of land use where that would penalise those who are seeking improved land management. Such regulation is likely to reduce the willingness and cooperative spirit with which farmers currently are addressing the problem. Strong regulation is also risky where the nature and causes of the problem are not well understood as this can precipitate further problems in land use because the incentives created do not correctly match the solution needed.

Problems have been experienced with the South Australian land clearing legislation introduced in May 1983 because it ignored factors associated with farm viability and social dislocation. The legislation also fails to address the need to manage natural landscapes for vermin, noxious weeds and wildlife. There was no consultation with those who are affected by the legislation. As a result the legislation has created administrative problems. This may be contrasted with the Western Australian case where consultation has gained the active support of farmers for land clearing legislation. Many farmers feel that the South Australian legislation has set back conservation ideals developed carefully in recent years and has encouraged many farmers to become law breakers. Farmers agree that there is a need to control wilful blatant offenders against sound land management. It is felt that this can be adequately achieved by attaching covenants to leases and maintaining a right of sanction against the few offenders.

Farmers in Western Australia have supported the forfeiture of several leases in the Kimberleys in recent years because of the failure of the leaseholder to properly manage the land. The conversion of term leases to perpetual leases is also supported to remove the uncertainty of tenure and its consequent adverse affect on land management. Conversion to perpetual lease is one example of the type of incentive which should be placed before farmers to improve land management. In general such changes should improve the incentives to conserve the land resource. United States Legislation on fragile land specifically removes benefits and farmer assistance measures where individuals fail to abide by land use recommendations.

Education is an important aspect of improving land management as has been shown in the 'Trees on farms' program in Victoria. The program has been successful because there is a receptive attitude among farmers to improving the aesthetic and productive condition of the land. The program is administered within the farmer organisation and this strongly influences farmer perception of the program and raises its credibility and acceptance.

Coordination of the research and presentation of land management information can be significantly improved. Research must have a more objective task-oriented and multi-disciplinary approach. There is a need for greater communication between scientific disciplines and between scientists, economists and sociologists in the formulation of strategies for land management.

NFF is concerned at the progressive weakening of the policy development process within the political parties. In both the major parties, policy development originating from the branch level and proceeding by debate through the hierarchy of the organisation has declined in importance in favour of executive decisions by the parliamentary wing of the parties and even by the Cabinet or senior office bearers.

The process of adoption of policy by government has in recent years become very much more dependent on input from pressure groups and from government departments. Input from pressure groups may be very beneficial. The pressure group is usually a collection of people who are close to a problem and have given greater consideration to it and the possible policy solutions than other sectors of society. In some cases a great deal of expertise and research effort will have been assembled and coordinated by the pressure groups. Such groups therefore have something to offer the policy process and should be encouraged to contribute.

The increasingly political context in which decisions are being taken leads to emotive debate of the issues and less rational consideration of policy alternatives. Improvements are needed to the policy-making process to allow more reasoned consideration of policy alternatives and to provide the opportunity for the expertise and views of pressure groups, academics and individuals involved, to be considered.

Non-government organisations: a synopsis by the Australian Conservation Foundation

Geoff Mosley

Non-government conservation groups are bodies set up by members of the community to promote the adoption and implementation of conservation policies. The groups address their advocacy both to the general public and to political institutions. In the case of land degradation non-government groups urge the deployment of funds and effort to the treatment of existing degradation and to the adoption of measures which will avoid future degradation; they oppose unwise land use decisions. Ballard refers to the educational role of non-government organisations. I will try to give some insight into the kind of contribution that conservation groups can make to the resolution of land degradation problems.

Conservation organisations provide the avenues for people interested in conservation to play a part in the decision-making processes. It is interesting to note, incidentally, that country people join conservation groups more readily than the city people. Conservation bodies also provide a means of drawing on a vast range of expertise. Scientists and other experts are usually willing to help conservation groups by providing knowledge as well as their opinions and views. These people give their services in a voluntary capacity and are thus not a burden on the public purse.

Conservation groups have some problems with their image. Initially small groups without a great deal of influence they became by necessity very heavily involved in defending the areas they perceived to be under threat — mainly nature conservation areas around the periphery of Australia. This issue is basically a conflict between incompatible land uses. Conservation groups, by trying to prevent something happening, were caught up in confrontation. Preoccupation with this type of activity meant that the groups had little time to devise suggested solutions; involvement with land use planning was minimal, especially in the early years. It has also typecast them as being fairly hardline confrontationists, which is not a totally accurate and productive characterisation. This has given the conservation movement particular problems in dealing with the rural environment and the inland. I believe the movement has a great capacity to help with the incorporation of conservation objectives into land use planning, especially, as Robb points out, if the various groups are involved at an early enough stage. This involves a greater degree of cooperative effort.

Some examples of the Australian Conservation Foundation's (ACF) work in the various fields are given below.

Public education

In 1983 the ACF conducted a 'Save the Earth' campaign which was directed first to the ACF's 12 000 members. Within 12 months from being voted a low priority by ACF members it became the highest priority (it is still in the top four).

The ACF next produced a TV community service announcement, directed primarily towards the urban population, and a leaflet was sent to all those responding to the announcement. Using funds raised by these efforts a soil conservation publicist was employed, who did extensive media work, much of it in the field. The job involved liaison with farmer organisations, government soil conservation groups etc. The timing of this initiative was critical since it came when a government commitment to a national soil conservation program was under strong attack. The public education work was therefore combined with a specific separate campaign within government aimed at first restoring and then expanding and implementing the program. A soil conservation analyst is now employed, whose work is closely related to the ways and means of gaining further public acceptance for the program.

Opposition to unwise land use

Another example of ACF work was opposition to what was considered to be unwise land use—the land clearance program in the south of Western Australia, roughly in the region between Ravensthorpe and Kalgoorlie, where for a number of years (about 20 years or more) it was planned that these lands be made into crop lands. In 1980 a firm proposal was made to clear 100 000 hectares a year until a total of 3.1 million hectares had been cleared. Activities in this area are various and involve close cooperation with local conservation groups, several of which include farmers. The poor economics of the proposal and the likely land degradation problems were exposed.

In a sense the ACF was fortunate because it could point to problems in the adjacent Jerramangup area that had already been cleared and where there were severe problems of soil erosion and salinity.

The ACF organised a lot of publicity via conferences, media releases etc. Alternative land uses were proposed and the role the area could play in conservation, if maintained in a natural state, was stressed. This work, including national park studies, is continuing. The ACF work was important because it developed a credibility and a series of cogent arguments on the lack of wisdom of the course embarked on. The Western Australia State Labor Party (then in opposition) was attracted to this point of view and at the election in 1983 promised a review of the land clearance program. In government it quickly placed a moratorium on clearing, but some further work was necessary to make sure the promise was kept. Subsequently the government decided the moratorium would be extended indefinitely. These battles are, of course, never won completely and the situation is still being monitored very closely.

Voluntary conservation groups clearly have a very important role to play in any work for integrated solutions. Some experience of how that might be achieved was gained through the meeting the ACF held at Broken Hill in 1982 — the National Arid Lands Conference (Messer and Mosley 1983). The participants of that conference were drawn fairly evenly from pastoralists, researchers, administrators and conservationists, and there was a sprinkling of politicians from New South Wales and South Australia. Six of the seven workshops were asked by ACF to work on a strategy for conservation of arid lands and they all came up with similar ideas about what should be done. This is an indication that once there is a suitable framework for cooperation and discussion a lot more progress can be made towards dealing with the problems discussed in this book. ACF is trying to get the right institutional framework. The suggested strategies which came out of the Broken Hill Conference have been considered by the Foundation's Council and incorporated in ACF policy. A strategy for the arid lands was proposed on which unfortunately not enough follow-up work has been done. However, it is interesting that what emerged was a suggestion for two new bodies, a National Arid Zone Council and an Arid Lands Institute.

The strategy called for a wide-ranging and periodic survey of the different parts of the arid lands to determine their capability for existing and projected forms of land use and to determine whether existing management measures are effective in attaining their objective. It was felt that this review would lead to the framing of land use policies and conservation plans for each region and would result in monitoring procedures. The conservation plans would provide for adjustments of land use, including the withdrawal of some areas from grazing, and specify the conditions which would apply until the next review. It was envisaged that the Arid Lands Institute would help with the research and assessment associated with the review and would also be responsible for gathering relevant knowledge and ensuring that it flows more rapidly and effectively to resource managers. The importance the ACF attaches to short-term leases as against freehold or perpetual leases should be seen in the context of its particular position and strategy.

With regard to John Ballard's proposal for an Industries Assistance Commission type of institution for land degradation issues this may be a useful model but it could produce excessive emphasis on economics. Land degradation has a large social component. It would not be helpful if land degradation was to be regarded simply as an economic problem.

Finally, the ACF welcomes the invitation to be on the Soil Conservation Council and hopes to be on many more such bodies because I believe it has an important role to play. Any successful strategy for dealing with land degradation will, I believe, incorporate the citizen groups and the industry groups.

14 Policy agents: their interaction and effectiveness

Robert Junor and Warwick Watkins

Historical development of our administrative structure

The early settlers, in their efforts to supply food for the new colony, developed land use practices by trial and error, using European agricultural practices which proved to be totally unsuitable for the Australian conditions.

Initially the settlers had considered the Australian bush to be a harsh, cruel environment incapable of sustaining adequate production and they therefore set out to clear and 'develop' the land as quickly as they could. Even in the early stages of the colony, it was not long before this development brought problems to the environment. As early as 1803, Governor King proclaimed that the removal of trees from the river banks in the Sydney region was prohibited.

By the late 1890s the combination of inappropriate land use management, the spread of the rabbit, and a severe drought had seriously degraded large tracts of the arid and semi-arid regions, to the extent that many landholders were forced off the land. The subsequent Royal Commission of Inquiry into the condition of Crown Tenants in the Western Division of New South Wales, completed in 1901, was perhaps the first real attempt to deal with land degradation in Australia. The Western Lands Commission of New South Wales was formed shortly afterwards as a direct response to the Royal Commission's findings.

The Constitution of the Commonwealth of Australia came into being in 1901 and the responsibility for management of the land and its resources was effectively retained within individual states.

Through the following years the need for specialist management of natural resources by state governments led to the establishment of

The valuable assistance provided by Mr Algis Sutas, Soil Conservationist, Soil Conservation Service of New South Wales with the research and preparation of the draft manuscript is gratefully acknowledged.

numerous departments. The philosophy of the time was to create single-issue or single-purpose departments each with its specific area of interest.

The first soil conservation agencies were formed in the late 1930s and early 1940s in response to the serious soil erosion problems that had occurred in Australia and the experiences in American agriculture which culminated in the 'dust bowl' conditions.

In 1942 the Commonwealth took over from the states the exclusive role of collecting income tax. The states were thus dependent on the Commonwealth for financial assistance and the Commonwealth secured a voice in resource management because it could direct funding to specific areas.

In 1946, a Standing Committee on Soil Conservation (SCSC), comprising representatives of the Commonwealth and of each state soil conservation agency, was established to assist the states in coordinating their work. Cooperation and assistance from trained personnel became available and special research projects were undertaken in consultation with the states (Aust DEHCD 1978a).

Community awareness of the environment was increasing rapidly in the 1950s and 1960s, with controversial issues such as sand mining, woodchipping and hydro-electric schemes focusing attention on the use and management of land.

The Lake Pedder controversy in the early 1970s must be considered a turning point in attitudes towards land use planning and decision making on resource management issues in Australia (Australia. Senate Standing Committee on Science, Technology and the Environment, 1984). It was the final report of the Lake Pedder Committee of Inquiry (Aust 1974) which put forward the argument for more effective coordination of land use policy in Australia. A recommendation was made for the establishment of an Australian Land Conservation and Natural Resources Council to coordinate land use, resource management and environmental management on a national level.

Most of the major environmental issues that have become the subject of public controversy in Australia over the past decade have essentially arisen from conflicts over land use. Unfortunately, there has been a concentration on preservation rather than conservation in the majority of these issues. Harmonisation of conservation and development of our land resources is in our view the desirable goal, but we recognise that some land features in Australia are unique and worthy of preservation.

Identification of the need for change

In 1974, the Report of the Committee of Inquiry into the National Estate

(Australia. Parliament of the Commonwealth of Australia, 1975), pointed to the complexity of structural relationships between agencies responsible for land management. The committee found that there was '. . . an incredible complexity and overlapping of authorities, boards, commissions, councils, government departments etc . . .' The committee stressed the need for some supervisory and coordinating authority.

In the same year, the working group preparing a Green Paper on rural policy in Australia made the point that many decisions on land use or land management were made by individuals or by government bodies who were ill-informed or whose interests were too narrow or short-sighted. (Harris, et al 1974).

That working group also considered that in the management of land, (particularly Crown land), the control of land use by a number of separate authorities tended to discourage, or in some cases exclude, alternative land use. The working group suggested that there appears to be a good case for making these divisions less inflexible and to encourage greater multiple use of Crown land.

Following the 1972 Premiers' Conference, an interdepartmental committee (IDC) was formed in 1974 to examine the request for financial assistance to the states for soil conservation programs. The IDC proposed that based on environmental, social, economic and financial considerations, there is a case for Commonwealth involvement in a long-term accelerated soil conservation program, and that the program should be planned and carried out in the context of an integrated approach to land management (Australia. Department of the Environment and Conservation, 1975). The IDC had also proposed that the Commonwealth and states undertake a collaborative study with the aim of providing a basis for coordinated land resource management.

The Commonwealth and State Government Collaborative Soil Conservation Study was carried out from 1975 to 1977 and involved the cooperation of the Commonwealth and all state governments. It produced a number of reports which provide a source of information for the formulation of policy on financial and other elements of soil conservation and associated land and water management programs. The major report, 'A basis for soil conservation policy in Australia' (Aust DEHCD 1978a), is basically a summary of policy issues, and made the point that as many government policies give rise to, are affected by, or concerned with land degradation, there is a need for good liaison, communication, and coordination between government activities and the responsible agents. The report recommended that soil conservation in Australia should be intensified, extended and more closely integrated with policies for rural industries and planning for coastal, urban, recreational and mining areas. It recommended further that complementary action should be taken by governments in research, cooperative planning, institutional changes and more effective liaison and coordination. The report went on to recommend that the roles and functions of each Commonwealth department interested in soil conservation be more clearly defined so that adequate arrangements may be made to express their interests and to facilitate liaison and coordination on soil conservation matters within the Commonwealth and among Commonwealth, state and local governments.

This need for improvement in coordination of activities between Commonwealth and states arises in all sections of natural resource management. For example, a report from the Senate Standing Committee on Science and the Environment (Aust Senate Standing Committee 1977) dealing with the environmental impact of the woodchipping industry, stressed that Commonwealth-state coordination should be the aim, with the Commonwealth assuming a role supportive of the states.

In 1982, a working group report to the Minister for Primary Industry (Balderstone, et al 1982), believed that the Commonwealth government should adopt a substantially greater role in soil conservation, through specific-purpose funding and assistance with the coordination and evaluation of a national soil conservation program.

Also in 1983 'A strategy for the conservation of Australia's arid lands' published after the National Arid Lands Conference (Messer and Mosley, 1983), stated that there had been poor coordination between researchers, administrators and land managers. It was also made clear that inefficiency resulting from the division of administrative responsibility between too many government departments would be an obstacle to achieving the objectives of the strategy.

Developing a coordinating structure

In 1983, the report 'Land degradation in Australia' (Woods 1983) presented the detailed information on land degradation provided from the Collaborative Soil Conservation Study. The report presents extensive data on the extent of land degradation and highlights the need for prevention and rehabilitation. In discussing organisational problems and deficiencies which may hinder the control of land degradation, or may actually contribute to it, Woods cites divided responsibility and gaps in responsibility for land matters, multiplicity of authorities involved and inadequate communication and coordination.

Woods suggests that improved communication, coordination, and a review of organisational responsibilities and functions, are the main avenues for improvement.

The publication of the World Conservation Strategy in 1980 prompted the development of a National Conservation Strategy for Australia, in 1983. The strategy considers the relationship between conservation and development, and discusses the causes of environmental problems, not just their symptoms, (Anon 1983). The strategy also suggests objectives and guidelines for future policies and actions, placing great emphasis on conservation for sustainable development. It recommended that to improve the capacity to manage, coordination of action within the Commonwealth, and cooperation between the Commonwealth and the states, and among the states, should be strengthened.

The main report of the steering committee studying Australia's water resources to the year 2000 (Australia. Department of Resources and Energy 1983) had also indicated that land and water management are closely interrelated and that the use of these resources requires the involvement of all relevant government agencies. The report suggests that only a framework of effective collaboration between water, agriculture and soil conservation authorities and the land users will ensure that management of land is consistent with its long-term productive capability and with the maintenance of quality and quantity of water resources.

However, it was not until 1984, when the Senate Standing Committee on Science, Technology and the Environment inquired into land use policy in Australia, that the key issues were concisely stated. The committee said that the natural resources which humans seek to utilise always occur in, and form part of, ecological and environmental systems which are not independent, but interrelated, and that changes introduced to one resource can have repercussions on many others. It should be realised that natural *resource* management is natural *systems* management and that therefore the management of natural resources must be viewed from an ecological systems approach. The committee considered that an evident lack of coordination led to duplication of effort by agencies, uncoordinated information on which to base land use decisions, and inconsistency in methods of management. The two major obstacles to the achievement of coordination are that land degradation is a complex, multi-disciplinary problem and that the existing structure of management agencies is not able to achieve the necessary interrelationships and coordination.

Interactions between policy agents

In discussing existing linkages and levels of interaction between policy agencies with responsibilities for land degradation, it may be advantageous to outline briefly why it is necessary for interaction to occur and the levels at which responsibilities for land degradation may lie.

In the outline of objectives for this workshop it was stated that the application of the term 'land degradation' should include all those adverse effects that land uses may have on services provided by land. The causes of land degradation are widespread and may involve nearly everything that people may do in conjunction with the land. Almost all uses of the land may be considered as degrading to the land, it is just a matter of what level of land degradation is socially, economically, agriculturally, and ecologically acceptable. However, it is major land degradation, such as soil erosion, salinity, tree decline, loss of native and unique land components, desertification, urbanisation, contamination by wastes etc, which have the more permanent effects on the land as well as on people. An examination of any of these elements of land degradation reveals many causes and effects. Soil erosion results primarily from removal of vegetative cover by one use or another, or by physical disturbance of the land surface. Soil erosion not only adversely affects the area where soil loss occurs, but also where the soil is later deposited. As water is a prime vector for soil movement, then the management of water also affects soil erosion. It is obvious that the causes of land degradation involve a number of factors, and that the consequences of land degradation affect a variety of land uses.

The question of acceptance of responsibility for prevention and mitigation of land degradation is of paramount importance if a coordinated attack is to be made on the problem. The level of commitment by the various sectors of society has been varied and, on the evidence of the current level of land degradation, is in our view inadequate.

Land users have the primary responsibility for the use and management of land and hence for implementation of preventative and corrective strategies. Therefore, the cost of such measures should, in the first instance, be borne by them. However, while there has been a growing recognition of the land-ethic principle within the community and a corresponding willingness to strive for measurable results in land conservation, the ability of the vast majority of land users to achieve this has been hampered by external influences.

The primary influences affecting the use and management of land are economic, social and technological, with market factors and seasonal conditions acting as modifying factors. The Australian farmer is endeavouring to survive in restrictive economic conditions and is usually unable to influence markedly the economic climate in which he has to survive. He has however, demonstrated an ability to survive through his own innovative actions and his responsiveness to changing market forces and physical resource pressures. This struggle for survival through maintenance of productivity has resulted in increased pressures on the land, and in many cases it has been used well beyond its inherent capability. The methods being employed to beat the cost/price squeeze, especially disinvestment, will generally see a decline in the potential of land for sustained productivity.

Possibly the largest factor contributing to continued expansion of land affected by degradation is an apparent inability to learn from past mistakes ('One thing that man has learned from history, is that man does not learn from history'). While they may have attained other social goals, the closer settlement schemes of past years have in many instances contributed substantially to the current level of land degradation. The expansion of agricultural enterprises and particularly the conversion of grazing land to cultivation, especially in the 'marginal' areas, has predisposed the land to accelerated degradation. Governments at all levels need to take a more responsible attitude to land use issues and to consider the interdependency of land and the fabric of society when evaluating policies which impinge on the use and management of land.

Management of land resources has traditionally been divided so that the various components of the land may be managed as individual disciplines. This single disciplined approach to natural resource management is not in keeping with the systems or multi-disciplinary approach needed to combat land degradation. Therefore to manage land resources effectively with our present management structure, there must be interrelations between the various disciplines.

Role of the Commonwealth

In Australia's federal system of government, legislative authority is divided between the Commonwealth and the states. The Commonwealth has a primary interest in the management of the economy, efficient allocation of resources and the equitable distribution of national income. Under the Constitution the states retain the responsibility for natural resource management because that responsibility is not directed to the Commonwealth.

The economic prosperity of Australia is dependent on the protection and maintenance of a productive and stable land resource. The Commonwealth government should therefore take an active role to ensure that the most appropriate land use policies are adopted in the national interest. It should contribute to the harmonisation of land use policies between various Commonwealth government departments and, where possible, between states. The National Soil Conservation Program (NSCP), National Tree Program (NTP) and the National Water Resource Program highlights the need not only for harmonisation of policies and programs but also for rationalisation to ensure the most efficient and effective allocation of funds through the targeting of programs.

Through its monetary and fiscal policies, particularly those on import/export and taxation, the Commonwealth government has the means to achieve an indirect but substantial effect on land use and management. The most direct and possibly the most important role for the Commonwealth government is in the provision of financial support to the states in matters of national interest. Land degradation is such an issue and NSCP is a very constructive contribution by the Commonwealth and reflects its level of responsibility in the area of land degradation.

The Commonwealth as a signatory to various international agreements (Declaration of the United Nations Conference on the Human Environment, Ramsar Treaty for the Conservation of Wetlands, World Heritage Agreement), as well as a participant in the World Conservation Strategy is strongly committed to the effective management of natural resources.

The Commonwealth therefore, is seen as the agency which should provide the necessary coordination between its responsibilities and those of the states. As a coordinator, the Commonwealth should take a positive role and through its membership on interdepartmental committees and ministerial councils, especially the newly created Soil Conservation Council, influence the formulation and adoption of land use policies and strategies for land degradation prevention and control.

Role of the states

The states have accepted the responsibility for land use and management and jealously protect their role which results from lack of direction in the Constitution. The states should therefore ensure that they have adequate statutory provision to implement strategies for land conservation.

Tasmania remains the only state yet to enact soil conservation legislation. A legislative framework needs the support of adequate financial and human resources to ensure that a concerted program of activities can be mounted. Most importantly, land degradation should be treated on a multi-disciplinary basis.

Restrictive departmental barriers and the desire to protect 'one's ground' need to be replaced with a more open combined sense of purpose in tackling the land degradation problems crying out for attention. In most cases the framework for action exists, what is required is the political and departmental will to ensure the adoption of the desired strategies. There must be more widespread adoption of preventative strategies to ensure that Australia advances beyond its current, predominantly corrective phase of land degradation control.

As most of the state government organisations responsible for the management of natural resources originated in the early part of this century, at a time when the nature of natural resources systems was not fully understood, the present day structure consists of a number of agencies, each with the responsibility for a single resource issue (ie soil, water, forest, land etc). In some cases, their particular functions and the reasons for their establishment have disappeared or faded in priority (Victoria. State Conservation Strategy Task Force 1983) or their current legislative powers or functions are inappropriate to today's circumstances (Musgrave 1984c).

The management of the main components of the land resource (soil, water, forest and land) are currently fragmented between a number of departments and ministries within each state, and these divisions lack consistency between states (Table 15). The inconsistency of division of responsibility between states indicates that difficulties in coordination would naturally exist, and emphasises differing and outdated approaches to natural resource management. This not only causes difficulties in coordination within states, but also between states.

In most states, soil conservation is included under the banner of agriculture, because early soil conservation was predominantly agriculturally orientated. However, soil conservation is vital in mining, urban and coastal, environmental and other major state development projects and to place soil conservation within the responsibility of conservation-orientated ministry would appear more appropriate.

A soil conservation agency placed within a department of agriculture is seen as unnecessarily restrictive and likely to cause conflict and compromise because of the production bias of agriculture. Policy conflicts and confusion over priorities for allocation of resources to deal with soil conservation issues could also arise under such a structure.

Soil conservation should therefore be placed with the other natural resource areas of water, forest and land. In Victoria the recently created Ministry for Conservation, Forests and Lands has drawn together the management of soil, forest and land; however water is still not incorporated within this system. Through its integration of departments with responsibility in land degradation, the new system in Victoria is aimed at increasing liaison at both administrative and policy levels (Cahill, Department of Conservation, Forests and Land, Victoria, pers comm).

This new structure has improved avenues for coordination in land degradation policy, and follows the approach set down in the World Conservation Strategy. The structure, whereby most natural resource agencies are combined under the same ministry, will need to be

Table 15: Departmental and ministerial division of management responsibility within individual states

State	Soil	Water	Forest	Land
New South Wales	Soil Conservation Service Minister for Agriculture	Water Resources Commission Minister for Natural Resources	Forestry Commission Minister for Natural Resources	Crown Lands Office/Western Lands Commission Minister for Natural Resources
Victoria	Soil Conservation Authority Minister for Conservation, Forests and Lands	Rural Water Commission Minister for Water Resources	State Forests and Land Service Minister for Conservation, Forests and Lands	Division of Crown Lands Management Minister for Conservation, Forests and Lands
Queensland	Department of Primary Industries Minister for Primary Industries	Department of Water Resources Minister for Water Resources and Maritime Services	Department of Forestry Minister for Lands, Forestry and Police	Department of Lands Minister for Lands, Forestry and Police
South Australia	Department of Agriculture Minister for Agriculture	Engineering & Water Supply Department Minister for Water Resources	Woods and Forests Department Minister for Forests	Department of Lands Minister for Lands
Western Australia	Department of Agriculture Minister for Agriculture	Minister for Water Resources	Forest Department Minister for Forests	Department of Lands and Survey
Northern Territory	Conservation Commission Minister for Conservation	Department of Mines and Energy Minister for Mines and Energy	Conservation Commission Minister for Conservation	Department of Lands Minister for Lands
Tasmania	Department of Agriculture Minister for Primary Industry	Rivers & Water Supply Commission Minister for Water Resources	Forestry Commission Minister for Forests	Lands Department Minister for Lands

Source: This table was compiled from information obtained from Government Information Centres or their equivalent in each state. For states where no such facility exists, the relevant departments were contacted directly.

carefully monitored by the other states and its effectiveness evaluated.

As the responsibility for natural resource management in the states is fragmented there is scope for conflict and duplication. Conflict may also arise from differing objectives of agencies, for example, between agriculture (being productivity motivated) and soil conservation (being environmentally motivated), which may provide differing advice on land use, such as in use of marginal lands. Duplication of effort also occurs when agriculture and soil conservation may both provide advice on agronomic and soil conservation matters. This highlights that there must be cooperation and coordination of policy regardless of the ministerial or departmental placement of agriculture and soil conservation.

As mentioned previously, the original framework for natural-resource management agencies was single-issue orientated, and their internal management and legislation is also thus orientated. There is insufficient legislative machinery for coordination of activities or policies; thus, many agencies are not willing to interact for fear of losing their individuality or identity, and therefore their power and status. In some cases, land resources information held by one agency, and which may be of obvious value to another, may not be exchanged because the existence of the information is unknown or unrecognised. (Australia. Department of Environment, Housing and Community Development 1979).

In some states, there are avenues for coordination of activities, for example, the Soil Conservation Advisory Committee in Western Australia, which aims to coordinate the services from the departments involved in the committee, as well as all other public authorities, to achieve soil conservation. In Tasmania, the Standing Committee on Soil Conservation, which comprises representatives from resource-based departments, provides a framework for effective natural resource policy formulation, but one which is yet to be fully used.

Some state soil conservation statutes also incorporate provisions for giving advice and settling disputes on the disposal, care or use of lands, such as in section 38 of the *Victoria Soil Conservation and Land Utilisation Act*, 1958; section 37 of the same Act also provides for the Soil Conservation Authority to implement directives and to coordinate services of appropriate government departments and public authorities.

In other instances of land degradation, for example, tree decline, there is a similar single-issue approach and lack of coordination. Numerous legislative measures provide restrictions on the cutting down of trees, but the application of controls is inconsistent, irregular and incomplete.

Interstate interaction

In areas of similar land use and land type, such as arid lands where states may have mutual interests, or in areas where natural resources cross state boundaries, such as the River Murray, coordination and cooperation are needed for the application of similar land management techniques and interchange of technical information and mutually benefiting standards for maintenance of natural resources.

The prime channel for interaction between states is through a number of Commonwealth-state ministerial councils (Table 16). However, as no council has a specific interest in land degradation an adequate concentration on land degradation is not achieved. Even though land degradation matters may be raised in all these councils, there is little machinery for liaison and coordination between the councils. (Aust DEHCD 1978a).

Channels for coordination of activities between states do exist, such as the River Murray Waters Agreement, but in light of the major problems which still arise, it is reasonable to say that such channels are not effective. Although Queensland controls a significant part of the Murray-Darling Basin, it is not a party to the agreement. The River Murray Waters Agreement also fails to acknowledge the ecological interdependence of parts of the system irrespective of political boundaries, as well as the principles of multi-disciplinary planning (Jackson, 1982).

Commonwealth-state interaction

As already stated, interaction between the Commonwealth and states on land degradation matters can occur at several Commonwealth-state ministerial councils (Table 16). The Standing Committee on Soil Conservation (SCSC) currently reports through the Standing Committee on Agriculture (SCA) to the Australian Agricultural Council (AAC). As such, the AAC is the major channel for coordination. However, as soil conservation issues are not solely associated with agriculture, and in view of the differing philosophies of agriculture and conservation, AAC is not an adequate forum for soil conservation. AAC has also been criticised for its slow procedures and lack of decision making (Balderstone et al 1982). The restricted power of SCSC, such as in the lack of a national focus for arid land management and land resource surveys, (Bunker 1984) has also been obvious.

The representation of state soil conservation agencies at less than departmental head level on SCSC (especially from those States where soil conservation is a branch of a larger department of agriculture) has been

a major impediment to its effectiveness. Because of less senior representation the committee has addressed administrative issues rather than major policy developments. The opportunity for policy formulation and implementation is further restricted at the AAC as the ministers responsible for soil conservation are not, in all cases, those responsible for agriculture.

The formation of a new council, the Soil Conservation Council (SCC) in February 1986, is expected to resolve the current problems, by providing the much needed focus on land degradation. Even though the new council's terms of reference will be directed towards soil conservation, land degradation in general will no doubt be a major topic.

The SCC will also largely dissolve the need for a National Land Use and Resource Management Council (NLURMC) as initially recommended by the Lake Pedder Committee of Enquiry, in that many of the matters which would have been addressed by such a council will now be undertaken by the SCC. This also solves the problems of establishing and maintaining a NLURMC, which in itself would have been a large and possibly unwieldy body due to the number of ministers needing to become involved, as well as the overlapping upon some of areas of the existing ministerial councils.

In addition to the formation of the SCC, an increase in communication and coordination between all ministerial councils with an interest in land degradation and land use would be extremely beneficial.

Other areas of interaction occur through programs such as the NSCP and NTP. Through the NSCP, funds are provided by the Commonwealth to the states for specific soil conservation programs. The Commonwealth, by administering NSCP funding, can coordinate activities by the states, as well as ensure conduct of work in areas of its own interest or where there are mutual interests. Early in 1985 the Commonwealth Parliament passed the *Soil Conservation (Financial Assistance) Act*, 1985, which puts the NSCP on a statutory basis and includes the setting of guidelines on which programs will be funded. Further, the Act creates the Soil Conservation Council to advise the minister on research into land degradation problems and funding priorities for the NSCP.

Table 16: Commonwealth/state ministerial councils with interests in land degradation

- Australian Agricultural Council
- Australian Environmental Council
- Australian Water Resource Council
- Australian Forestry Council
- Australian Minerals and Energy Council
- Council of Nature Conservation Ministers

The objectives of the NTP include the promotion of coordinated activities by governments, community groups and individuals in arresting tree decline, and has provided a sound framework for effective cooperative action, despite limited levels of funding (Thatcher 1984).

Blockages in the system

The lack of interactions among policy agencies with responsibilities in land degradation occurs at all levels of government (Australia. Senate Standing Committee on Science, Technology and the Environment 1984). The lack of interaction has resulted in duplication of effort, uncoordinated and inadequate information, inconsistency of attitudes on land degradation, and an incomplete approach to land management.

Uncoordinated and inadequate information exists in dealing with the problem of tree decline with many landholders in New South Wales finding it difficult to obtain information and advice on the planting and general management of trees. This in part arises from the fact that no one government agency has the specific function to deal in those matters. The Department of Agriculture generally does not possess the necessary expertise, the Soil Conservation Service may provide information on tree planting vis a vis soil conservation, but functions and expertise of their field staff is also limited. The Forestry Commission, which possesses the greatest expertise in this area, concerns itself primarily with crown lands and production forestry.

One of the major findings of the workshop held in Melbourne during the 1980 Focus on Farm Trees Conference was that government departments involved with trees on the farm present an inadequate, understaffed and fragmented image (Oates et al, 1981). The workshop agreed that greater coordination and collaboration was required between government departments to avoid the current fragmented approach to the problem of tree decline.

The facts that tree planting in the rural sector calls for expertise from a number of disciplines, and that the agencies involved are limited in their scope by their specific responsibilities, means that only a coordinated effort will succeed. In New South Wales, a combined trees-on-farms project involving the Soil Conservation Service, Department of Agriculture and Forestry Commission, is remedying that situation.

Inconsistency in attitudes and responsibilities to land degradation also flows from lack of interaction. Agencies such as those responsible for providing roads and service corridors, and for administering mining and similar resource uses, may not be as conscious or concerned about land degradation as the prime natural resource management agencies are. Certain anomalies in legislative powers aggravate this situation.

In New South Wales, under section 21C of the Soil Conservation Act, the Catchment Areas Protection Board has the power to control the damage and/or destruction of trees in areas of protected lands (lands which have been mapped as having a slope in excess of 18-degrees). However, as the Crown is not bound by the controls and the controls do not, by specific reference, extend to areas of state forests the Forestry Commission could, therefore, log areas with a slope in excess of 18 degrees. Agreements and standards have been made so that this does not occur but these agreements are not policed.

Conditions are similar between the Commonwealth and states. Section 109 of the Constitution states when a law of a state is inconsistent with a law of the Commonwealth, the Commonwealth law shall prevail. This may apply where a Commonwealth department or authority, working under Commonwealth legislation, undertakes activities which may interfere with those of a state. This could occur under the Telecommunications Act, which gives the Telecommunications Commission the power to cut down timber and excavate soil from any land in the course of construction and maintenance of service lines, or it may occur under the Defence Act in relation to fortifications and other defence works. If these works contravene state powers or activities, such as where a soil conservation agency may prohibit or control certain works, the state has no power. (Aust DEHCD 1978a).

These inconsistencies can be settled, not merely by legislation, but by forming mutual agreements and guidelines, and through interactions among the relevant agencies.

Developing the working structure

To be realistic, radical changes by governments in their administration of natural resources could not be expected to proceed rapidly or uniformly throughout Australia. However, gradual changes could be expected as the elements of the natural resource administrations come to realise that they cannot achieve workable programs within a time frame acceptable to community interests in isolation from other areas of resource management.

The time has arrived for administrations of the single-issue resources of forests, water, soil and land to recognise that if they are to continue to remain effective it is essential to implement natural resource management policies that have multi-disciplinary objectives.

From a national perspective, support for the establishment of policies on land degradation will gain the acceptance of most agencies when a broader focus to encompass all elements of environmental management is applied.

One approach to establishing national policies for the treatment of degrading lands would be to concentrate on the land degradation problems confronting the principal component areas of land use in Australia. These component areas would include the land degradation issues associated with arable, grazing, irrigated, semi-arid and timbered lands, and arid, coastal and alpine environments.

All Australian states have to develop programs and policies for at least five of these eight component areas of the Australian environment affected by land degradation. In the face of this, the Commonwealth government could more readily bring together the technology and expertise held by individuals, authorities and state government departments to develop policy, coordinate research effort and prepare programs for each component area of land use that could be integrated within the policy framework of individual states.

As an example, recent conferences on the management of semi-arid lands have demonstrated that the technology is available that would allow:

- determination of lands unsuited for agriculture or grazing purposes;
- identification of programs that could be applied to restore or allow edible shrub regrowth;
- coordination of national programs aimed at feral animal control;
- coordination of research activities such as the use of fire in rangeland management or the control of inedible shrub invasions.

With the recently established Soil Conservation Council and a restructured Standing Committee on Soil Conservation an administrative structure has been established which should provide the direction for the development of land degradation policies in Australia. This structure would now comprise:

Soil Conservation Council
(Council of Ministers)
Determines national policies and priorities.

Standing Committee on Soil Conservation
(Committee of state representatives of soil conservation agencies)
With the aid of the Advisory Committee on Soil Conservation, researches and develops national policies and programs and recommends priorities for allocation of expenditure under the National Soil Conservation Program; coordinates research in Australia.

State Soil Conservation Agencies
Determines state priorities and programs; coordinates the preparation of multi-disciplinary plans by state resource management agencies; coordinates the implementation of programs via state agencies, local government, land user committees or individuals.

There is now a structure which will allow national policies and programs to be established. What is required is creative and imaginative thinking by soil conservation agencies to develop the programs to capture the participation of other resource management agencies and gain the ready acceptance of the community.

National policies need to be developed for the following:
- the restoration of eroded and degraded land;
- the sustainable development and use of the soil, water and vegetation resources of the nation;
- promotion, education and increasing the levels of understanding between community, industry groups and governments in the management and conservation of soil resources; and
- determination of strategies that will achieve a balance between development and conservation of catchment areas, the coastline, aquatic environments, prime soils and unique lands.

It is imperative that the Soil Conservation Council be supported in a positive and constructive way through the Standing Committee on Soil Conservation to develop imaginative and practical policies for each of the component land use areas mentioned above.

The Commonwealth government has an important role to play in coordinating the development of these policies and programs for implementation through the various state agencies. Funding provided to states under the National Soil Conservation Program can provide the incentive to ensure that national policies become more closely integrated among states.

At the state level in 1984, the New South Wales government took a major initiative by establishing an interdepartmental committee (IDC) for total catchment management. Membership of the IDC comprises the Soil Conservation Service (chairman), Forestry Commission, Water Resources Commission and the Department of Agriculture, National Parks and Wildlife Service, Crown Lands Office, Public Works Department, State Pollution Control Commission and the Department of Environment and Planning.

The principal objectives of total catchment management are to encourage effective coordination of policies and activities of relevant departments, authorities, companies and individuals which impinge on the conservation, sustainable use and management of the land resources of the state's catchments, including soil, water and vegetation; and to ensure the continuing stability and productivity of soils, a satisfactory yield of water of high quality and the maintenance of an appropriate protective and productive vegetative cover; and that land within the state's catchments is used within its capability in a manner which retains as far as possible, options for future use.

The IDC will provide a forum for the development of a more fully integrated approach to the management of natural resources in New South Wales. Its success will depend wholly on the participation and commitment both of the individual members and of the organisations they represent.

Discussion

In achieving a coordinated approach to land-management policy formulation and land management practice, it is essential to first define a working structure to allow incorporation of the various agencies involved. The various levels of government maintain differing constitutional, legislative and community responsibilities and possess different levels of practical expertise. While the states maintain the direct responsibility for managing natural resources, and possess the necessary legislative and practical structures to manage, they do not actually implement land practices; that is the responsibility of the landholders, for it is only the landholder who can implement land management practices. The Commonwealth has an indirect responsibility with natural resource management, but is able to supply monetary incentives and should act as the supervisor, coordinator and motivator.

Although land degradation control is basically the responsibility of those who own or manage land, it is also the concern of all present and future generations who will consume food and other services provided by the land (Roberts 1985b). As the landholder is the day-to-day decision maker on land management, he is the one able to adopt conservation practices and able to ensure that use is compatible with conservation. The landholder uses the land to earn an income, which is dependent on the condition of the land, and so the landholder has the responsibility to himself as well as the community to conserve land resources.

Local government agencies also have a major responsibility in implementing land conservation practices through their role of providing and maintaining public amenities and services.

The role of facilitating the implementation of land management falls to state governments who, while having the legislative background, technical expertise and departmental structure, also have the responsibility to act on behalf of the community, and therefore should be heavily committed to programs of land management (Downes 1971). While not having the role of actually implementing land management, the state government must be able to provide those who implement (landholders and local government) with the necessary technical knowledge, provide them with assistance in meeting the cost, as well as provide the basis for management.

In natural resource management there is also a need to provide the necessary coordination, supervision and motivation to those who facilitate and implement land management on a multi-disciplinary basis. This task falls within the responsibility of the Commonwealth, as it is concerned with community benefit on a national scale. By having the power to provide financial incentives, the Commonwealth can motivate as well as indirectly specify land management. This can be seen from the 'catalytic' value of the National Soil Conservation Program, as well as other national programs, which through the injection of Commonwealth funds, also encourages the input of state and landholder funding. As the national programs can specify areas of financial input, the Commonwealth may also coordinate work within and between states.

Other agencies which can act as motivators are the Commonwealth/ state ministerial councils as well as their various standing committees. These already possess avenues for coordination, and are in an ideal situation to provide incentives and motivation.

Conservation and community groups also motivate by gathering community support and applying it in political forms. As governments are acting on behalf of the community, community and conservation groups can play a major role in influencing government action.

Conclusion

There has clearly been sufficient discussion and documentation on the need and direction for change in the formulation and implementation of policies for the use and management of land. It is unfortunate that so much comment has inspired so little constructive effort in the administration of land resource issues.

It is now time for implementation of known technologies, initiation of problem-orientated and strategically directed research and rationalisation of programs at a state and national level to ensure that available resources are directed towards identified areas of need and not perceived areas of importance.

The working structure exists and while some infrastructural changes, especially within state administrations, would increase effectiveness, the predominant missing link is the political and departmental will to achieve discernible results. The ground swell of public concern in the area of resource management, which has been predominantly preservation-orientated so far, may well be the catalyst which achieves the necessary infusion of interest and action at all levels in the community.

Commentary

John Paterson

The chapter by Bob Junor and Warwick Watkins reviews the position of the soil conservation fraternity in New South Wales, and reflects the position in which, I think, most other states see themselves at the present time. It is the perspective of the insider which, of course, is quite different from the perspective given in Chapter 7 by Bradsen and Fowler. But both are pondering, from quite different vantage points, on where we stand and where we go.

The chapter describes the interests of people in the field, the history of public attitudes and major events over the past four decades, as the authors see them. The stance of the professional soil conservator is apparently that almost all land uses constitute degradation and the issue is to define what is acceptable. They go on to the steps required for the assignment of responsibility in the process of degradation and its amelioration.

The authors make some comments on the limitations of landholder action. Acceptance of responsibility of the landholder can be prejudiced by economic conditions, ignorance, and by lack of understanding of the wider ramifications of what the individual landholder does in his primary pursuit: that is; earning a living.

On the other hand, they see the acceptance of bureaucratic responsibility as being hampered in different ways; by imperatives of territoriality, by the slow transition from corrective measures to preventive measures, by a single-resource orientation both in soil conservation and in neighbouring agency territories with which they rub shoulders. They query the appropriateness of soil conservation being stabled with agriculture, as it is in most states, although not in Victoria. They note a variety of existing standing arrangements for communication in the field.

The chapter places very great emphasis, or perhaps hope, on a lead from the Commonwealth and on the outcome of initiatives by John Kerin to bring soil conservation into a national forum. This move has been reflected in the standing committee which has been operating for two years and is to be reflected in a Ministerial Council in 1986.

Parenthetically it might be worth noting that the effort to get soil conservators together in a national forum under ministerial auspices is happening at the same time as similar things are happening, though with a broader ambit, in relation to the Murray-Darling Basin. Initiatives have recently been taken to organise a Ministerial Council on the Murray Darling Basin, to include the three southern states and the Commonwealth, each fielding a team of three or four ministers.

I now touch on some structural issues in the soil conservation field, and deal with the ramifications of some of those for other fields of research management activity. I differ with the authors in some of their conclusions. My differences would arise fundamentally from having less faith than the authors do in the curative power of communication and coordination.

Communication and coordination are very fine things. Unless, however, we resolve some structural underpinnings of the evident problems, I believe we can communicate our heads off without making any great progress. Let me approach it this way. First, I will present my own quixotic review of progress in resource management agencies, in soil conservation and incidentally in some other fields. Second, I will look at where the remaining sources of discontent stem from.

There is no doubt in my mind that soil conservation activity over the past 40 years has not been a waste of time. The soil conservators cannot claim sole responsibility for the disappearance of the rabbit, of course. A wide popular perception of dangers of land degradation processes as they were occurring up to the 1930s legitimised ameliorative measures for stabilisation of the grosser forms of erosion. Forms of cultivation changed and reduced erosive tendencies, leading to marked improvements in the management of rural land. Any memory of a 1940s and 1950s agricultural landscape is a sufficient basis for a conviction that the present situation is greatly different from what it was 30 or 40 years ago. There is absolutely no doubt in my mind that organised state activity in the soil conservation field, aided and abetted by the Commonwealth, had a major impact. It has substantially achieved the *immediate* goals of soil conservation activity that were envisaged when the legislation came into force in the various states during the 1930s and 1940s.

In this field, as in any other, the achievement of *first order* objectives simply places one in a position of confronting more fundamental and, inevitably, more subtle and complex issues. Soil conservators are now facing *second generation* issues. They are not alone in this. The same kind of phenomenon has arisen in neighbouring fields of endeavour. Take my own field of water management. There is no doubt that the *first imperative* was to organise reliable sources of water supply. Something like a hundred-fold increase in the volume of stored water has been achieved on this continent over the last 70 or 80 years. Reliability and availability have greatly improved as a result; water management has thus met its first order objective. Secondary objectives of the water management agencies included drainage, protection of rivers and streams, and protection of water quality. These were, however, subordinated in the drive to increase regulated supplies of water. In the water portfolio, as in the soil conservation portfolio, those first order objectives have largely been achieved at the expense of secondary effects and interactions with the larger environment, which were, under statute, of little concern to the agencies responsible for water resource development.

In the pollution control field, the first order objective was to stop point sources of discharge from identifiable sources of pollution. These too have been very effectively mopped up over a period of two decades by legislation on clean waters and environment protection. A universal system of point source discharge control has had a measurable, demonstrable and invaluable effect on water quality in many rivers and streams.

The issue then is what to do next. The pollution control people are now looking at non-point or diffuse sources of pollution. These include sedimentation and the entrained nutrients and poisons involved in sediment loads. Simultaneously the soil conservators, having dealt with gross phenomena of gully and sheet erosion and so on, are now looking at the more subtle issue of long-term soil loss. Second order objectives bring those two sets of agencies into territorial conflict, because from different primary viewpoints they are fighting for the responsibility for the long-term loss of soil cover.

In the water field, the regulation of rivers and streams has been achieved to the extent that further advance now largely turns on achievements in the stabilisation of catchment hydrology. The stability of rivers and streams and the life of storages is extremely vulnerable to discharge rates which are in excess of those with which they are able to cope. As a result of having achieved their first generation objectives the various bureaucratics have wandered slap-bang into each other's territories, because the way ahead involves control of the interaction of their main functions with the main functions of others.

Before we pursue what might be done, there are a few other subtleties in the operation of these agencies which need to be considered and which are not discussed by the authors. I have made some unkind remarks about the cosy relationship that tends to exist between soil conservation authorities and their clientele. They will step in where the farmer is cooperative but tend to avoid the hard case. The hard cases are horribly visible because they are where the remaining gullies are. When you ask the soil conservation people 'What are you doing about that?' they will tell you that farmer Brown is a hard case. Let me illustrate with a specific instance. The Avon River Improvement Trust in Victoria is doing its best to rehabilitate the Avon River, which is one of the most degraded rivers on the entire continent. At one point the trust had problems with a major erosion gully. Much of their effort downstream is rendered ineffective by the gully but the soil conservation people will not touch it because the farmer is a difficult character.

It is is not only the soil conservators who rejoice in that kind of comfortable relationship with their client group. Irrigation authorities stand in fear of their clients and the clients stand in fear of the authority. They step around each other very politely and often avoid hard decisions. The authority avoids taking firm measures even where these are apparently necessary, and even where there is some popular support for such measures. We are now confronted by the necessity of dealing with the outstanding cases. Voluntary cooperation has taken us a long way in many respects, but in the face of a 'stand out', or what economists call a 'free rider', it no longer enough.

Some of you might be familiar with Anthony Downs' book called 'Inside Bureaucracy' (Downs 1967). 'Downs Theorem' of public regulation is that the regulator inevitably becomes the protector of the regulated. We can see that expressed in water management and in soil management. We see it emerging in pollution control, though to a lesser extent, because regulator and regulated have not had as long to get to know each other.

The cosy relationship has major strengths, which should be retained, but we must recognise that it also has shortcomings. If we are not on occasion prepared to wield the stick, there is no way of dealing with the 'free rider' problem. The 'free rider' problem is fundamentally destructive of cooperative measures and it limits the effectiveness of voluntary action. More particularly it is a fundamental barrier to the pursuit of second order objectives and the more subtle *interaction* effects of first order objectives. Effective action requires a degree of subtlety and careful operation which has not been vital in the first round.

So I am forced to dissent from the view emerging from the Junor/ Watkins chapter that the essence of the way ahead is communication and coordination. They are necessary but not, in my opinion, sufficient.

Water authorities now have a vital interest in catchments. If catchments are not well managed there is not much more we can do about rivers and streams, and drainage management. Equally though, unless broader powers are available to effectively control cross-country flows in catchments, then the soil conservators are substantially lacking the machinery required to make significant advances in *their* prime

field. One could of course extend discussion of interactions into issues such as control of cultivation, and of forestry—which can be a gross offender—and so on.

Underlying the structural problem, of course, is the complexity of the cause-and-effect relationships. They are fairly obvious in first generation problems and in dealing with first order objectives, but not nearly so obvious in the second generation and in the second order of interaction efforts. More particularly, the law is not well equipped for dealing with other than the simplest manifestations of *systems*.

When one is simply providing for the taking of water from one place and its reallocation elsewhere, or for arresting gross destruction of territory by erosion, then the law is able to describe the phenomenon, give it a name and specify how it will be dealt with. Like the rest of us, lawyers are not much given to embracing complexity in systems other than their own. Complex systems are reduced to absurd simplicities to permit them to be described in statute and adjudicated by judicial persons. There are fundamental difficulties in the mapping of the physical and social systems that we wish to regulate into legal systems. What was simple when the object was first order objectives and first order causes and effects, becomes anything but simple when we attempt to describe a complex physical system in legal terms. It is something we have not thought much about in a systematic sense.

Let me then suggest a solution to this conundrum which, as I have said, cannot be left purely to goodwill, communication and coordination.

First, management by objectives is something that can be readily accommodated under statute. The courts can be encouraged to look at the objectives of legislation, and to adjudicate where necessary by recourse to properly stated objectives. Incorporation of objectives in statute has become fashionable in some circles, but implementation has been unspectacular to say the least, and difficulties have been created for the courts because objectives have tended to be specified as motherhood, rather than being stated in instrumental terms.

Second, the assignment of rights and duties is simply done when it is a matter of vesting a resource such as water in the Crown, or vesting public responsibility for soil erosion control measures. It becomes much more complicated when one is dealing with more complicated systems, requiring more sophisticated management techniques to regulate those systems. Happily much of what is required is already present in statute law. The practical problem arises because within the cosy relationship, the assignment of responsibilities has rarely been enforced. Under water legislation there have always been powers to prevent people polluting. Those powers have not been read as being applicable to the case where a catchment is detached from a hillside and dumped in the river.

Similarly, the powers to influence the flow of waters is available within soil conservation legislation but has not been applied systematically. This interface is one of the classic areas for territorial combat between water conservators and land conservators. Regulation of *small farm dams* is the business of the water authority; regulation of *gully control structures* is the business of soil conservators. These structures are often indistinguishable. The enforcement of existing powers would take us a significant distance down the track but this would call for substantial statutory reassignment of powers and obligations.

Third, the agencies involved in these overlaps still have first order objectives which they must continue to pursue. The problem lies in the definition of boundary conditions when overlap arises. If looked at from a systems viewpoint each of the overlapping Acts will be seen to give powers, now in disuse, over the others. Water authorities *do* have the power to take action against landholders who pollute rivers. These are rarely used if the 'territory' involved is seen as belonging to the soil conservators. The same self-imposed constraints can be found in pollution control, forestry and so on. The overlap areas, instead of being the focus of attention, more often become vacant, a bureaucratic no-man's land that is too dangerous to enter.

A long tradition of hostile bureaucratic relations has been part of the problem. This is changing to some extent in some states. Convention now is turning to dictate an effort by bureaucracies to talk to each other. In fact there is a daunting number of interdepartmental committees seeking to improve communication and coordination. They consume an inordinate amount of time. Even when goodwill abounds they tend ultimately to resort to territoriality as the contemporary expression of goodwill. By and large, formal coordination efforts have achieved much less than hoped for, and tend often to blur the issues.

Why do they do that? Largely because boundary assignments are not clear in statutes. They can often be clarified. Water legislation should not include powers to monkey with catchments. It must, however, include the power of veto over any activity in a catchment which is under the jurisdiction of some other body (forestry, soil conservation etc) and is capable of dumping sediment in the water system. Conversely the effective operation of powers to regulate overland water flows requires a power of veto for the soil conservators. This will permit them to say 'that is not on because it is having an unacceptable effect on our first order land management objectives'.

A formal system of *separability* can achieve a result that coordination cannot (Paterson 1985). It can be written into the statute. Forms of words can be devised which describe first order interaction effects. Even though one cannot describe a system holistically in legal terms, it is possible to use formal separability conditions to achieve a good part of the same result. Most machinery of government issues can be substantially resolved into assignments of primary responsibilities and a structured set of priorities and veto powers to define the boundaries.

So, to conclude, I suppose that I would agree with the authors of the chapter that to live among people of goodwill is a good thing but I would not place too much faith in it; poorly specified structure places excessive strain on goodwill.

VI

Towards more effective policies for controlling land degradation: an overview

15 Contributions from the physical and biological sciences

Graeme Robertson

The use of land often promotes land degradation and this use is in turn a function of the economic, social, political and technical environments. The economic, political and social environments and their influence on land degradation have been widely canvassed in preceding chapters. The purpose of this chapter is to assess the capacity of the current technical knowledge on biophysical aspects of land degradation to contribute to solutions to the problem.

Definitions

It is perhaps late in the proceedings to define what should be meant by land degradation. As Chisholm notes in Chapter 12, land degradation is a complex phenomenon. In Chapter 3 Wasson notes that it is a change that makes land less useful for humans.

It is important to view land degradation in terms of use, either active or passive. Hence land clearing should not be considered per se as a land degradation process, as Burch, Graetz and Noble have inferred in Chapter 2, unless one of two preconditions is met. That is the land must previously have been more useful in a passive role, for example wildlife conservation, or the clearing must instigate other land degradation processes that detract from the capacity of the land to continue supporting its desired use and reduce its capacity for future uses or options.

The land use environment

Land degradation is fundamentally the product of human decisions, as noted by Quiggin in Chapter 10. These decisions are made in a socio-political-economic environment unique to present-day Australia. This environment will affect the state of biophysical knowledge and the application of that knowledge.

While agriculture is not the only land use in Australia causing land degradation, it is by far the major land user. Moreover, the characteristics of agricultural land use that constrain the application of solutions to land degradation are similarly held by most land users in Australia.

These important characteristics are:
• Large area of land per unit of output.
• Low labour input per unit of output.
• High capital investment on machinery per unit of labour.
• Low input (fuel, fertiliser and pesticides) per unit area.

Ruttan (1982) produces a series of international comparisons that clearly demonstrate the unique position Australian agriculture holds with respect to the above characteristics when compared to all other agricultural producers.

In 1985, a fifth factor must be added, that is, rapidly declining terms of trade and profitability (Table 17).

The effect of these characteristics on land degradation is that the land user has little access to labour or management to identify, investigate or monitor land degradation problems. Moreover, given the extensive nature of production there is inadequate economic capacity to add intensive inputs designed to ameliorate land degradation. This has always been the case in Australian agriculture, but is further exacerbated by current low profitability. To intensify management to reduce land degradation would be to fail to recognise the historical basis for the undoubted economic success of Australian agriculture: that is, its technology of low input per unit area, exploiting Australia's natural advantage in large land area (low cost land) and avoiding the disadvantages of limited labour and capital.

The low levels of labour and other resources per unit of land managed have similar implications for government land management agencies as they do for farmers. Except in unusual situations, resources to conduct research, extension, regulation or control will not be available in levels

Table 17: Index of declining terms of trade for agricultural producers in Australia*.

Comparison of quantity of goods that could be purchased by 1 tonne of wheat	1974–5	1984–5
1 tonne wheat (average return less freight, $/t on farm)	92.50	124.85
Superphosphate (tonnes)	6.29	1.28
Urea (tonnes)	1.06	0.43
Chemical ('Spray Seed', litres)	26.4	17.0
Fuel (diesel, litres)	1 592.0	227.0
Freight and handling (per tonne)	7.34	3.96
Tractor power (75 Kw tractor, Kw)	0.63	0.22
Tractor power (130 Kw tractor, Kw)	0.48	0.18

* Data compiled from Farm Budget Guides produced annually by Western Australia Department of Agriculture and Western Farmer and Grazier Newspaper.

adequate to duplicate intensive systems of land degradation control, as applied in, say, USA or Western Europe.

The above point has been illustrated by the biophysical papers presented in this forum. Burch et al have highlighted the lack of knowledge of the processes causing land degradation and particularly emphasised the lack of monitoring of the technical, economic and social processes. Wasson has also promoted the idea that current monitoring is grossly inadequate and, indeed, Wasson infers that in many instances the appropriate type of monitoring is still a question of debate.

Processes of land degradation

Land degradation includes wind erosion, water erosion, salinity, waterlogging, soil compaction, structural decline, acidification, development of non-wetting, flooding and vegetation decline. Although there are linkages between some of these processes, most of them work independently of each other. Consequently, designing and implementing a technical package to minimise these processes as a group is difficult. Amelioration of one process may often adversely affect another. For example, earthworks to control water erosion can, under Western Australian conditions, lead to increased groundwater recharge and salinity. Similarly, improved legume pastures can restore soil structure but promote soil acidification and non-wetting. As it is necessary to consider the non-technical aspects in application and adoption, the scenario becomes more complex and the integration of knowledge and the assessment of the system becomes at least as important, if not more important, than knowledge of the components.

Biophysical causes

As previously noted, clearing of land or intensification of agriculture are not in themselves causes of land degradation. Rather they can predispose land to degradation. Just as examples of land degradation associated with intensification of land use can be identified, equally it is possible to find examples where the converse is true. For example, in Western Australia intensification of cropping will certainly reduce recharge to groundwater and hence reduce salinity hazard (Nulsen and Baxter 1982).

Doubtless the intensification of cropping on recharge areas, while reducing salinity, will result in other land degradation problems unless management practices are appropriate. Minimum tillage, stubble retention and other practices not yet conceived may be required if stable land use under continuous cropping is to be achieved in the situation outlined.

The causes of land degradation are the use of management systems inappropriate to the continued stability and maintenance of that use, not the use per se. It may be, for some land, that there does not exist within the current context a management system capable of achieving desired land use without degradation. This raises another issue, that of land capability assessment and the prevention or regulation of use beyond the assessed capability of the land. This approach, favoured by soil conservationists, is often argued against by economists concerned at the removal of market influences from land use decisions.

Potential solutions

The chapters dealing with the biophysical aspects of the problem were all critical of the level of understanding of the main land degradation processes. Wasson suggested that there was indeed a considerable amount of knowledge available, but poor ordering of priorities has resulted in much of the available knowledge not being able to be applied.

The social scientists have re-emphasised this perspective. In his commentary Greig highlighted the lack of relevant functional relationships and Davidson made similar observations.

In view of the pessimism about the level of understanding and even the identification of the problems, there must be doubt about the availability of solutions to land degradation in Australia. Undoubtedly there is inadequate documentation of land degradation processes, rates, and economic and social implications on which to develop sophisticated national policies to avoid or repair land degradation. Moreover, the resources required to obtain the desired information within a reasonable time frame are unattainable.

The policy issue, given the acceptance that land degradation is a national problem, albeit unquantified (unquantifiable at this time?), is a choice between whether the knowledge constraints are sufficiently serious to prevent movement towards solutions, and whether there is an approach possible despite the constraints.

The latter is the only acceptable choice and it then befalls the pragmatist to endeavour to conceptualise solutions and indicate necessary policy measures.

Although it may be hyperbole, it is possible to assert that there are known technical solutions to all recognised land degradation problems observed in Australia, although some of these solutions may contain elements that are unacceptable from an economic, social or political viewpoint. A solution involving non-use would often contain such unacceptable elements. Hence the existence of solutions does not indicate that we are close to being able to implement these solutions in any but a few situations.

In addition, there may be some problems not yet encountered or recognised because of the constraints to knowledge as discussed. Past disregard for the importance of the soil and land resources probably still conceals land degradation problems. The recently identified traffic compaction of sandy soils in WA (Henderson 1985), its widespread occurrence and effect on productivity is a sobering lesson on the limits of our knowledge of problems that may be present.

However, despite this concern we can follow to a large extent the United States Department of Agriculture (USDA) and assert that technology is adequate to reduce much soil erosion to acceptable levels (USDA 1981). Unfortunately, in most cases the fact that technology exists does not mean that it will be adopted. Unless the technology is fully integrated into the management system in a way that results in a more economically efficient production system, it is unlikely to be adopted. Strip cropping and stubble retention in northern NSW and southern Queensland may be one of the few examples of successful technology in this respect.

There are examples from Western Australia and probably elsewhere that serve as illustrations of how readily adopted a soil conservation strategy can be if it is integrated into the management system and is profitable.

In the eastern wheatbelt of Western Australia soils dominated by sodium and magnesium have declined severely in structure since clearing. Cultivation, loss of organic matter and dispersion have resulted in a very poor plant growth medium that has low infiltration rates, puddles and waterlogs, surface seals and in fact makes life very difficult for a plant in a low rainfall environment.

Minimum tillage, a one pass direct drill operation with a combine, has resulted in higher wheat yields over time (Table 18). This has been associated with improved soil structure, better germination and better water relations (Jarvis 1985, Jarvis et al 1985). As a consequence, direct drilling has been widely adopted in the eastern wheatbelt, but its adoption was related to the benefits the system had to offer farming in the area, not the fact that it avoided a serious soil degradation problem. Direct drilling was easily integrated into the farming system in the area, it was suited to a low labour/large area production system and was a lower cost alternative. As a bonus, in the long term it improves soil structure and yields.

Minimum tillage has similar, if not greater, benefits to offer soil conservation on the extensive areas of sandy surfaced soils, particularly with its effect on reducing wind erosion hazard. However, research data and farmer experience have shown that crop performance with minimum tillage on sandy soils is inferior to that with conventional systems (Table 19). Moreover, in the higher rainfall areas of the south coast, the major wind erosion area, weed control is less certain and often more costly. Other problems such as root diseases and seedling-damaging insects appear enhanced under minimum till conditions. Hence to date minimum tillage does not offer an attractive alternative to the farmer and consequently, despite the severe wind erosion hazard on the sandy soil types and the known conservation benefits of minimum tillage, it is not adopted widely.

Research needs

Too many eminent people and groups have indicated that more research and knowledge is needed to overcome land degradation in Australia for the suggestion to be dismissed. Noted recent examples include the Queensland Planning Committee for Soil Conservation (QDPI 1983), the Australian Soil Science Society (1984) and the Standing Committee on Soil Conservation (1984).

In his chapter, Wasson has indicated the large areas of inadequate knowledge of one element of land degradation, soil erosion. Moreover, he implicitly suggests that there is a bias of research effort and hence knowledge towards simpler, better defined systems that do not necessarily have much applicability on a practical scale.

If soil conservation practices are not being adopted in favour of land

Table 18: Wheat yields from continuous wheat. Red sandy clay loam; Salmon gum, Gimlet soil—Merredin 1977–84*

| | Wheat yields (kg/ha) | |
| | Continuous wheat | |
	DP	DDC
1977	199	306
1978	2 100	1 691
1979	1 647	1 777
1980	No yield—drought	
1981	568	1 190
1982	582	1 002
1983	602	810
1984	1 072	2 021
Average	846	1 100

DP = District practice (conventional cultivation)
DDC = Direct drilled combine
* Results from R J Jarvis (1985).

degrading practices, there must be major constraints present. In this forum it has been suggested that lack of technical knowledge may be the major constraint.

Chisholm and Kirby and Blyth have highlighted the difficulty, if not impossibility of regulatory agencies having adequate knowledge to implement various regulatory approaches to minimising land degradation. Bradsen and Fowler have in turn argued that lack of information similarly restricts the use of economic incentives and in many cases the use of regulation may require less information. Given the difficulty of using current knowledge efficiently for regulation or development of economic incentives for conservation, or disincentives for land degradation, it is clear that more knowledge and hence more research are required. The question is what type of research is most likely to be cost effective in achieving the desired goals?

Certainly there is a need for continuing basic research. However, the role and potential contribution from such research must be placed in perspective. While it is desirable to develop mechanistic descriptions of land degradation processes, in the medium term it may not be possible. Moreover, it may not help solve a water erosion problem to know the chemical and physical mechanisms of clay dispersion, although that knowledge may assist prediction of such events. Also, knowledge of mechanisms may lead to solutions, for example water erosion may be reduced by chemical modification of the soil. However, equally, a management practice such as strip cropping may be an

Table 19: Wheat yields from two cropping methods under two rotations on sandplain soils—1977–84*.

Wongan Hills and Esperance

	Wheat yields (kg/ha)					
	Esperance		Wongan Hills			
	wheat/clover rotation		continuous wheat		wheat/clover rotation	
	DP	DDC	DP	DDC	DP	DDC
1977	1 836	1 770	582	616	1 117	1 242
1978	3 006	2 601	1 786	1 130	1 968	1 931
1979	1 252	1 134	1 609	1 465	1 028	1 014
1980	2 072	1 510	1 203	984	804	549
1981	2 223	1 972	1 916	1 690	2 964	2 488
1982	1 686	1 344	1 900	1 779	2 324	2 566
1983	1 063	832	1 682	1 524	2 874	2 179
1984	1 340	1 126	1 799	1 597	2 741	2 343
Average	1 804	1 536	1 559	1 348	1 977	1 789

DP = District practice (conventional cultivation)
DDC = Direct drilled combine
* Results from R J Jarvis (1985).

adequate solution to water erosion in an area and there may be no need for mechanistic-based research in such a case. This is the dilemma of the research administrator. Is the best approach to solutions to a management problem a reductionist research approach or is it a systems based approach? Wasson suggests that the development of practical solutions will require a halt to reductionism, a suggestion fortunately being increasingly propagated.

It is appropriate at this point to re-emphasise that land degradation is a problem of land use and the management system being applied. Only the land user—the farmer, the forester or the miner—can really apply solutions to land degradation and hence solutions must be attractive to the land user. This attractiveness must usually, not always, be expressed in terms of increased economic efficiency.

When attempting to develop modifications to a management system that will lead to both increased economic efficiency and reduced land degradation, a management-orientated research will be essential. By its very nature such research must be multi-disciplinary and consider the economic, social and political components as well as the biophysical. In the main the research will be applied and have a regional or district focus. There will remain a need to gather detailed process information that can be interpreted locally. However, the major research input needs to be at the local level. This has two implications in the context of this forum. Firstly there is a need for economists to provide input and direction at the local or micro model area. Far more economic resources are required at the local level than at the esoteric, perhaps more intellectually satisfying, yet less relevant, macro model and theoretical resource economic areas. At the moment the majority of the scarce resources is in the latter area.

Similarly the biophysical research area currently has a biased focus towards a reductionist approach, or at best a mechanistic approach. Since the Birch Inquiry (Aust Independent Inquiry into the CSIRO 1978), CSIRO has apparently avoided management-orientated research on the grounds that it is not in line with its policy objective that its research must be of national significance. Management-orientated research is usually of local significance although in aggregation such research would have a profound effect nationally, at least in the area of land degradation.

It is appropriate to compare the research needed into land degradation with research common in agricultural production. Both must be applied, systems-based research with the integration and adoption by the land user being the ultimate objective.

Agricultural research has several interesting characteristics. It has very high rates of return, often in excess of 50 per cent per annum (Ruttan 1982, Marsden et al 1980). However, of equal importance is that the return often accrues mainly to the region where the research is done (Ruttan 1982, Fels and Quinlivan 1978). In the case of the technological base for sheep production in WA studied by Fels and Quinlivan, the converse held also. That is, most of the technology used was developed locally.

It is supportive of theoretical approaches to problem solving to identify type studies where the advocated system has been successfully applied. Fortunately there is one excellent example of such an approach. Nelson and others (Nelson 1985) identified a serious recurrent wind erosion problem in a continuous cereal cropping area on the sandplain around Geraldton in Western Australia. The problem was not only a land degradation problem but farm income was being severely affected.

Nelson and others using traditional management-orientated research produced a new system of farming, the use of a trash retention cereal/ lupin rotation specifically aimed at reducing wind erosion. The new system was both agronomically and economically more efficient and was virtually 100 per cent adopted in the region within three years. The wind erosion problem is now a memory in this particular region.

Perhaps unfortunately the Nelson system is not readily transferable to other regions. This reinforces the hypothesis that the application of any mix of biophysical solutions to land degradation will be very site-specific and will depend on the nature of the land, the way it is being used and the total economic and social milieu in which that use is occurring. It is likely that solutions to the management problems that result in land degradation are most likely to be found in regionally based systems research projects. Little gain will be made by researching interesting problems. There are many interesting research problems, some of them are important.

Policy implications

The application of management-orientated research as a major contributor to solving land degradation should be the prime policy objective. Of course the efficient use of such a policy requires at least some ranking of problems within and between regions and hence some resources need to be allocated to monitoring on a macro scale. However at this stage a high level of precision is not needed in this field.

The use of local management models is likely to be an efficient way to focus expertise from various disciplines on problems and potential solutions and in this area economists have a vital role to play. Such systems models should play a significant role in ensuring that the research is constrained by the need to produce a management system that reduces land degradation and at the same time improves, or at least maintains economic efficiency, within the social and political environments. If this is achieved, adoption will be rapid. In many cases technology transfer and local adaptation of known principles will contribute more to solutions in the medium term than will the achievement of new knowledge.

Conclusions

The biophysical sciences have a major role to play in providing the basis for solutions to land degradation in Australia. These solutions will be most readily achieved if current biophysical knowledge is applied, together with social and economic information to management-orientated research. This research by its nature must be applied and be focused at the district and regional level.

16 Contributions from the social sciences

Robert Dumsday

Introduction

How are we able to assess the impact that land degradation has on Australia's environment and the welfare of Australians? What do we believe to be the extent and severity of land degradation problems? Do we have a problem? If we have a problem, what options are available to solve or ameliorate it? These were central questions which the workshop organisers set down and to which participants frequently returned.

I have chosen to deal with these central questions directly in this chapter, in an attempt to distill the views that were presented in previous chapters and in subsequent discussion, and to assess the progress made towards the objectives. In addition, I consider a number of secondary questions that were put before the contributors in this section.

However, before tackling these questions it is necessary to review what is meant by land degradation. The organisers suggested that land degradation includes all those adverse effects that land uses may have on the services provided by land. They identified as the major forms of land degradation, soil erosion and salinity of land and streams, rural tree decline, loss of native and unique habitats for flora and fauna, damage to land through recreational use, and desertification through loss of vegetative cover in arid regions.

In his commentary on Section II, Peter Greig observes that there could be no absolute definition of land degradation—a rural allotment in the process of reverting to natural bush might be 'degrading' from a commercial farmer's point of view but 'rejuvenating' if the owner seeks to live on a 'bush block'. Greig adopts the definition that degrading land is land that is losing its productive capacity, measured against current land use. However, he does not suggest how this definition might account for potential or intended land use.

In Chapter 1 Chartres also refers to land degradation as potentially the cause of loss of the land's productive capacity. He focuses attention on water and wind erosion, salinisation and waterlogging as the major forms of land degradation in Australia. In Chapter 3 Wasson defines land degradation as a change experienced by land that makes it less useful for humans.

In discussion, Gibbons suggested that, ranked from most important to least important, the main forms of land degradation in Australia were water and wind erosion and their onsite effects on cropland; water erosion and structural decline on grazing land; offsite effects of erosion on public utilities; irrigation salinity; and dryland salinity. He did not include rural tree decline per se or some of the other forms[1] listed by the workshop organisers; however, his main point at the time was that the important forms of land degradation mostly occurred on privately operated freehold and leasehold rural land and that policies should therefore be focused on private rather than public decision makers.

In summary, no single definition of land degradation is likely to satisfy all those interested in the subject. This certainly was the case during discussion when participants sometimes became frustrated at the lack of consensus on the matter. Cooke and others suggested that even the term 'land degradation' was unsatisfactory and that 'land deterioration' would be a better term.

The lack of consensus on the definition of what constituted land degradation was probably more apparent than real. There were no substantive differences between most of the definitions proposed during the workshop and that proposed by the organising committee at the outset. There also seemed to be agreement on the major forms of land degradation, at least as measured in biological and physical terms. Blyth and McCallum point out in Chapter 4 that the physical existence of land degradation is not necessarily evidence of an economic or social problem; it is in this area that the real disputes arise.

Central questions

How are we able to assess the impact that land degradation has on Australia's environment and the welfare of Australians?

There are significant scientific and technical problems in assessing or measuring the physical impact that land degradation has on the environment. There is a vast array of physical and biological measures such as the soil chemical factors of pH, salinity, nutrient levels and levels of aluminium and manganese oxides; soil physical factors such as pore size, bulk density and moisture-holding capacities; vegetative factors such as proportion of cover, plant diversity and plant productivity; and other measures such as soil loss per unit area of land, sediment loads in streams, and rates of tree decline.

Some people attach fundamental importance to the physical and biological measures of land degradation in terms of their relationship to human welfare. Thus, soil loss is often seen by conservationists as something to be minimised or eliminated in the belief that benefits to society in the long run will inevitably exceed any costs.

Greig draws attention to a related set of arguments advanced by 'deep' ecologists and others that the common view of land degradation was excessively anthropocentric. These people argue that all living and non-living things have rights and that whether or not the long-term benefits of controlling land degradation exceed the costs to humans, there are ethical grounds for ensuring that degradation is reduced or eliminated.

Economists, especially those of the neoclassical school, believe that they have the best methodologies for assessing the impact of land degradation on human welfare. They tend to be unashamedly anthropocentric on the grounds that all decisions on the environment will continue to be made by people and that whether or not other living and non-living things have rights, like the unborn, their rights must be represented by franchised humans. Passmore (1980 p188) is pessimistic about the likelihood of persuading populations to adopt new moral principles on the environment. He suggests that 'new modes of behaviour are much more important than new moral principles'. In Chapter 10 Quiggin suggests that it would seem unwise to wait until a conservation ethic is firmly established before mounting an attack on the problem.

The workshop organisers took the view that while ethical considerations should not be ignored in environmental policy making, emphasis should be placed on understanding and working with the behavioural causes of land degradation, given existing value systems. Those who are not satisfied with this view are referred to the writings of Passmore (1980), and may take additional solace from the fact that several social scientists regard ethical concepts as important to the economic analysis of environmental issues and are actively working in this area (Sagoff 1981, Pennock and Chapman 1982, Sen and Williams 1982, Wunderlich 1984).[2]

Chisholm (Appendix A) outlines the economic framework within which the welfare effects of degradation or conservation should be assessed. While the framework has a number of weaknesses it does not appear to have any serious competitors from other disciplines and is likely to be increasingly applied to evaluation of environmental issues.[3] However, it is important to note that while the conceptual apparatus for this evaluation is available it has seldom been applied in practice. In Australia there have been no comprehensive empirical evaluations of land conservation programs from a national perspective. There have been isolated case studies made in Allora, Queensland, and Eppalock, Victoria, as part of the Commonwealth and State Government Collaborative Soil Conservation Study (Aust DEHCD 1978a, Abelson 1979, Chapter 4). There have also been regional evaluations of salinity control in Western Australia (Bennett and Thomas 1982) and Victoria (Lumley 1983) but we are yet to see national studies of the type done by Heady's group in the USA (eg Wade and Heady 1977).

What is known about the extent and severity of land degradation problems?

Several local, regional and state surveys in Australia have attempted to tackle this question. However, the only survey providing a national perspective on land degradation problems is the 1975-77 collaborative study referred to above (Aust DEHCD 1978a). Detailed information from this study has recently been published by Woods (1984). A broad conclusion from the study was that 'Fifty-one percent of the total area used for agricultural and pastoral purposes in Australia was assessed as needing some form of soil conservation treatment under existing land use'. (Aust DEHCD 1978a, p135).

The collaborative study was useful in increasing community awareness of the nature and extent of land degradation problems. It probably contributed to the heightened interest of community groups such as the Australian Conservation Foundation (ACF) and to higher political profiles for land degradation problems at national and state levels. It also serves as a valuable source of information on land degradation and the policies for combating it. However, the information was subject to a number of limitations, as described by Woods (1984 p2). These limitations included inconsistencies between states in the way land degradation was assessed and in the corrective measures advocated, overestimation of the extent of degradation[4], and the joint occurrence of several forms of degradation leading to difficulties in classification. More importantly, because this was the only study of its kind, there is no reliable assessment available of how, from a national perspective, the extent and severity of various forms of land degradation have changed over time.[5]

Apart from the two benefit cost studies mentioned above, the collaborative study reported economic information from across Australia on the costs of structural works and management practices 'required' to maintain land in its present use, and the capital value of the land which these measures would protect. Given the restraints under which the study was conducted it was not possible to assess whether the benefits from implementing the 'required' measures justified the costs involved, or whether land should be maintained in its present use in the first place. Unfortunately, some people have chosen to ignore these important qualifications in making widely publicised claims (eg Burton 1983, p83) that the cost of carrying out 'essential' soil degradation control is of the order of $1.6 billion.

In summary, while these surveys have been valuable as a source of information and in increasing public awareness of land degradation they do not help significantly in formulating economically efficient policies for dealing with the problem. However, nearly a decade has passed since the last comprehensive national survey and it is probably time to repeat the exercise, but with more emphasis on case study benefit cost analyses and less emphasis on estimating the costs of 'required' practices.

Do we have a problem: are rates of land degradation excessive in some sense?

It is clear that many people think that land degradation is a problem; most governments find it politically beneficial to maintain substantial departments of land protection, soil conservation, etc and there is also substantial private investment in land conservation measures (Aust DEHCD 1978a). The question is, are these levels and types of public and private investment in control of land degradation inadequate, excessive, or about right? I have already suggested in answer to previous questions that we have very little useful information on which to base formal answers to this question. The chapters by social scientists do not even consider this question on 'in principle' grounds.

Some participants were of the view that land degradation problems were not nearly as serious today as they were in the 1930s and 1940s before controls were first introduced. In discussion, Paterson and Davidson noted that the Australian landscape improved dramatically following control of rabbits and other pests like prickly pear. Substitution of clover leys for fallowing in wheat production led to similarly dramatic improvements in land quality and productivity (eg see Burch, Graetz and Noble, Chapter 2).

There is an interesting parallel in land management with developments taking place in water management in Australia. Watson and Rose (1980) and Randall (1981) have suggested that the Australian water industry has entered a new phase in the past few years. They suggest that the water industry went through a period of rapid development with the opening of new irrigation areas. Then, as suitable sites for these works began to dwindle, the 'development phase' was left behind and greater emphasis was placed on allocative mechanisms to improve the efficiency of water use from existing systems—the 'management phase'. Paterson's comments on Chapter 14 brought attention to this argument and in my opinion it is useful to view land degradation problems in a similar manner.

In my view we are no longer in the 'exploitative phase' of land management in Australia which was characterised by widespread ignorance of the factors leading to degradation, and rapid depletion of the services provided by land. Australian landholders, scientists, and extension officers have developed and applied new knowledge of land use which, as Davidson suggests (Appendix B) has resulted in less frequent catastrophic environmental events and a trend towards rising or stable productivity. The Minister for Primary Industry in his introduction to this book, does not see the need for a massive injection of funds and resources to control land degradation. Instead he advocates greater coordination of the activities of state and federal governments, with which I concur but to which I would add increased emphasis on developing and applying new policy and managerial instruments, at both private and public levels, in order to improve the efficiency with which land is managed.

One aspect of the management phase of land use which was not an important element of the exploitative phase is the competing uses of land. Agriculture can no longer be regarded as the pre-emptive form of land use. Following a long period of rising real incomes in the economy, and given the likely continuing impact of the EEC Common Agricultural Policy on Australian agricultural exports, we can expect to see increasing demand for non-agricultural uses of land for recreation, national parks, flora and fauna reservations, and so on. Some people might suggest that it was a good thing that we farmed the land when we did.

The question of controlling land degradation in the management phase could be thought of in two parts: degradation *within* land uses and the effects of different *classes* of land uses.[6] Rather than trying to control degradation within, for example, agricultural land uses, it may be economically more efficient to remove land from agriculture or not to allocate it to agricultural use in the first place. Of course, such actions have already been taken in practice, although it has not always been clear that they were economically efficient actions. Preventing the allocation of land to agriculture in the Little Desert region of Victoria was probably economically desirable (eg see Lloyd 1970)[7], whereas I question the economic justification for removing agriculture from parts of Western Australia in order to control dryland salinity, or the banning of land clearing in parts of South Australia.

In summary, it does not appear that land degradation problems are as severe and widespread as they were in the early stages of European settlement. Many current problems tend to be local in nature, although soil acidification problems may become widespread and salinity levels in the Murray-Darling Basin are causing growing concern. We have entered a management phase where costs and benefits of controlling land degradation need to be compared at both public and private levels. This, of course, represents one of the main contributions to be made by social scientists in helping to decide whether there are any actions to be taken to improve the efficiency of land use over time. A result of these analyses, which some people would undoubtedly find unpalatable, is that, given alternative uses of public and private resources, there may already be over-investment in conservation in some areas.

What are the policy options for controlling land degradation?

In Sections III and IV the social scientists have carefully evaluated a range of policy options available to governments for controlling land degradation. They include *economic incentives* in the form of charges or pollution taxes—on emissions of salt or soil etc (or on some factors related to the output of these pollutants), subsidies—including concessional finance and income tax concessions on investments in conservation measures, and transferable rights (for example on tree clearing); *regulations*—including zoning land for permitted uses and setting management standards; *provision of information* through research and extension programs; review of private *property rights* associated with freehold or leasehold land tenure; and review of other *government policies* for their unanticipated effects on land degradation.

It may be useful to group policy options for controlling land degradation into two types: those that should find application in the short run versus those that may be applied only in the long run. It was clear at the workshop that many participants were not willing or able to grasp some of the policy proposals that were advocated, especially those advocated by economists. There is a well-known expression in the economics profession that the long run is made up of a series of short runs, so some proposals for the long run may never be implemented. For example, economists have been advocating pollution taxes for many years, yet there are few cases where such taxes have been implemented in practice, even for the relatively simple case of point-source pollution. Some economists believe that the political and other costs of administering these systems have been underestimated.

On the other hand, to the case of water resources, it seems that there is a lag of something like 20 years between policy proposals and implementation. More rational pricing policies and transferable water entitlements, for example, have been advocated for many years by economists, but it is only since water use entered the management phase that such policies have been implemented in practice. We cannot expect that participants will leave workshops fully armed with new ways of going about their business when they return to their respective offices. Adoption of ideas is probably more often a process of osmosis than some threshold event.

Secondary questions

Turning now to some of the more detailed questions, I wish first to deal from a policy perspective with some of the questions faced by biological and physical scientists.

Are the processes of land degradation well understood?

Most people at the workshop, including the biological and physical scientists, felt that the answer to this question was 'yes'. The more common forms of degradation, such as wind and water erosion and salinity, have been studied for many years and a substantial body of information on the processes involved has been accumulated. However, this does not mean that new forms of degradation, or the future severity of old forms, will be accurately anticipated. In addition, there was agreement that it was not possible at present to quantify many of the processes involved in a manner which would be useful to policy formulation.

Are the predictive models which relate land use to land degradation adequate for policy purposes?

There is no doubt that the development of models to represent degradation processes is growing, both in Australia and internationally. There are good reasons why this should be so. First, the processes involved are usually complex and it is difficult to analyse the relationships involved without such models. Even the fairly simple USLE model is valuable in disentangling the relationships between the effects of land slope, vegetative cover, management practices etc on soil loss and, indirectly, on productivity. Second, as often claimed for such models, they have significant tautological value—they assist decision makers to compile information in an orderly fashion and to build their understanding of the systems involved. Third, the practical contributions that social scientists might make are severely constrained in the absence of such models.

Several participants held the view that there was a need to continue to develop predictive models, either to improve on the capabilities of existing models, or to describe relationships for new situations. Some research groups are, of course, already developing models with managerial and policy objectives, as described by Blyth and McCallum in Chapter 4.

The scientific acceptability of most available models is a source of serious concern to many researchers. For example, the USLE has been adopted by the Australian Agricultural Council for development and adaptation to Australian conditions. This decision has been criticised (eg see Stocking 1984) on the grounds that the USLE was developed for soil and climatic conditions in the USA quite unlike those to be found in this country. It should be noted that Stocking does not offer an alternative to the use of the USLE. Other criticisms of the use of the USLE are to be found in Edwards and Charman (1980) and Chisholm, Chapter 12.[8] However, given that there have already been some 40 years of research in Australia into degradation processes, decisions need to be made about the costs and benefits of embarking on projects to build uniquely Australian versions of models like USLE, CREAMS and EPIC. From a policy point of view it should be recognised that the ability to predict relationships between land use and soil loss accurately, for example, represents only one component of the task of administering control programs efficiently.

As something of an aside, we are repeatedly told that the Americans have the advantage of 10 000 plot years of data on which to build their models. This argument has never greatly impressed me. The New South Wales Soil Conservation Service (NSWSCS) has been running plot experiments at six research stations for up to 30 years to give a total of 5 200 plot years (G Cunningham pers comm 1985). It is true that most other states have not conducted similar experiments and that even the NSWSCS experiments were often run more for extension than research purposes. However, it is also likely that the USLE relationships for many USA locations have been determined by groups of experts at workshops drawing on their collective experience rather than by rigorous statistical analysis of carefully designed field or laboratory experiments. For policy purposes this process of introspection may often be sufficient. To be *acceptable* for policy purposes the models should be capable of contributing to an *improvement* in decision making, while not necessarily being able to represent real world processes and systems *accurately*.

What is known about onsite versus offsite effects of land use?

It is apparent from the contributions by Upstill and Yapp (Chapter 5) and Woods (Chapter 6) that little is known about the economic consequences of the offsite effects of various land uses. While there are some data on these costs for the USA, they appear to be almost non-existent for Australia. Most of the attention in both countries has focused on the onsite effects of degradation. However, while I agree with Upstill and Yapp that there is a need to collect more information on the offsite costs, I do not believe that they are likely to assume the importance that Upstill and Yapp ascribe to them. The data from the USA do not necessarily apply to Australian conditions, primarily because of our lower population densities in rural areas. A large proportion of the US population lives in provincial towns of small-to-medium size so the total utility derived from controlling offsite effects is likely to be larger than in Australia where, for example, the pollution levels of rural streams may not be of concern to many people. If the city of Adelaide were not involved, even the salinity levels of the Murray River would probably not attract much political interest.

Do the concepts of rates of time preference, intergenerational equity, externalities etc, help our understanding of the causes of land degradation and suggest whether anything further should be done about them?

There was substantial agreement at the workshop that land degradation was a social phenomenon, caused primarily by human actions rather than the result of random natural forces. For these reasons it is important to understand the behavioural forces at work in order that they may be influenced, if necessary, to change rates of land degradation. It is no longer sufficient to view the problem as one that can be solved simply by directing more public funds to conservation programs.

In Sections III and IV the social scientists discuss a range of behavioural phenomena that may contribute to problems of land degradation and which have potential for government intervention. However, there is some disagreement among the authors about the relative importance of some of the behavioural factors, or even in which direction their effects are felt. Table 20 is an attempt to summarise the relationships between behavioural factors and their effects on the direction of deviation from optimal resource usage.

It should be noted that Kirby and Blyth explicitly exclude irreversibility and market interest rates as important factors and criticise intergenerational equity and market failure arguments for adjusting social rates of discount. I am not completely convinced by their arguments on irreversibility as they seem to be too 'agrocentric', to coin a term. Many people see the irreversibility argument as including the possibility that agricultural land use will lead to irreversible loss of other land use options for future generations. Further, I am not convinced that their arguments on interest and discount rates rebut Quiggin's concern about strip-mining the soil (Chapter 10).

Table 20: Behavioural causes of land degradation[9]

	Direction of deviation from optimal resource usage	
Behavioural factor	Under-exploitation	Over-exploitation
Discount rates or imperfect capital markets		ACEQ
Irreversibility		ACQ
Intergenerational equity		ACQ
Uncertainty	Q	Q
Lack of knowledge	E	ACEK
Property rights		E
Freehold tenure		B
Long-term leasehold		K
Fixed-term leasehold		KQY
Government policies	CEKQY	CEKQY
Externalities and public goods	CD	ACDEK
Perception		R
Economies of scale		A

A = Chisholm, Appendix A
B = Bradsen and Fowler, Chapter 7
C = Chisholm, Chapter 12
D = Davidson, Appendix B
E = Edwards, Commentary on Section IV

K = Kirby and Blyth, Chapter 11
Q = Quiggin, Chapter 10
R = Rickson et al, Chapter 9
Y = Young, Chapter 8

In my view the concepts discussed by the social scientists do help in understanding and solving environmental problems. However, some of the concepts may best be viewed as having long-term potential—it seems that many decision makers in the land management area either do not understand the concepts or see them as being irrelevant at the present time.

On the question of rates of time preference, discount rates, etc, at least one participant remarked that economists did not seem to know which rate should be applied to evaluation of environmental projects. In my view this is not the case, although there is certainly no universal agreement among economists on this important issue. Most economists hold the view that there is no justification for setting special discount rates for environmental projects and recommend the use of market rates of interest or the rate adjusted for taxation effects etc as recommended by the federal treasury. Some Australian state treasuries (for reasons best known to themselves) apply to all public projects rates that are significantly lower than both market and federal treasury rates. In these cases there appears to be even less justification for setting special rates for environmental projects.

In the long run there may be acceptance of arguments, such as those advanced by Quiggin, that special discount rates (including zero rates) should apply to the economic evaluation of some public projects.

However, until now there does not appear to have been official acceptance of these arguments in any country. For example, the British studied these arguments in detail in the economic evaluation of options for disposing of radioactive wastes and decided (rightly or wrongly) that even here, there was no justification for employing lower or zero rates of discount (eg see Pearce 1982).[10] Given the relative importance that most people attach to land deterioration as compared to environmental pollution by radioactive wastes, that result would appear to offer little hope to British soil conservationists.

Despite the above observation that no treasuries appear willing to support low or zero rates of discount for evaluating environmental programs, even those involving irreversibilites and long time horizons, it will often be the case that governments are willing, in practice, to make available 'low interest' loans for private investment in, say, soil conservation structures. It appears that this is just another case where formal economic analysis does not capture all the political costs and benefits of different levels of intervention. To quote (out of context) from one of the doyens of the economics profession:

> It is sometimes argued that finance by taxes burdens the present, and finance by loans the future; and, therefore, that the choice of method should depend on how far the present and the future respectively benefit from the expenditure. But this is not so. The issue is not one of justice between generations, but of what is technically convenient and politically feasible. (Pigou 1947 pviii).

Chisholm (pers comm 1985) quickly reminded me that some of Australia's worst environmental (and economic) disasters like the Ord River scheme and other irrigation and closer settlement schemes were introduced following reasoning of this kind.

What importance should be attached to the 'unpriced' benefits of environmental improvement versus the efficiency effects as expressed in markets?

This question concerns some matters that are close to the hearts of many conservationists who generally feel that economic analysis plays down the importance of values that are not expressed through markets. Unpriced values may include the aesthetic attributes of landscapes and the utility that is derived from preserving options for future generations. Sinden and Worrell (1979) provide a comprehensive review of the values which fall into this category and the ways in which they may be assessed.

Despite the sometimes ingenious work of researchers like Sinden and Worrell it is my view that it is often better to assume that where there are no prices, there are no prices. Economists usually face significant difficulties in analysing important problems even when price information is available through the workings of markets. The methods for eliciting 'prices' in the absence of markets are sometimes suspect and may serve only to increase the unreliability of recommendations which are based on such methods.

The foregoing does not mean, of course, that we should not at least list or describe the unpriced benefits and costs that may be associated with a given environmental program. Descriptions of the type provided by Woods in Chapter 6 are valuable to decision making and may often represent the extent of useful analysis for some items in benefit-cost appraisals. It should also be noted that leaving evaluation of 'unpriced' benefits and costs to political processes represents less than an ideal situation. Political assessments are subject to substantial imperfections and may have little to do with advancing aggregate human welfare. There is a strong case for economists to continue to work in this area.

Who owns the rights to environmental quality?

This is an important question having serious practical implications for the administration of environmental programs. It is clear that most conservationists believe that governments hold, or should hold, the rights to environmental quality and that even where there is freehold ownership of land, the bundle of rights that attach to that ownership should not include the right to pollute or degrade the environment. However, in practice, landholders have either possessed legal or de facto rights to clear the land, lose soil etc and governments have generally found it necessary to allocate subsidies rather than apply penalties to landholders to achieve environmental objectives. In other words, at least until recently, landholders have won the political battles which determined who should pay (in the first instance) for environmental quality.

Most economists would take the view that landholders should not be seen as holding rights to degrade the environment. Agriculture should be seen in the same light as other industries which emit pollutants. There are several practical reasons for taking this view, as discussed by Chisholm in Chapter 12. In brief terms, the polluter-pays principle suggests that, if pollution is to be controlled, it is more efficient to tax the polluters rather than subsidise them because, among other things, the latter approach leads to the attraction of more resources into the industry than is desirable from a welfare maximising point of view.

Of course, many farmers do not see themselves as having the right to degrade their properties, whether or not it appears to be a profitable practice. Further, it is likely that many, if not most, farmers are motivated to avoid degradation actively and peer group pressures are often brought to bear on those who may otherwise lack the motivation. But the fact remains that there will always be cases under such voluntary systems where private and public interests diverge, owing to the behavioural factors summarised in Table 20. For this reason economists would argue that penalty (eg pollution tax) or transferable quota systems should be considered.

Despite the above arguments, participants often argued strongly against the substitution of penalties or regulations for subsidies in programs aimed at controlling land degradation. The main reason for their stance was that the difficulties (costs) of monitoring and policing programs based on penalties or regulations would be prohibitive. It is clear that there is substance to this argument. Despite the existence of spy-in-the-sky technology and the practical difficulties of concealing various forms of land use, the landholder is in a position to bypass land use controls in many ways. After all, even for the highly specialised case of field production of marijuana, control programs have not been completely successful.

While the polluter-pays principle could be seen to apply to forms of degradation like salinity, soil erosion etc, it may be more appropriate to argue for the user-pays principle to apply in cases such as tree decline where the degradation may be confined to loss of aesthetic attributes and be independent of other forms of degradation, as appears to be the case for parts of South Australia. Of course, there are again practical difficulties, for example, in charging passers-by for the privilege of admiring pleasantly treed vistas. Instead, the proponents of trees on farms programs usually manage to persuade governments to meet the costs from consolidated revenue. The additional stratagem of persuading farmers that it is profitable to retain or replant trees on farms has been recently challenged by Tisdell (in press) who points out that many of the studies on which such claims are based have ignored important considerations like the opportunity costs of land under trees and the time it takes to establish plantations.

In summary, it appears that society will increasingly see governments, on behalf of the general public, rather than landholders as holding the rights to environmental quality and we can expect to see increased use of coercive measures to control land degradation. However, there are risks that this approach will lead to a worsening of environmental problems owing to difficulties associated with monitoring and policing. It may be more efficient to exclude agriculture from some areas rather than to attempt to administer a control program in the presence of agricultural land uses.

What efficiency gains are potentially available from adopting taxes rather than regulations to combat degradation?

This question involves assessing the worth of implementing some of the policy measures advocated, especially by Chisholm in Chapter 12. Chisholm quotes instances of 50 per cent reductions in program costs through adopting pollution taxes rather than regulatory measures in control of industrial pollution. These savings are likely to be lower for the case of land degradation in rural areas because of the relative complexity of the systems to be monitored and policed. Taylor and Frohberg (1977) claim cost reductions of about 30 per cent for soil conservation programs based on taxes.

Even if the cost savings from applying economic concepts to environmental control are not large, this does not mean that they should be abandoned. The concepts described by Chisholm and others provide a benchmark against which other policy proposals may be compared and evaluated. It is useful to know that because of the complex informational requirements of non-point pollution control for both tax and regulatory systems, there may be little advantage in economic efficiency terms for the pollution tax systems favoured by economists. There are likely to be political costs of implementing a tax system which, for any given level of pollution control, will usually impose much higher costs on polluters than will a regulatory system.[11] This is one reason, of course, for paying serious attention to the suggestions of Bradsen and Fowler in Chapter 7.

Are land degradation problems best viewed in national, state, regional or local contexts for policy purposes?

This question was frequently discussed during the workshop where many people favoured policy formulation at the state, regional or local level and suggested that, largely because of the highly localised nature of degradation problems, policy formulation at the national level was not likely to be of great assistance. However, it must be appreciated that the bulk of public funding for degradation control programs is obtained through federal sources and that the federal government should be expected to take an interest in the competing claims of the states for such funds.

Policy makers at the national level need to determine, first, what proportion of the federal budget should be allocated to national and state programs for degradation control and second, whether the proposed expenditures by states appear to be rational from a national perspective, especially given the interdependence of some of these expenditures in regions such as the Murray-Darling Basin where four states are involved.

Even within states it is necessary, of course, to have an overview of degradation problems across the state in order to allocate research and extension resources efficiently among competing localities. So while local knowledge is critical to the solution of degradation problems, the policies which build on this knowledge will usually need to be formulated at state and national levels.

Does the available evidence favour any particular form of land tenure in terms of controlling land degradation?

Bradsen and Fowler make the claim in Chapter 7 that '. . . There is no clear evidence that land degradation has occurred at a greater rate overall on pastoral lands held under fixed-term lease than on agricultural lands held under the more secure forms of tenure, ie, freehold or perpetual leasehold'. Young in Chapter 8 cites evidence from a number of public inquiries and his own research to support the view that '. . . if left to free market forces New South Wales pastoral lands, its pastoralists and its consolidated revenue may all be in better condition.' Young appears to favour strongly the freehold system of land tenure, but taking cognisance of political realities, advocates perpetual leasehold with attached covenants for the pastoral zone. He is clearly opposed to the concept of fixed-term leases, as is Quiggin in Chapter 10.

One of the main advantages claimed by Bradsen and Fowler, and Mosley, for fixed-term leasehold is that it permits a more flexible form of land use than that provided by freehold systems. Their argument is that once land is alienated to freehold it becomes politically very difficult to change the form of land use from, say, agriculture to national parks.

I found difficulty, in the face of available evidence, in accepting the propositions that fixed-term leasehold would lead to lower rates of land degradation, or more flexible land use, even in the arid zone. However, it may help to view the problem of land degradation again in two parts — the aspects within industry (eg pastoral) and the aspects concerned with allocating land to different classes of use (eg pastoral versus national parks). The proponents of fixed-term leases have concentrated on the latter. Even here I find it difficult to accept that political opposition, say, to removal of land from pastoral use, would be any less weighty whether the tenure system was fixed-term lease, perpetual lease or freehold. The fiscal costs to governments of transferring land from pastoral to public uses also need be no different under the different forms of tenure — the forgone rents should be about the same.

The question of Aboriginal use of land was not explicitly considered by the organisers before the workshop. Perhaps we had empathy with H C Coombs' view (pers comm 1985) that Aborigines had little to learn, especially from us, when it came to the subject of land husbandry.

However, it emerged during the workshop that, under new land rights proposals, Aborigines would soon control, under inalienable freehold title, approximately ten per cent of Australia's land. This area is equivalent to the total area of land currently allocated to freehold title and is therefore of some significance irrespective of its economic potential. It also emerged that some of this land was already being leased to developers who, in some cases, immediately set about clearing it. In addition, there can be no guarantees that the freehold will remain inalienable. In my view there were inconsistencies between the arguments for fixed-term leases and the treatment of tenure for Aborigines in the arid zone. The same environmental controls should presumably apply to all land in the arid zone. Barker also addresses these issues in his commentary on Chapter 7.

Given all these considerations it would seem reasonable to advocate more radical policies for the arid zone than those which retain the status quo or which move in the direction of fixed-term leases. The introduction of (alienable) freehold title for most land in the arid zone, accompanied by appropriate environmental programs, should lead to greater efficiency in the use of resources and less land degradation than most current proposals. The main qualification I would add to this proposition is that cross-subsidisation of infrastructure in the zone (or in all rural areas for that matter) should be re-considered, with a view to phasing it out. This may mean the demise of pastoralism in some parts of the arid zone with the land returning to something like its native state — giving a result that is consistent with the preferred strategy of Bradsen and Fowler, and Dillon (1985) but by a different route.

Is the database adequate to formulate efficient policies for controlling land degradation?

In physical terms, the environmental data base for rural land use in Australia is grossly deficient compared with, say, that of the USA (despite comments above about plot years of soil conservation data). We have probably under-invested in national data systems because of our constitutional reliance on state control of land, in contrast to the USA. There is little doubt that lack of federal coordination of data collection and analysis has severely hampered efforts to obtain national perspectives on land degradation and the rational allocation of public resources to conservation programs. However, it is also true that construction of improved databases could represent a bottomless sink for public resources given the wide range of information which many would feel should be included.

Whatever is done in this area at this time should involve both physical and social scientists. Cullen and other participants were able to point to costly data collection exercises that ended in futility owing to lack of clear objectives for the uses to which the information would be put. To a significant extent, physical scientists in Australia are evaluating the availability and applicability to Australian conditions of models like CREAMS and EPIC and databases from Canada and the USA like the Environmental and Technical Information System (ETIS) developed by the US Corps of Engineers. However, these evaluative exercises have not usually involved social scientists so I have doubts about their usefulness from a policy perspective.

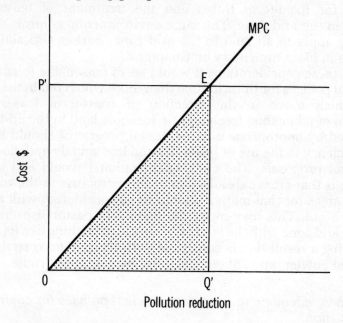

Figure 21 Comparison of the cost to polluters of a tax versus a regulation.

Concluding remarks

It occurred to me before the workshop, and before Ballard (Chapter 13) proposed the establishment of a national conservation commission modelled on the Industries Assistance Commission (IAC), that the IAC itself should be asked to hold an inquiry into assistance measures for the control of land degradation. The IAC does not provide the continuing forum proposed in Ballard's model for conservation issues but, at least in the short term, we are more likely to see an IAC inquiry established than a new commission.

The advantage of an IAC inquiry of the sort proposed is that, unlike the outcomes of workshops, the policy connections are more explicit.

There are formal procedures for considering IAC recommendations and the bureaucratic and political stakes are higher than they are for most workshops. Many of the other advantages advanced by Ballard for his model would also apply to such an inquiry.

The contemporary question of whether or not proposed capital gains taxes should apply to agricultural production has important implications for land degradation. In the short run the imposition of these taxes may lead to increased rates of land degradation as land holders find it less profitable to invest in conservation works and practices. In the long run capital gains taxes may lead to reduced rates of degradation in areas where agricultural uses of land become unprofitable, following imposition of the taxes, and the land is permitted to revert to its native condition. Given the cumulative nature of land degradation processes, some empirical work would be needed to assess the net impact of the short-run and long-run effects.

Endnotes

1. Soil acidification through the use of leguminous pastures is yet another form of land degradation that has been recognised in Australia (eg see Burch et al, Chapter 2) but which was not given prominence in the workshop. Some researchers believe that it will become a serious and widespread problem in the next five years.
2. These references were selected from a reading list prepared for a workshop on Ethics and Economics organised by Gene Wunderlich for a joint meeting of the American Agricultural Economics Association and the Association of Environmental and Resource Economists, Florida, August 1985.
3. See, for example, the international conference on The Economics of Dryland Degradation and Rehabilitation organised by the Australian Department of Arts, Heritage and Environment and the United Nations Environment Program, Canberra, March 1986.
4. Because of ascribing degradation to the whole of a region or area when in fact only parts of it are affected.
5. Rural tree decline is perhaps the only phenomenon for which we have such information.
6. In my view most of the economists' contributions to this monograph have focused too narrowly on optimising rates of land degradation within agricultural land uses, while neglecting the potential for government intervention between different classes of land use. This is understandable given historical patterns of land use in Australia but will need to be re-examined in view of likely future conflicts.
7. Among other things, Lloyd claimed that the low rates of return to capital of 2.5 to 5 per cent, ignoring tax savings, did not justify destruction of a valuable conservation and recreation area. It seemed that the major attraction to private developers was the income tax savings that would have accrued to them.
8. I take issue with some of Chisholm's more detailed comments. For example, the USLE can be quite easily made to contend with stochastic and catastrophic characteristics of water erosion processes. The forcing function for these events, erosivity of storm rain, can be calculated on a daily basis for long periods of time

(eg using the computer package RAINEI described by Dumsday et al 1984) and passed to appropriate models. The USLE has also been used to predict satisfactorily *relative* differences in soil loss rates for management systems in the wheat belt of southern Queensland and northern NSW (Dumsday 1973). Finally, because of the availability of new computing software, it is not necessarily costly to apply the USLE, even at the individual farm level (eg see Dumsday and Seitz 1985).

9. Some of the categories should possibly be grouped, eg uncertainty and lack of knowledge; irreversibility, intergenerational equity and discount rates.

10. I am reading this into Pearce's paper. Pearce does not advocate the treasury position; he comes to no firm conclusion on the matter. In addition, he provides a useful discussion of the effects of assumptions concerning ethical rules on the outcomes of benefit cost analysis.

11. This is easily seen in Figure 21 where the cost of a pollution tax P' is compared with that of a regulation setting pollution reduction at Q'. MPC is the marginal private cost of reducing pollution.

17 The practicalities of policy solutions

Bruce Davis

In analysing any public policy matter, a number of interrelated questions must be considered (Jenkins 1978, Ham and Hill 1984):

- is there a problem — if so, what is its scale and priority, relative to other issues on the public agenda?
- who has articulated the problem and what are the various perceptions and attitudes of interested or affected parties?
- what is the political culture (milieu) within which the issue will be debated?
- which agency or authorities should have administrative jurisdiction over it?
- how well is the machinery of government equipped to handle the problem? (There are various elements of organisational structure and process which must be considered, as well as necessary resources such as time, money, expertise and information).
- what is the feasibility of policy formation and implementation, given the likely impact of political expediency, precedent and external influence, as well as questions of timing and accidents of history?

Not all of these aspects were discussed in detail during the workshop and they cannot be analysed to any substantial extent within the limits of this attempted overview; nonetheless a few observations may be recorded to highlight some practical difficulties in achieving policy solutions.

As far as land degradation in Australia is concerned, is there a problem? Contrasting views are expressed by authors in this book and the evidence collated seems rather ambiguous. Clearly there are many individuals and groups within the community who genuinely believe land degradation to be a substantial and insidious problem likely to cause major environmental and economic difficulties in the future. Equally, there are some who argue that existing administrative arrangements and land conservation programs are adequate. The weight of evidence perhaps tends to favour the former (alarmist) perspective, but even here disputation arises about the severity of the problem and feasible remedial measures.

Such ambiguity of goal has important implications for policy making (Anderson 1979), not least some prospect that the issue may not attract the political attention it deserves. There are three principal reasons why this is so. First, the issue is difficult to articulate since it involves complex ecological factors and long-term impacts which are hard for the politician or layman to grasp. Second, there is no identifiable public interest group to enunciate the problem clearly or to press for its consideration by government. Whether we regard the Australian political culture as elitist, corporatist or pluralist (Lucy 1983), attention tends to be focused principally on issues projected by the media, following demands by powerful interest groups. In the case of land degradation, who *are* the advocates for substantial and integrated action? Third, even if the problem is comprehended, there are many political and administrative constraints, arising both from limitations of government (Hood 1976, Passmore 1981) and competition between the states for development (the impact of federalism).

In addition to these barriers to agenda-setting (Jones and Matthes 1983), there are four fundamental paradoxes in recent attempts to improve environmental policy and natural resources management:

- We constantly preach the interrelationships of ecosystems and unity of the biosphere, but for administrative convenience divide natural resources management among a myriad of narrowly specialised functional agencies, creating considerable coordination and conflict-resolution problems within the public sector.
- We erect elaborate mechanisms of administration, involving statutes, policies and procedures, but then discard rationality in favour of outright political expediency, by granting ministers excessive discretion within Westminster-style government.
- We advocate a conservation ethic, but immediately undermine it by the countervailing activities and power plays of resource development agencies and private corporations; or else provide substantial financial incentives and subsidies devised by the public sector but employed for purely private gains.
- We assume that resource agencies should develop their thinking on ecological principles, without realising that the latter are only a means towards societal ends. Natural resource policies reflect *human* ends and wants; in the end *all* environmental policy must be based on *human relations* among interest groups within the community.

If politics is 'the art of the possible', it follows that 'governing' is not always a neat and orderly process, but sometimes imprecise and haphazard, hence public expectations about achievements or anticipated outcomes are often somewhat unrealistic (Richardson and Jordan 1979). There are many practical limitations to what the bureaucracy can achieve. This is particularly so in such complex fields as environmental management, where different opinions exist about the relationship of man and nature (O'Riordan 1976); where natural resources are vied for as sources of wealth and power (Sandbach 1980); and where knowledge and prognostication are often imprecise (Felveson et al 1976). In the matter of collective public choice, awkward trade-offs are required among conflicting and complementary objectives; thus land use management is a field where consensus or concerted action is difficult to achieve (Thompson 1972, Gilpin 1980 a and b).

The thrust of comments thus far may appear unduly pessimistic. Surely some constructive policies and effective administrative arrangements can be devised? There *are* prospects of improved performance in land use management, but particular problems and limitations attend each option. As Young (1982) has pointed out, governments are simultaneously obliged to perform three basic functions in carrying out natural resources management. These are custodial activities (conservation); promotion activities (development); and regulation (intervention). Hence policies must be devised which conform to this framework.

The chapters of this book canvass some of these prospects, noting the necessity to have appropriate information for decision making; the need to inform politicians and the lay public better about implications of land tenure systems and public-private sector cooperation; the important role of incentives and regulation, but with some doubts expressed about effectiveness of the latter mechanism.

The issue of administrative jurisdiction must be considered. In Australia, natural resource management is primarily a function of the states, with the Commonwealth role limited to policy coordination (often through ministerial councils), persuasion towards common standards and funding of essential research programs. But there are sometimes international pressures for Commonwealth intervention in environmental issues; moreover, there is widespread public expectation that the national government will act if states default on major issues or act against the perceived public interest (Scott 1980, Davis 1981). In this respect, competition between the states for economic development is often a cause of political expediency and environmental attrition, so that Commonwealth action may be in prospect to press the national interest or ensure more equitable treatment of affected parties. There is little doubt that if the Commonwealth chooses to act, it has powerful means of exerting leverage, principally through its control of public expenditure and grants-in-aid, but also via constitutional powers arising from recent High Court decisions, such as the Tasmanian Dams Case (Sornarajah 1983). There *are* political costs involved in federal intervention and thus far the Commonwealth government has largely left land conservation measures to the states and territories. This may not remain the situation in future, given increased public pressure for improved practice in natural resource management.

One of the principal difficulties in tackling land degradation in Australia is the wide variation in policies and practice within and among the myriad of agencies involved in land use management (Bates 1983, Porter 1985). Individual state soil conservation services may be achieving effective results within their own limited spheres of operations, but far too often those agencies are small in scale, mere appendages of larger portfolios, and have to compete within a budgetary framework where they are granted low priority and limited resources. Unless the land degradation problem becomes more visible and acute as a public issue, current programs will remain largely a token gesture.

This is not to denigrate the efforts or achievements of many dedicated professional officers or the valuable contribution of voluntary conservation groups and some agricultural producers, but rather it reflects the political reality as far as land degradation in Australia is concerned. In short, unless the national government provides leadership and some funding of an integrated action program to foster land conservation, the scale of the problem (and potential treatment costs) are likely to escalate. There may be prospective grounds for a Commonwealth ultimatum that unless the states put their own houses in order within a specified time, the Commonwealth will intervene in what is essentially a national problem. But where is there a visionary political leader, firm and dedicated enough to tackle this daunting prospect?

Whichever level of government takes up the running, a mild warning is in order about the data base on which decisions will be based. Not only are there significant gaps in relevant information, but it would be foolish to suggest improved decision making will automatically result within the public sector, if we introduce more comprehensive and sophisticated forms of land use information. This is not to reject the notion of additional research or information exchange, where clearly scientifically useful, but rather to note that attempts at a 'technological fix' through massive information systems are doomed to failure, as numerous case examples show (O'Riordan & Sewell 1981, Clark et al 1984, Hawkins 1984). The reality is that politicians rarely use the comprehensive data made available to them; decisions are generally based on values, beliefs and perceptions, selective advice offered by key support staff and diverse pressures arising from the electorate (Spigelman 1972, Tribe et al 1976, Nethercote 1982). The form and content of policy advice is therefore crucial; unless technical and public service staff know how to project options and implications to key decision makers, sub-optional choices are likely to be made.

One must be equally sceptical that improved land tenure systems will act as an incentive for improved pastoral practice or vegetation retention and thus obviate a need for regulation. This has not been the case in the past and human nature being what it is, with economic pressures abounding, overstocking and other poor agricultural practice is likely to continue. Indeed, the fundamental questions are: do Australians really understand the marginal nature of much of their continent and its uneven climate and do they truly comprehend how to cope with it? (Bolton 1981).

A good case could be made for limiting or preventing certain pastoral or forestry activities in large tracts of some states or territories, given the remedial costs that ultimately accrue either in the form of agricultural subsidies to private interests or public charges to treat land degradation, flood prevention, irrigation and other environmental problems. Evidence from other nations suggests that development controls over private land in rural locations are likely to *increase*, rather than decrease, in future years, as a result of diverse community pressures (Davis 1976, Cherry 1976, Aust Senate Standing Committee 1984). What is really missing from current land conservation measures in Australia is a more diverse mix of socially acceptable intervention methods (regulation) and financial incentives. Existing controls and much of the taxation system tend to foster land abuse, rather than land care. We need to become more innovative in devising incentive packages and anticipatory programs, rather than remedial measures to correct past errors of judgment or land abusage.

Property rights also imply some community obligation. We should simply reject the notion that every man's home is his castle, if it can be demonstrated that private actions are imposing subsequent costs on the remainder of society. Nonetheless, cooperation is likely to prove more productive than regulation and in an era of invasion of privacy, excessive secrecy within government and some rather unethical business practice at times, it would be wise to ensure that whatever governmental policies are adopted to deal with land degradation, adequate review and appeal provisions exist to cope with the concerns of aggrieved citizens. Ideally the starting point should be community participation in identifying land conservation measures, rather than judicial redress when technical measures fail to solve social problems.

The foregoing commentary implies the task ahead is one of considerable complexity and magnitude. Legislatures and laymen are not yet well attuned to understand land degradation; existing institutional arrangements do not facilitate coordination of effort or clarity of policy direction and there are doubts about which level of government should have responsibility for initiating action. If these appear formidable obstacles, they are probably not much worse than in many other functions of government. This is not an excuse for inaction, but it does imply the need for clear thinking and increased discourse about land conservation, as well as increased research effort.

Seminars, drawing on a diversity of expertise and experience, are a useful catalyst. Another admirable guide to future action lies in priorities and recommendations recorded in the National Conservation Strategy (Anon 1983), with its acceptable theme of 'conservation within sustainable development'. Grand designs run some risk of failure, but modest and well-publicised demonstration effects could be achieved via regional projects. It is to be hoped that the workshop that produced this book will serve as a reminder to government to increase the diversity and accelerate the pace of land conservation measures. Australia's Bicentennial Year falls in 1988: there could be no better gift to future generations than a continent better managed, productive and with a firm commitment to a new land conservation ethic.

Appendix A

Rational approaches to environmental issues

Anthony Chisholm

Preamble

In these background notes I will first attempt to highlight some of the basic assumptions and value judgments underlying economic rationality and then develop and extend some ideas in the context of the 'benefit-cost' approach. Benefit-cost analysis is relevant to the problem of land degradation and it also provides a useful vehicle for highlighting the main aspects of rational approaches to environmental issues.

My comments focus almost exclusively on mainstream (also termed neoclassicial or rational) economics which is the approach, by and large, taken in the economics contributions to this book. Of course, not all the economists who have contributed chapters to the book would agree with the particular selection of points I choose to emphasise, or perhaps with some of the specific points I attempt to make. But I have done my best to outline a few parts of mainstream economic thinking which I think are important, together with some aspects of benefit-cost analysis.

The later part of my commentary, on benefit-cost analysis and land degradation, includes some ideas which have their origins in a paper I presented to a conference on Soil Degradation: The Future of Our Land? held in Canberra in November 1984.

Scarcity and choice

Society is endowed with a limited supply of a wide array of resources: land, skilled people, clean air, water, non-renewable resources, time, and so forth. A fundamental economic and social problem faced by society is how best to use the scarce resources. If the resources were not in limited supply, there would not be an economic problem; everything that anyone wanted could be provided here and now, and an infinite amount of resources would still be left over to satisfy everyone's wants completely in the future.

Resource scarcity implies that any specific pattern of resource use involves opportunity costs; for example, if more resources are allocated to land conservation, fewer resources will remain to be allocated to health care, education, defence, symphony orchestras and other goods and services.

While people differ in their particular ideas for the best use of scarce resources there is general agreement that resources are too precious to waste. Resources are currently being wasted if it is possible to change the pattern of resource allocation so as to improve the well-being of one person or groups of persons without reducing the well-being of any other person or persons. That is to say, in the words of the Italian economist Vilfredo Pareto: *An efficient allocation of resources is one from which no person can be made better off without making another person worse off.* Any allocation of resources which is not efficient (Pareto optimal) involves waste and is called inefficient.

The concept of efficiency refers to individual people being either better or worse off. Individual preferences count. This is not to say that the welfare of small and large groups is not important, but rather that groups count only insofar as they represent an aggregation of individual wants and preferences. Certain individual desires may, of course, be widely shared by members of a small group or by a society as a whole.

The practical application of the concept of efficiency requires a procedure for deciding whether someone's well-being has improved or deteriorated following a change in the pattern of resource allocation. In western societies, the principle of *consumer (individual) sovereignty* represents a value which most people share: *each person is considered to be the best judge of his or her own welfare.* And the welfare of society is judged to depend on the individual welfare of its citizens.

The term 'consumer' in the widely used term 'consumer sovereignty' is something of a misnomer. The principle is intended to apply to *all* economic decisions which affect an individual's welfare, including, for instance, the value an individual attaches to a particular job, as well as the value they place on the consumption of particular goods and services. From now on I will use the term *individual* sovereignty. By and large, mainstream economists take tastes and preferences of individuals as a given (exogenous parameter) for the purposes of most economic analysis. This is not to deny, of course, that environmental philosophers, scientists, sociologists, political scientists, lawyers, and perhaps even economists, may have some influence on moulding individuals' attitudes towards, say, the environment and thereby influence the amount people are prepared to pay to attain a cleaner and better environment. It's just that economists usually take the above influences and the myriad of other forces that ultimately mould individuals' tastes and preferences as essentially given and hence beyond their realm of analysis.

Although the principle of individual sovereignty implicitly underpins most of the economic analysis contained in the chapters of this book, it is not the only way of implementing a concept of efficiency based on the well-being of individuals. An alternative procedure, for example, is to let some other person (eg a monarch), or a democratic socialist or communist government, be the judge of each person's welfare and determine the allocation of all resources accordingly.

While western societies give fairly free rein to individual sovereignty, there is a number of situations in which strict adherence to the concept is commonly thought to be inappropriate. These situations typically arise in circumstances where individuals have incomplete or erroneous information or for various reasons are unable to process the available information properly. Young children are an obvious example and parents usually substitute their own judgments for those of their young children. Similarly, for both children and adults, most governments permit individual sovereignty to operate only within a restricted (regulated) market framework for the supply and consumption of medical and other drugs. This should not be construed as an issue of market processes versus political processes. The essential point is that in some areas government restricts individuals' freedom of choice; eg, few (if any) governments are prepared to hold a referendum to decide whether or not to permit free distribution and consumption of the so-called 'soft' drugs.

Economists themselves have advanced arguments for governments overriding individual preferences (individual sovereignty) for certain types of goods/bads. In the economist's jargon these are referred to as merit and dismerit goods. Some may argue, for example, that education has some characteristics of a merit good while soft and hard drugs are a dismerit good. And government action should expand consumption of the former and constrain consumption of the latter. There is considerable controversy in the economic literature both about whether adult market preferences can be 'wrong' or 'distorted' in some meaningful sense, and about whether political processes can be reasonably expected to 'correct' such faulty preferences. For a recent survey and analysis of the issues see Brennan and Lomasky (1983).

The benefit-cost framework

Benefit-cost analysis formalises a commonsense concept of rationality, namely that a rational individual, group, or society will weigh up the advantages (benefits) and disadvantages (costs) of a particular action. It is squarely anthropocentric insofar as resources are valued in terms of their contribution to human welfare. The underlying presumption is that to have more goods and services that contribute to human well-being is a desirable thing. The measure of human well-being is largely based on the previously outlined concept of individual preferences and sovereignty. It should be noted that the term goods and services is used here in a very broad sense. It includes, for instance, the joy humans experience from visiting a rare natural environment, and also the joy they derive from the mere knowledge of the existence of such creatures as blue whales, as well as the joy some individuals derive from eating a big Mac or driving a Porsche.

People who advance arguments for environmental rights, and claim that the benefit-cost approach is unduly anthropocentric, are obviously not proposing that humans should hand over the decision and policy-making process to animals and plants. Ultimately there is simply no alternative but to have the decisions relating to the environment made by human beings. The concern and outright abhorrence expressed by some people relate to the nature of the decisions reached through this process and what they perceive to be an undue weighting given to the 'baser' wants of humans.

There is a rapidly growing literature on philosophy which considers the difficult issues of ethics and environmental policy concerns by proposing that strict adherence to humanism be abandoned and moral considerability be explicitly extended to non-human species. Other philosophers have argued that once a start is made on the 'slippery slope' of abandoning strict adherence to humanism, there is no logical point to draw a line short of according moral considerability to everything in existence. Kneese and Schultze (1985) provide a recent review by economists of this literature. Their major conclusion is that the naturalistic ethical ideas are presently 'much too abstract, or insufficiently formed, to mesh tightly with actual public policy issues or with economic concepts' (p200). In the remainder of this Appendix we return to the 'humanistic fold', but it is important that economists who are concerned with environmental issues keep abreast of the main ideas emerging from the naturalistic ethical literature. Most economists, including this writer, have been remiss, at least in the past, in generally paying scant attention to this area of literature.

We now consider, in a little more detail, the meaning of individual preferences and an 'economic vote' and compare it with a 'political vote'. First, it is true that benefit-cost analysis uses the convenient measuring stick of money (dollars), but it is not attempting to measure benefits in any absolute sense. Rather, benefit-cost analysis is concerned with relative benefits and it assumes that relative prices reflect the rate, at the margin, at which people are prepared to give up some consumption of one good or service (eg health care) in order to get more of another (eg entertainment).

Benefit-cost analysis involves the appraisal of all costs and all benefits, relating to a proposed course of action or investment project, from the viewpoint of society as a whole. As we shall later see, the appraisal of an investment project by a private investor and the benefit-cost appraisal of the same project from the viewpoint of society as a whole will, in some circumstances, result in an identical evaluation. Benefits are measured by willingness to pay, measured in dollars, which in turn reflects willingness to sacrifice other goods and services. If the dollar value of an individual's willingness to pay for a particular action (project) exceeds their share of the cost, a positive economic vote would be recorded for that individual. For most actions (projects), some individual valuations would show net benefits and some will show net costs. Benefit-cost analysis essentially involves the adding up of dollar votes in such a way that the concept of *net social benefit* is defined by the difference between the dollar votes in favour and the dollar votes against a proposed action. A positive net social benefit indicates that the benefits to the gainers from a project exceed the amount the losers would require in compensation in order to be no worse off.

To economists, one of the major appealing features of an economic vote is that it permits an expression of *intensity of preference*: it enables an individual (subject to their income constraint) to express clearly how deeply he or she wants or does not want a particular project or good. By contrast, the conventional political vote through a ballot box permits the recording of only a 'yes' or 'no' vote. A majority referendum 'yes' vote may result in a project going ahead when any reasonable weighting of intensity of preferences would have precluded the project. Moreover, voters usually express a political vote on a whole package of issues represented by party political platforms. It is simply not feasible to hold referenda on every social decision. The chances of an individual's vote being decisive are extremely slight and the corresponding incentive for a self-interested individual to be well-informed on the issues is also small.

With economic voting, significant costs and benefits are very explicitly borne by individuals. Rational self-interest thus provides a strong incentive for individuals to be well-informed. The *willingness to pay* for benefits concept which underpins benefit-cost analysis deals with dollar votes as expressed in the marketplace or in 'proxy' (shadow) markets for goods for which markets are non-existent. There are, of course, some considerable practical difficulties involved in estimating willingness to pay for, say, environmental goods and services, for which markets are usually non-existent (see Freeman (1979) for a comprehensive survey and analysis of economic procedures for evaluating the benefits of environmental improvements).

The first major value judgment underlying the use of the willingness-to-pay concept in benefit-cost analysis; that individual preferences should count, has been discussed. The second major value judgment is that individual preferences should be weighted by dollar votes, and thus to some degree by an individual's income.

Economists have differing views on how distributional considerations should be handled in benefit-cost analysis. One view is that the money value of costs and benefits be added together without regard to who incurs them or to whom they accrue (Harberger 1971). An alternative view to the 'dollar is a dollar' principle is that some set of distributional weights should be employed in adding across different groups in society. If a benefit-cost analysis is being undertaken for government, the government's distributional values are presumably the relevant ones. But Parish (1976) suggests that it would be naive to expect governments to give explicit instructions on distributional weightings and, to support his view, quotes from the United Nations Industrial Development Organisation (UNIDO) Guidelines:

> ...the qualifications of political leadership are many and varied, but for better or worse a grasp of the tools of economics has never been high on the list. Yet without an understanding of the over-all methodology of benefit-cost analysis...it is difficult to imagine that the value judgments necessary for calculation of national parameters [of which distributional weights are an example] could be elicited from the political leadership. The most important reason why the political process discourages rather than encourages explicit quantification of political value judgments with respect to the goals underlying calculations of national economic profitability is that political leaders rely on the support of distinct interest groups that are partially (at least) in conflict with one another. In such circumstances ambiguity has obvious advantages. One can hope to appear all things to all men. The corollary is an understandable if deplorable reluctance to take explicit positions that indicate the precise extent to which the political leader values one objective (with its particular lobby) over another objective (with its particular lobby) (UNIDO 1972, p137).

It is clear that deriving an appropriate set of weights for a benefit-cost analysis is not easy. As a consequence, there are those who firmly advocate that benefit-cost analysis should go no further than seeking to answer the question; are the net social benefits positive? That is to say, how much money would the gainers from a government action be prepared to pay for their benefits and does the aggregate sum exceed the amount the losers would require in compensation.

Economists who hold this view reject the admixture of political values into economc analysis that distributional weighting implies (eg Mishan 1982, Parish 1976). On the other side are those economists (eg Pearce and Nash 1981) who argue that, in practice, compensation of losers rarely takes place. They then see 'conventional' benefit-cost analysis as making unduly sacrosanct the prevailing distribution of income.

The above views represent differences in value judgments. In addition, subjective judgments of facts are inevitable in applied benefit-cost analysis, particularly in securing and processing information on 'willingness to pay' for certain environmental benefits.

Benefit-cost analysis and land degradation

Land conservation may be viewed as an investment, both from the perspective of the individual land user and of the community as a whole. In the broadest sense, the economic goal is to use our land in a manner that maximises the well-being of the community both now and over time. Underlying this goal is an assumption that the values and preferences of all individuals count. Furthermore, the term 'land-use' embraces all possible uses of land, including agriculture, forestry, National parks, and plant and wildlife sanctuaries. Generally speaking, the above goal will not be met when some pure biological or technical target, such as minimising soil loss, is used for decision making.

The most satisfactory economic interpretation of the goal of maximising community well-being is that it maximises the present value of assets in land use. The approach requires the appraisal of alternative land use patterns and soil conservation practices and measurement of the associated net income (benefits minus costs) flows produced through time. In order to compare differing streams of net income deriving from alternative land use patterns it is necessary to convert them to a lump sum value at a particular time. The most natural procedure is to convert net income flows to their discounted present value. Discounting requires the choice of an interest rate so that a set of weights may be attributed to dollars of net income received at different times. It is well known that $1000 here and now is worth more to an individual, and to society, than a certain $1000 in, say, ten years time because between now and then the money may be used for profitable investment or desired consumption. With a positive rate of interest, the more distant the future benefits, or outlays, the greater the discount factor used to convert their future value to a value in the present.

Once all net incomes associated with a contemplated investment have been discounted to their value in the present they then can be summed to obtain a net present value. The present value of assets in land use will be maximised when investments with only non-negative net present values are undertaken. When there are two or more possible courses of action (investments), and taking any one precludes all other, the course of action yielding the highest net present value should be chosen. Roughly speaking, the optimal land use pattern is that which maximises the price that could be paid for a block of land while still earning a normal rate of return on the capital valuation of the land.

Embodied in the goal of maximising the present value of assets in land use is the important notion that current land use practices must take into account their impact on the future physical state and value of land. Generally speaking, the adoption of land use practices today which substantially degrade the land available for use tomorrow will lower the present value of assets in land use. However, the economic goal, in contrast with most technical targets, does not rule out the possibility that sometimes it may be in the interests of both farmers and society as a whole to adopt land use practices which cause a rate of land degradation greater than the natural rate. An important related economic notion is that the existence of degraded land today does not in itself provide a prima facie case for its repair. Previous land degradation is a *sunk* cost. Whether or not existing degradation should be repaired is a question of the present value of the costs and the benefits. (For other discussion of the issues in this section, see Blyth and Kirby 1985, Dumsday and Edwards 1984, and Musgrave 1984a.)

To this point, I have deliberately not distinguished between individual land users and the community as a whole in the goal of maximising the present value of assets in land use. A helpful point of departure is to note that, under certain conditions, the land use pattern which maximises an individual farmer's present net worth will be identical to that which maximises the present value to the community. In a properly functioning land and capital market, the value of assets is determined by discounting the net income stream produced through time when the assets are put to their highest value use. Farmers seeking to maximise the present value of their net worth will usually avoid farming practices which cause excessive land degradation and a consequent decline in asset values and future income-earning potential. Under ideal free enterprise conditions, farmers seeking to maximise their net worth by using sound commercial accounting and investment practices will be guided by an 'invisible hand' to manage land in a way which maximises community welfare. However, note John Quiggin's (Chapter 10) observation that farmers with the most optimistic expectations of the stocking potential of land, ceteris paribus, will be prepared to bid the highest price to acquire land.

Farmers will adopt farming practices and invest in levels of soil conservation which are optimum for both them and the community when, in the economist's jargon, there is no *market* failure or *political* (government) failure. When these conditions are met, investment appraisal from the viewpoint of an individual farmer and social benefit-cost appraisal are one and the same thing. However, when these conditions are not met what counts as a benefit or a cost to a particular individual, or group, does not necessarily count as the same benefit or cost to the community as a whole. With benefit-cost analysis we are concerned with the welfare of the community as a whole.

The necessary conditions for the private pursuit of profits (wealth) in a free enterprise system to bring about an efficient allocation of resources can be boiled down to two: that markets be perfectly competitive and that all goods and services relevant to the well-being of all individuals be properly priced in the market. When these conditions are not satisfied some degree of market failure occurs. Farmers will then base their land use and soil conservation decisions on distorted price signals and the resulting allocation or resources will be sub-optimal for the community as a whole. In these circumstances there may be opportunities for governments to intervene in market (free enterprise) processes to attain a socially more desirable pattern of land use.

The major areas of potential market failure through land degradation are: imperfect information, externalities (discussed in Chapter 12), economies of scale, imperfect capital markets and inter-generational equity issues and irreversibilities. The essential features of these areas of market failure, as the author sees them, are now summarised.

Imperfect information

It is well known that knowledge about land degradation has 'public good' characteristics. Unlike normal goods, the use of a piece of information relating to a profitable soil conservation measure by one land user does not diminish the supply of that information available for use by other land users. Furthermore, it is difficult to exclude any land user who does not offer payment for the information from benefiting from it. Consequently, in the absence of government action, the free enterprise system will invest too little research in acquiring knowledge. A clear role therefore exists for government-sponsored research into, and extension of, viable soil conservation practices.

It appears that many of our past mistakes in land degradation can be attributed to imperfect information on the part of both farmers and governments. Since the late 1940s, governments in Australia have actively allocated resources to soil conservation research and extension, but it is only in very recent years that research into economic aspects of land degradation in Australia has really begun. It is claimed by some that the major information gaps on the physical nature and extent of land degradation in Australia have now been closed. Is this true? I believe that those attempting to apply benefit-cost analysis to problems of soil conservation will be often frustrated by imperfect information, despite the great advances that have been made in our understanding of soil degradation problems over the last quarter of a century. Land productivity is influenced by many factors and the causes of soil degradation are often slow-acting and subtle. There is some evidence, both in Australia and from overseas, that many farmers currently have very little idea of the amount of soil loss (tonnes per unit area) from their properties. The task ahead of us is clearly a multi-disciplinary one. The benefit-cost approach does no more than attempt to provide a systematic framework for thinking about problems of social choice. The usefulness of the results is utterly dependent on the quality of the physical and economic data. Crucial questions are raised, and as yet remain unanswered, about public (government) investment in land/ water degradation research and extension. Is more public investment required? Is the allocation of research funds between the natural and social sciences (and within research areas in these two broad groupings) about right or not? What can be done to improve upon the existing land conservation/degradation institutional structure for research and extension in Australia?

Economies of scale

Market failure occurs when economies of scale in particular methods of combating land degradation (eg large mitigation projects) preclude the use of these techniques by individual land users. In some situations (eg if there is a small number of land users who benefit from such a project and no benefits to other parties) land users may voluntarily cooperate to share the costs and benefits of a large mitigation project. However, most often these types of degradation abatement projects will not take place without government intervention. In such circumstances, the alternative courses of public action should be evaluated with the aid of benefit-cost analysis.

Benefit-cost analysis has sometimes been criticised on the grounds that because subjective judgments of fact are inevitable it is open to bureaucratic/political manipulation. This criticism may be even more telling if distributional weightings are incorporated in the analysis. I indicated in Chapter 12 that some criticism had been made of likely excessive investment in large mitigation projects (and too much reliance on technical criteria) and too little investment in preventive land use management practices. Some bureaucratic/political incentive structures appear to have characteristics which point in this direction and there is a strong likelihood that if benefit-cost analysis were more widely used to evaluate such projects it would be open to manipulation by self-interested public agencies. A possible solution to this problem would be for state Treasury experts (and, where appropriate, Commonwealth Treasury experts) to audit carefully benefit-cost analyses carried out by various public agencies charged with the responsibility of evaluating (and possibly constructing and/or administering) large mitigation projects.

Imperfect capital markets and intergenerational equity issues

Imperfect capital markets will cause a divergence between the weights attached by farmers to income received at different times and the 'ideal' weights attached by the community. In particular, it is widely believed that farmers use a set of weights (discount rates) which attach too small a weight to the future from the viewpoint of the community. The choice of a social rate of discount is one of the most conceptually complex and controversial areas of benefit-cost analysis. There is no clear evidence that the author is aware of that Australian farmers pay higher market rates of interest for medium to long-term loans than other borrowers, or that they are unduly rationed in the amount they may borrow at prevailing market rates. The issue then appears to revolve around the adequacy of investment funds (credit markets) for medium to long-term loans which are applicable to long-lived investments like soil conservation. In this respect, it may be argued that the aggregate level of saving and investment by the community is too small which implies, of course, that present consumption is too high. A shortage of investment funds would force farmers and other investors to be unduly myopic and thus discriminate against investments of longer life.

Two separate forces operating in this direction may be identified, one of which is imposed by government and the other a form of market failure. The first is the existence of the income tax which discriminates against saving (future consumption) relative to present consumption. The second is a type of market failure which exists for a community that desires to save for future generations. It can be shown that the amount provided for future generations, if everyone acts in an individual capacity in the market place, is less than all individuals, taken together, would collectively desire. These two forces provide a case for giving some sort of subsidy to both public and private long-lived investments in general. They do not provide a case, though, for singling out soil conservation for special treatment.

A further complication arises when it is recognised that there is a form of market failure analogous to that of inadequate saving in charitable giving to the relatively poor for present consumption. People acting in an individual capacity will make a smaller aggregate charitable gift to the poor than is collectively optimal. Insofar as one form of market failure points towards too little saving for future generations, and the other towards too little charitable giving for consumption by today's poor, we cannot unambiguously conclude that there is too little of *both*. If there is a trade-off between people giving for future generations and making charitable contributions to today's poor, it is possible that in the absence of government intervention, there may be too much of one and too little of the other. So far as the author is aware, this point has not been made in the very substantial theoretical literature on optimal social discount rates. These trade-offs between the present and the future are perhaps most starkly and tragically represented in some of the developing countries—most notably in Africa—where extreme land degradation is occurring in many areas (with obvious implications for future generations), as millions struggle to get sufficient food and shelter to live.

On balance, I do not believe that a strong case exists for using a rate of discount to evaluate investments in soil conservation which is lower than the market rate of interest. It is important to recognise that if all benefits and costs of contemplated investment in soil conservation are measured in *real* terms the discount rate used should also be a real rate. The appropriate discount rate is the expected real rate of interest which is equal to the observed market (nominal) interest rate minus the expected rate of inflation. For example, if the observed market rate of interest is 11 per cent, and the expected rate of inflation is six per cent, the real rate of interest is five per cent. Particular applications of benefit-cost analysis should include sensitivity testing by presenting results for a range of discount rates.

Irreversibility

Irreversibility occurs when some natural resources disappear or when some valuable uses of resources are for all practical purposes lost forever. The loss of animal and plant species, natural waterways, ecosystems, gene pools, and many forms of land degradation are examples. The notion of irreversibility assumes particular importance when the resource is unique or rare (eg Yellowstone National Park or the Franklin River). If a resource is abundant (eg coal) or has close substitutes the concept of irreversibility is not very relevant. In land degradation in Australia, the larger the area of land threatened by degradation the smaller is the remaining area of non-degraded substitute land and the more significant is the concept of irreversibility. There is also an important question of how irreversible is soil degradation. An irreversible outcome may be said to occur when it is either technically impossible or prohibitively costly to restore a resource to its original state. While there are some striking examples of irreversible soil degradation in Australia there are also examples which show that land can be remarkably resilient to past bad management. For example, there appears to have been a considerable improvement in the state of the land in much of the NSW pastoral zone as a result of a change of tenure system from short-term leasehold to perpetual leasehold (Richard Condon, pers comm).

The significance of truly irreversible land degradation is most clearly seen when we imagine an uncertain world in which the passage of time brings new information about the social benefits of alternative land use patterns. The new information can be taken into account only to the extent that land has not been used in a manner causing irreversible soil degradation. Once irreversible degradation has taken place, new information that indicates it would be a bad mistake cannot affect the outcome (see Krutilla 1967, Chisholm 1973, and Fisher 1981). Accordingly, there can be some value of preserving an *option* to use land in certain ways in the future by refraining from an irreversible action that otherwise looks profitable. Option value may be thought of as what society would be willing to pay as a kind of insurance premium to retain the option of being able to use non-degraded land in the future. The notion of option value should be distinguished from two related concepts. Firstly, there is the concept outlined in the previous section, namely, that a community may desire to conserve resources for future generations. This concept is commonly referred to as *bequest* value. Secondly, there is the concept of *existence* value. Whereas option value refers to a willingness to pay to retain the option to use land at some time in the future for a particular form of production, existence value refers to a willingness to pay for the knowledge *alone* that, say, Australia's semi-arid lands exist in a non-degraded state. The concepts of option value, bequest value, and existence value are extremely difficult to quantify. Their importance depends critically on uncertainty as to society's future demand and/or supply for irreplaceable natural resources.

It seems not unlikely that uncertainty surrounding the determination of the expected rate of social discount, over the long term, overshadows the combined quantitative importance of option, bequest, and existence values. However, in an uncertain world in which new information emerges through time, irreversible land degradation does carry some additional cost to society which is attributable to the above factors. These considerations may help to explain why many people appear to believe that, in affluent countries at least, severe soil degradation should not *knowingly* be permitted to occur.

Another focus on the issue of land degradation may be obtained by posing the following question: from economic and ethical perspectives what differences, if any, are there between human exploitation of non-renewable energy resources (eg fossil fuels) and depletion of the non-renewable components of land? As a point of departure, let us first assume that there is only one form of utilisation for each resource. The mining and utilisation of the non-renewable energy resource effectively reduces, for ever, the stock of that resource available for use by future generations. Also, we assume that human usage of a particular area of land will, over time, irreversibly reduce the productivity and social value of the land to zero.

For both land usage and mining of non-renewable energy resources, the question is whether the resources should be exploited (and effectively destroyed) or left unused. Let us also assume that we know with certainty that we will never acquire any more knowledge about ways of exploiting and using the non-renewable energy resource and that we will never have a technology which would permit us to do anything other than 'mine' the land. That is to say, 'sustained yield' land usage is, and always will be, a technical impossibility. Such land is clearly just like a non-renewable resource and its usage poses precisely the same ethical and economic issues as does the mining of any non-renewable resource.

If we lived in a world where all resources had the above characteristics, it would be a very Hobbesian world indeed. Solow (1974) suggests that such a world may be described by the 'lollipop' model. The world is a lollipop; each lick leaves precisely one less lick for future generations and one day, if the world or the human species survives long enough, the lollipop will be completely consumed.

When human survival and sustenance requires some consumption of the lollipop, the commensense guide to a society that cares about future generations is like a parent's advice of moderation to a child: lick it slowly and don't bite chunks off. I should perhaps add that I recognise the dictum that matter cannot be destroyed—only its form changed. The presumption in the lollipop model of the world is that the change in the form of matter, following usage of the resource, renders it useless for mankind.

At the opposite end of the spectrum to the lollipop model is the Garden of Eden model. The earth is a garden which provides an abundant flow of goods and services that can be sustained forever without using up any resource stocks. The principle is maintained if we relax the Garden of Eden image and allow for lean years and fat years. So long as the earth effectively replenishes itself, the issue of depletion of resources does not arise. Some forms of subsistence agriculture where any depleted nutrients are replaced by organic manure, humus etc, collected onsite are fairly accurately described by this model. Apart from stochastic climatic influences, the land is naturally replenished and the future is a replica of the past, assuming that there is a stable population.

In modern societies, both of these extreme visions of the world are, of course, misleading, but I believe they highlight the crucial roles of information and uncertainty, substitutability among factors of production, and technological change. Modern agriculture contains some elements of natural replenishment of the land, but it is becoming increasingly like an industrial factory, with many inputs imported from offsite, including non-renewable inputs (eg phosphatic fertiliser and fossil fuels); soil erosion and the runoff from fertilisers and pesticides is the agricultural non-point pollution analog of industrial point pollution. This dimension of the production process is captured by the picture of the earth as a spaceship.

Returning to the lollipop model, we can now show why it is a misleading representation of the world. First, we do not really know the amount of stocks of phosphate, fossil fuels etc existing in the earth. It is much too conservative to assume that no new reserves of non-renewable resources will ever be found. As Solow says, at unpredictable times the lollipop will expand by unpredictable amounts. We should allow something for the likelihood of future discoveries even if it turns out that some anticipated new discoveries are not made and from an ex post perspective we over-consume some resources. Sharing resources with future generations should also include sharing risks with them. Second, the lollipop model is misleading because since it is possible to substitute labour and capital for scarce non-renewable resources, there is not a one-to-one correspondence between the finiteness of resources and the finiteness of output. Third, the technology of lollipop licking has not changed, but in the real world there is the possibility of technological change. The economic substitution of capital and new technology for land is well illustrated by figures for crop production in the United States. Over the period 1950-54 to 1975-79, United States crop production increased 63 per cent, while acres of crop land harvested declined two per cent. The aggregate productivity of all resources increased 58 per cent over this period (US Department of Agriculture 1982).

Clearly, to believe in the notion that appropriate technological advances will always occur is being unduly optimistic, but the notion that no new technology will ever turn up is equally unrealistic. However, many observers believe that the post-Second World War rate of technological advance in agriculture, particularly United States agriculture, was unprecedented with major breakthroughs like hybrid corn which they believe are unlikely to be repeated even with the research in genetic engineering now under way (Crosson and Stout 1983). It is interesting to note that natural stocks of nitrogen contained in topsoil threatened to be a severe constraint to sustained yield agriculture in parts of the United States and some other other countries until the development of cheap nitrogenous fertilisers. Technological advance converted an exhaustible stock resource into a virtually inexhaustible flow resource.

Land is still by far the most important single asset in agriculture and any decline in its productivity due to soil erosion, salinisation etc may increase the costs of production of food and fibre and impose higher costs on future generations. The most important point to emerge from the foregoing discussion is that our real obligation to future generations is 'capacity to produce'. In this respect the lollipop model of the world is extremely misleading. There is no clear limit to the substitutability of other inputs and new technology for non-renewable resources. Social judgments about the seriousness of soil erosion and other forms of land degradation depend crucially on the expectations of a society about the long-term prospects for new land-substituting technologies.[1] From an ethical and intergenerational perspective, depletion of some components of the land may be justified so long as the overall productive capacity of food and fibre producing systems passed on to successive generations is maintained.

Endnote

1. Uncertainty is pervasive in economic life, but more than the usual degree of uncertainty surrounds issues of long-term sustainability of biologically-based productive systems like agriculture. Even though in many situations we are unable at present to quantify the benefits of land conservation I believe that the preceding conceptual framework is a very useful one for thinking about issues of resource use and social choice in an ordered and structured way. In situations where it is not possible to quantify benefits, the framework of benefit-cost analysis can be used to determine the most cost-effective way of meeting a conservation standard or to rank alternative ways of proceeding with a large conservation project. A detailed discussion and analysis of the issues of sustainable resource use, safe minimum standards of conservation, and social choice under uncertainty is provided in Chisholm (1986).

Appendix B

Comments by Bruce Davidson
Department of Agricultural Economics
University of Sydney

**'Australia's land resources at risk' by Colin Chartres
(Chapter 1);**

**'Biological and physical causes of land degradation' by Gordon
Burch, Dean Graetz and Ian Noble (Chapter 2)**

The theme of these chapters appears to be threefold. Namely, that
Australian soils are naturally inferior to those of Europe and the United
States of America, that European farming in Australia has led to land
degradation, and that this is continuing.

No attempt is made to construct some index of initial soil fertility
which would make it possible to quantify the degree to which the total
fertility of soil in Australia is inferior or superior to that of Europe and
the United States of America (Leeper 1970, p23). It is even possible that
the area of good soil on a per capita basis is higher in Australia than that
in Western Europe or the USA. Many Australian soils, including the
black cracking clays of northern New South Wales and central
Queensland and the black soils of the Wimmera are more fertile than
most European soils in their natural state. It would be difficult to find
poorer or more eroded soils than the slopes of the Welsh hills where all
the top soil has been removed and farming is carried out on the sub-soil
(Robinson 1934). It is not recognised that many of the best soils in
Europe, including the wheat lands of East Anglia, the Polders of
Holland, and the fields of Flanders, are due to man-made drainage
(Summers 1976, Liddell Hart 1970, p330-331).

In addition, it is impossible to discuss soil quality without considering
other aspects of the environment and the purposes for which it is to be
used. The heavy clay soils are an advantage to wheat growers in
northern NSW and Queensland as they store large quantities of summer
rainfall for the wheat crop grown during autumn and winter. In the
southern Mallee areas where all rain falls in the winter, the sandy soils
are superior, as little water is stored by the soil and all is available to
the plant. Many vegetable growers favour sterile sands as this gives

them complete control of the chemical status of the soil. Vegetable growers at Moorabbin in Victoria bitterly objected to their forced removal from that area of sterile sand to the rich clay soils of the drained KooWee-Rup swamp in the 1950s for this very reason.

It is not recognised that in any case the initial status of the soil is only a short-term advantage. In time cropping will exhaust the chemical fertility of any soil and cultivation will destroy its structure. Chemicals required by plants must then be restored by fertilisation, and structure by adding organic matter in some way (Leeper 1970, p23). One might ask, if the poverty of Australian soils compared with those of Europe and the USA is such a disadvantage, why do the farmers of these two regions use heavier dressings of fertiliser on them than do Australian farmers, and why do they have to be protected from cheap Australian wheat and meat which has to be transported half way around the world? What matters is the productivity of all the resources used in agriculture and the soil is only one of these.

Even more important than the initial fertility of the soil is the failure to recognise that in most farming regions in Australia with a growing season of more than five months, the land is much more productive than it was when Europeans first arrived. It is in this region that 80 per cent of Australia's agricultural output is produced (Davidson 1969, pp111 and 120). The first step in improving the productivity of the region consisted of ringbarking large areas of forest, which increased the carrying capacity of the grazing land from one sheep to three acres to one sheep per acre (Hodgson 1846, p51, F C Crofts, University of Sydney, pers comm 1981). The introduction of improved pastures, including legumes and fertilisation with superphosphate led to a further doubling or trebling of the carrying capacity of pastures in this region (Davies 1958, Davies and Humphries 1965).

Annual legume pastures which were introduced in the 1940s made it possible to change the wheat rotation from bare fallow-wheat to a period under pasture followed by wheat crops. As the new rotation abolished the bare summer fallow, wind erosion was greatly decreased. The authors describe the dust storm in Melbourne in the summer of 1982-3. This was due to unsown wheat land being cultivated throughout the summer because the drought prevented sowing in 1982. With bare fallowing in the summer, dust storms occurred in Melbourne much more frequently. During the 1930s when bare fallowing was practiced four severe dust storms were recorded in Melbourne but in the 1960s and 1970s only two were recorded in each decade (Australian Meteorological Bureau, Melbourne, pers comm 1985). The pink snows caused by the deposition of wind blown soil in the Australian Alps and even on the snow fields of New Zealand which were fairly common in the 1930s are no longer found.

Even in the dry interior it is doubtful if the productivity of the land is declining. The graphs in Chartres' paper suggest that the average number of sheep in the Western Division of NSW has not declined since the early 1900s. The real question is, could a system of occupation have been developed for western NSW which, after allowing for the rabbit plague, would have permitted larger numbers of sheep to be carried in the region than at present? If this was not possible, the system adopted by the original graziers of reaping a large harvest in the first few years and then accepting lower returns, was probably the one which gave the highest net returns.

The clearing of land which later led to the dryland salting of some of it was also the correct decision, providing the returns from clearing and cropping were greater than the total costs involved. If the land was not cleared it could not be used at all. The only superior alternative would have been some form of development from which salting did not arise and from which net returns were higher than from the system used. In addition, the 130 000 hectares quoted by Chartres as being affected by dryland salting in Western Australia is a very small proportion of the 2 800 000 hectares developed for agriculture in the south-west of that state during the last 30 years (Fitzpatrick 1974).

Burch, Graetz and Noble claim it has taken over a century to adapt the agricultural traditions to the realities of the Australian environment. The real problem the European settlers faced was how to exploit the comparative advantage they possessed in having a large area of well-watered land per capita, and the disadvantage of a small population which provided only a small domestic market and a limited labour force demanding high wages. They had to produce commodities for which an export market existed, which could be profitably transported over long distances by land and sea to those markets, and to produce them using little labour. By 1820, only 30 years after first settlement, a wool industry was developed which satisfied these conditions (Davidson 1981, p77-90). Large cultivating machinery, the development of mechanical harvesting equipment in which Australia led the world, and the construction of railways, made it possible to export wheat profitably by the 1870s. The profitable export of meat and butter was not possible until the development of refrigeration in 1880. In short, the problem was one of developing techniques suited to Australia's economic conditions and location rather than an environmental problem (Davidson 1981, p175-221).

The same paper contains a curious contradiction. Initially we are informed that 'Australia has the paradox of having a detailed understanding of the process whereby our renewable (or is it "non-renewable") resources are being destroyed or degraded, but without a rigorous quantitative assessment of how much has been degraded and how severely or whether this degradation is accelerating or stable' (p 27). Later we are asked 'why has a wealthy nation and developed society allowed extreme rates of land degradation to continue?' (p 45). If we do not know how much has been degraded how can it be said that society has allowed extreme rates of land degradation?

One also wonders if the problem is as serious as is suggested. After all, Australia has more than doubled its agricultural production in the last 30 years (QRRE 1979, p172, 1985, p294). Almost all of this increase has come from land which has been farmed for over 30 years; very little new land has been brought into production. Could such an increase in production have been obtained from soils which were steadily deteriorating?

The Commonwealth-State Collaborative Soil Conservation Study (Australia. DEHCD 1978a) estimated that it would cost $675 million at 1975 prices ($1225 million at 1981 prices) to repair all degraded land in Australia. The net value of rural production in 1981-2 was $3736 million, and in the drought year of 1982-3 it was $1729 million, a decrease of $2007 million (QRRE 1985, p287). Thus the total cost of repairing all of the damage due to erosion after 200 years of European occupation was less than the loss caused by one severe drought.

It is also worrying that neither of the papers mention the possibility of any gains from erosion. The soil and fertility of some alluvial flats is constantly renewed by floods. Is it possible that wind erosion in the drier areas of the continent moves soils to areas with a better rainfall where they can be used more productively?

'Degradation pressures from non-agricultural land uses' by Lance Woods (Chapter 6)

Degradation from mining

It is disappointing that the economic aspects of rehabilitation of lands used for open cut mining is not discussed more fully. The author points out that this cost is often of the order of $10 000 per hectare. This statement is not correct: the total cost to the company of restoring the land is unknown. The $10 000 quoted by the author as the cost of restoration is the bond the company must deposit with the NSW government as a guarantee of restoration. It is known that the cost of sowing and fertilising pasture after restoration is approximately $800 per hectare. Restored land of this type in the Hunter Valley, one of the major open cut mining areas, is valued at only $500 per hectare (NSW Soil Conservation Service, Singleton, pers comm 1985). Is the social loss of a sum well in excess of $300 per hectare mined justified? Instead a company could be taxed for an amount exceeding $800 per hectare mined, and this sum used for purposes other than wasteful land rehabilitation.

In the same valley the top soil was eroded away long before mining commenced. Yet mining companies are still forced to replace surface soil on the restored land and this increases the cost of rehabilitation. In many cases better plant growth would be obtained at less cost if lower layers of the soil were placed on the surface (P. Charman, NSW Soil Conservation Service, pers comm 1985).

Prime agricultural land and urban development

It is stated that there are no good reasons other than location why prime agricultural land should be selected for urban development, rather than less productive types of land. On the contrary there are excellent reasons for doing so. Good agricultural land is often cheaper to build on than poor agricultural land, for example, Sydney sandstone is more expensive to build on than the clay loams of Camden because of the higher cost of foundations and installing underground drains. Even the greater ease of home gardening might make home building on good agricultural land preferable.

It is also difficult to understand why the author suggests that a shortage of good agricultural land is beginning to cause concern in the USA and Europe when both countries have unmanageable surpluses of agricultural commodities. Nor is it true that Australia does not have large areas of good agricultural land. If good agricultural land is defined as land capable of being cropped or supporting improved pastures, Australia, on a per capita basis, has twelve times as much good land as Europe and four times as much as the USA (Davidson 1981, p5-13).

'Land degradation: legal issues and institutional constraints' by John Bradsen and Robert Fowler (Chapter 7)

This chapter suffers from the same weakness as the first two chapters in this monograph in that it fails to explain how steadily increasing production is being obtained from allegedly deteriorating land. It is claimed that Britain, after hundreds of years, has a looming land degradation problem (page 136). In 1930 the United Kingdom imported 80 per cent of the wheat consumed and 67 per cent of the total cereals consumed. In 1980 a mere 12 per cent of both wheat and total cereals used were imported. This increase in crop production was achieved although the total human population increased by 22 per cent between 1930 and 1980. During the same period the number of livestock, many of which have to be fed on cereals, increased by the following proportions, cattle by 74 per cent, sheep by 27 per cent, pigs by 194 per cent and fowls by 89 per cent (International Institute of Agriculture 1931, FAO Production and Trade Year Books 1980). Do the authors think that the United Kingdom is likely to revert to a level of reliance on agricultural imports similar to that of the 1930s, so long as British farmers continue to be paid the high prices they receive for grains and livestock under the Common Agricultural Policy of the European Economic Community?

The authors suggest that appropriate land use planning must be based on appropriate land use planning studies, the aim of which should be to identify districts or areas and their needs essentially on an ecological basis. Such an approach implies that scientists know how land should be used. However, some of our greatest failures to maintain the environment followed the establishment of irrigation and its expansion after the Second World War. This was based on the best scientific advice. Is the scientific knowledge available at present any better? If not, the solution offered is inappropriate. In addition the irrigation areas based on such advice were an economic disaster (Davidson 1965, pp 147-201, Davidson 1974, Davidson and Graham–Taylor 1982).

'Land tenure: plaything of governments or an effective instrument?', by Michael Young (Chapter 8)

The major purpose of using long-term leases rather than freehold tenure for pastoral lands was the belief that at some future date such land could be used profitably in smaller units than at present. This might arise because of changes in technology or even because of long-term changes in market conditions. In these circumstances it is doubtful if the nine conditions listed are necessary for satisfactory leasehold tenure.

The aim is to encourage the lessee to maximise net returns from the land without damaging it. This can be achieved as in freehold tenure by simply granting a perpetual lease which can be passed on to heirs or sold. The decrease in the market value of a perpetual lease should deter the lessee from damaging the land just as effectively as it does with freehold tenure.

The question of the state wishing to repossess the land at some future date could be overcome by including a covenant in the lease that enables the state to repossess the land at its market value as a grazing lease at the time it is repossessed.

Appendix C

Comments by John Thomas
CSIRO Division of Groundwater Research

Blyth and McCallum (Chapter 4) claim that two recent studies of the costs of salinity degradation are of limited use for public policy, other than perhaps as providing measures of the magnitude of the degradation. The studies were by the Working Party on Dryland Salting in Australia (1982) and Peck, Thomas and Williamson (1983). In his commentary on Section II Greig agrees with Blyth and McCallum: A few grand scale evaluations of the costs of land degradation have been attempted, yet the results are of questionable significance for policy making. Greig continues 'the costs ought to be considered on a case-by-case, or catchment-by-catchment basis, beginning with the most rapidly degrading ones'.

These commentators equate policy-relevance with benefit-cost analysis, which I suggest is a narrow viewpoint. The desire for action to combat land degradation does not, as a matter of fact, begin with benefit-cost analysis, but with observation of degradation processes and (in broad terms) their social and economic impacts. Evaluation of alternative policies and particular projects comes later. But in any case the step from evidence to policy will be taken with or without benefit-cost analysis. Expositions of utilitarian purity may or may not be policy-relevant, depending on who the policy-makers are (Bennett and Thomas 1982 discuss this at length). Nevertheless, both benefit-cost analysis and broad-scale appraisal of the problem may influence policy if they are incisive enough. Peck et al produced a number of propositions about salinity which are relevant to policy, a point which seems to have escaped Blyth and McCallum, and Greig.

First, Peck et al derived estimates of maximum benefits which could have been obtained between 1971 and 1981 if land and water salinisation had been avoided, together with actual costs of attempted abatement, monitoring and research into the problem. Broadly, the study conclusions are consistent with Davidson's (Appendix B) observation that, in comparison with other causes of productivity loss, land degradation has been significant but not dominant (my interpretation of Davidson).

Second, Peck et al show that salinity may be expected to increase in extent and that the real costs per unit of physical damage are also likely to increase in future. Since no salinity abatement would be cost-free, it follows that *either* abatement expenditure must rise or the level of damage costs must increase, *or both*. Whether it is economically efficient to allow damages or abatement expenditures to increase involves, of course, a question for benefit-cost analysis and a judgment of who should (or is prepared to) bear the costs.

I agree with Blyth and McCallum, and with Greig, that it would be nice to see further benefit-cost studies in the area of land and salinity rehabilitation, where the technical and physical relationships are known.

Appendix D

Participants at Workshop on Land
Degradation and Public Policy
3-4 September 1985

Mr Andrew Arch
Economics and Marketing Branch
Department of Agriculture and Rural
Affairs (Victoria)

Professor Eric Bachelard
Forestry
The Faculties
Australian National University

Dr John Ballard
Political Science
The Faculties
Australian National University

Mr Michael Barker
Law Faculty
Australian National University

Dr David Bennett
Land Resource Policy Council
Premiers Department
Perth

Dr Jeff Bennett
Department of Economics and
Management
University of New South Wales
RMC Duntroon

Dr Michael Blyth
Bureau of Agricultural Economics
Canberra

Professor Jim Bowen
CRES
Australian National University

Mr John Bradsen
Faculty of Law
University of Adelaide

Mr Howard Briggs
Land Resources Branch
Queensland Department of Primary
Industries

Professor H Brookfield
Human Geography
Research School of Pacific Studies
Australian National University

Mr John Brumby, MP
Member for Bendigo
Parliament House
Canberra

Dr Gordon Burch
CSIRO Division of Water and Land
Resources
Canberra

Mr Dennis Cahill
Deputy Chairman, Soil Conservation
Authority
Department of Conservation, Forests and
Lands (Victoria)
Kew

Emeritus Professor Keith Campbell
Department of Agricultural Economics
University of Sydney

Dr Colin Chartres
CSIRO Division of Soils
Canberra

Dr Anthony Chisholm
CRES
Australian National University

Dr John Cooke
Department of Conservation, Forests and
Lands (Victoria)
Mildura

Mr Peter Cullen
School of Applied Science
Canberra College of Advanced Education

Dr John Dargavel
CRES
Australian National University

Dr Bruce Davidson
Department of Agricultural Economics
University of Sydney

Dr Bruce Davis
Department of Political Science
The University of Tasmania

Dr Diana Day
CRES
Australian National University

Dr John Dixon
Environment and Policy Institute
East-West Center
Hawaii
USA

Dr Robert Dumsday
School of Agriculture
La Trobe University

Mr Geoff Edwards
School of Agriculture
La Trobe University

Dr John Formby
CRES
Australian National University

Mr Robert Fowler
Faculty of Law
University of Adelaide

Mr Reg French
Agricultural Resources Branch
SA Department of Agriculture

Mr Frank Gibbons
National Soil Conservation Program
Department of Primary Industry
Canberra

Mr D. Gilbert
Division of Plant Industries
NSW Department of Agriculture
Sydney

Mr Jeff Gilmore
Assistant Private Secretary
Minister for Primary Industry
Canberra

Mr Gary Goucher
National Farmers' Federation
Canberra

Dr Peter Greig
Corporate Development Unit
Melbourne and Metropolitan Board of
Works

Professor David Griffin
Forestry
The Faculties
Australian National University

Dr John Handmer
CRES
Australian National University

Professor Diana Howlett
Department of Geography
Faculty of Arts
Australian National University

Mr Graham Hunter
Department of the Premier and Cabinet
Melbourne

Mr Lee Jackson
National Soil Conservation Program
Department of Primary Industry
Canberra

Dr Anthony Jakeman
CRES
Australian National University

Dr David James
Centre for Environmental and Urban Studies
Macquarie University

Mr Ken Johnson
Department of Geography
Faculty of Arts
Australian National University

Mr Robert Junor
Soil Conservation Service of NSW
Sydney

Dr Michael Kirby
Bureau of Agricultural Economics
Canberra

Dr P Laut
CSIRO
Division of Water and Land Resources
Canberra

Mr Brian Lees
Geography
The Faculties
Australian National University

Mr Andrew Lothian
SA Department of Environment and Planning
Adelaide

Mr Andrew McCallum
Bureau of Agricultural Economics
Canberra

Dr Geoff Mosley
Australian Conservation Foundation
Hawthorn VIC

Professor Warren Musgrave
Department of Agricultural Economics and Business Management
University of New England

Dr Ian Noble
Research School of Biological Sciences
Australian National University

Mr Lindsay Nothrop
National Soil Conservation Program
Department of Primary Industry
Canberra

Mr Warwick Papst
Land Protection Service
Department of Conservation, Forests and Lands (Victoria)
Kew

Dr John Paterson
Director-General
Department of Water Resources (Victoria)
Armadale

Dr Robert Pearse
Department of Agricultural Economics and Business Management
University of New England

Mr Doug Pearson
Commissioner
Western Lands Commission (NSW)
Sydney

Mr Ray Perry
CSIRO Division of Groundwater Research
Wembley WA

Mr Tony Plowman
Member, Soil Conservation Authority
Department of Conservation, Forests and Lands (Victoria)
Kew

Mr John Quiggin
CRES
Australian National University

Dr E W Radoslovich
CSIRO Division of Soils
Canberra

Dr Roy Rickson
School of Australian Environmental Studies
Griffith University

Mr Andrew Robb
National Farmers' Federation
Canberra

Dr Graeme Robertson
WA Department of Agriculture
Perth

Mr M Ryan
Department of Primary Industry
Canberra

Dr Rod Simpson
CRES
Australian National University

Mr O R Southwood
Regional Director of Agriculture
NSW Department of Agriculture
Dubbo

Ms Joan Staples
Australian Conservation Foundation
Canberra

Dr R Sylvan
Department of Philosophy
Research School of Social Sciences
Australian National University

Mr Geoffrey Thomas
CRES
Australian National University

Mr K E Thompson
Department of Arts, Heritage and
Environment
Canberra

Mr N J Thomson
Faculty of Economics
Adelaide University

Mr Garrett Upstill
Department of Arts, Heritage and
Environment
Canberra

Dr Brian Walker
CSIRO Division of Wildlife and
Rangelands
Canberra

Dr Pat Walker
CSIRO Division of Soils
Canberra

Dr Robert Wasson
CSIRO Division of Water and Land
Resources
Canberra

Mr Warwick Watkins
Soil Conservation Service of NSW
Sydney

Mr Bill Watson
Bureau of Agricultural Economics
Canberra

Mr Adrian Webb
Soil Conservation Research Branch
Queensland Department of Primary
Industries
Brisbane

Professor Douglas J Whalan
Chairman of the Board of the Faculties
Australian National University

Dr Lance Woods
Department of Resources and Energy
Canberra

Mr Timothy Yapp
Department of Arts, Heritage and
Environment
Canberra

Mr Michael Young
CSIRO Division of Wildlife and
Rangelands Research
Deniliquin NSW

Bibliography

Abelson, P. 1979. *Cost benefit analysis and environmental problems*. Southhampton: Saxon House.

Adamson, C.M. 1976a. Stabilization of hill areas under adverse grazing and climatic conditions at Wagga Wagga, New South Wales, Australia. In. Luchok, J., Cawthon, J.D. and Breslin, M.J., eds. *Proceedings International Hill Lands Symposium, 1976 October 3-9, Morgantown, West Virginia*. Morgantown: West Virginia University Books: 381-385.

Adamson, C.M. 1976b. Some effects of soil conservation treatment on the hydrology of a small rural catchment at Wagga. *Journal of the Soil Conservation Service of New South Wales*. 32(4): 230-249.

Adamson, C.M. 1978. Conventional tillage systems as they affect soil erosion in southern New South Wales. *Journal of the Soil Conservation Service of New South Wales*. 34(4): 199-202.

Alcock, B.S. 1980. *The costs of soil erosion*. Toowoomba: Economic Services Branch, Department of Primary Industries. (Miscellaneous bulletin; no.11).

Anderson, J.E. 1979. *Public policy-making*. 2nd ed. New York: Holt, Rinehart and Wilson.

Anon. 1983. *A national conservation strategy for Australia*. Canberra: AGPS.

Appleby, P.G. and Oldfield, F. 1978. The calculation of lead-210 assuming a constant rate of supply of unsupported 210_{Pb} to the sediment. *Catena*. 5(1): 1-8.

Arch, A.M.J. and Dumsday, R.G. 1981. Measuring the benefits of soil conservation in dryland cropping areas: paper presented at the 25th Annual Conference of the Australian Agricultural Economics Society, February 10-12, Christchurch, New Zealand.

Armstrong, J.L. 1981. Soil Conservation Service of New South Wales. Unpublished report.

Arrow, K.J. and Fisher, A.C. 1974. Environmental preservation, uncertainty and irreversibility. *Quarterly Journal of Economics*. 82(2): 312-319.

Australia. Bureau of Agricultural Economics. 1984. *Australian forest resources 1983*. Canberra: AGPS.

Australia. Commission of Inquiry into Land Tenures. 1976. *Final reports, February 1976*. Canberra: AGPS.

Australia. Committee of Inquiry into the National Estate. 1975. *National estate: report of the Committee of Inquiry*. Canberra: Government Printer. (Parliamentary paper no.195 of 1974).

Australia. Department of Environment, Housing and Community Development. 1978a. *A basis for soil conservation policy in Australia*. Canberra: AGPS. (Commonwealth and State Government Collaborative Soil Conservation Study 1975-77 Australia. Report; 1).

Australia. Department of Environment, Housing and Community Development. 1978b. *Economic evaluation of Eppalock catchment soil* conservation project, Victoria. Canberra: AGPS. (Commonwealth and State Government Collaborative Soil Conservation Study 1975-77 Australia. Report; 9).

Australia. Department of Environment, Housing and Community Development. 1978c. *Economic evaluation of a soil conservation project in Allora Shire, Queensland*. Canberra: AGPS. (Commonwealth and State Government Collaborative Soil Conservation Study 1975-77 Australia. Report; 10).

Australia. Department of Environment, Housing and Community Development. 1979. *Towards a national approach to land resource appraisal*. Canberra: AGPS. (Commonwealth and State Government Collaborative Soil Conservation Study 1975-77 Australia. Report; 2).

Australia. Department of Resources and Energy. 1983. *Water 2000: a perspective on Australia's water resources to the year 2000*. Canberra: AGPS.

Australia. Department of the Environment and Conservation. 1975. *Financial assistance for soil conservation: Interdepartmental Committee report*. Canberra: AGPS.

Australia. Independent Inquiry into the Commonwealth Scientific and Industrial Research Organization. 1978. *Independent inquiry into the Commonwealth Scientific and Industrial Research Organization, August 1977*. Canberra: AGPS. Chairman: A.J. Birch.

Australia. Industries Assistance Commission. 1983. *Rural adjustment: interim report*. Canberra: AGPS.

Australia. Lake Pedder Committee of Inquiry. 1974. *Final report, April 1974*. Canberra: AGPS for the Department of Environment and Conservation.

Australia. Parliament. Senate. Standing Committee on Science and the Environment. 1977. *Woodchips and the environment*. Canberra: AGPS.

Australia. Parliament. Senate. Standing Committee on Science and the Environment. 1978. *Woodchips and the environment: supplementary report*. Canberra: AGPS. (Parliamentary paper; no.334/1978).

Australia. Parliament. Senate. Standing Committee on Science, Technology and the Environment. 1984. *Land use policy in Australia*. Canberra: AGPS.

Australia. Standing Committee on Soil Conservation. 1971. *Study of community, benefits of and finance for soil conservation*. Canberra: AGPS.

Australia. Standing Committee on Soil Conservation. 1984. *Soil erosion and productivity*. Perth: WA. Department of Agriculture.

Australia. Working Group on All Aspects of Rural Policy in Australia. 1974. *The principles of rural policy in Australia: a discussion paper*. Canberra: AGPS. Convenor: Stuart Harris.

Australian Environment Council. 1984. *Guide to environmental legislation and administrative arrangements in Australia*. Canberra: AGPS. (Australian Environment Council. Report no.16).

Australian Soil Science Society. 1984. *Future research needs in soil science in Australia*. (Publication; no.7).

Aveyard, J.M. 1981. Effect of long-term erosion on crop production. *Proceedings of the Third Australian Soil Conservation Conference, Tamworth*: 229-302.

Aveyard, J.M. 1983. *Soil erosion productivity research in New South Wales to 1982*. Sydney: Soil Conservation Service of New South Wales. (NSW. Soil Conservation Service. Technical bulletin; no.24).

Aveyard, J.M., Hamilton, G.J., Packer, I.J. and Barker, D.J. 1983. Soil conservation in cropping systems in southern New South Wales. *Journal of the Soil Conservation Service of New South Wales*. 39: 113-120.

Bagnold, R.A. 1941. *The physics of blown sand and desert dunes*. London: Methuen.

Baker, V.R., Pickup, G. and Polach, H.A. 1985. Radiocarbon dating of flood events, Katherine Gorge, Northern Territory, Australia. *Geology*. 13: 344-347.

Balderstone, J.S., Duthie, L.P., Eckersley, D.P., Jarrett, F.G. and McColl, J.C. 1982. *Agricultural policy issues and options for the 1980s: Working Group report to the Minister for Primary Industry*. Canberra: AGPS.

Barfield, B.J., Moore, I.D. and Williams, R.G. 1979. Prediction of sediment yield from surface mined watersheds. In. Carpenter, S.B., ed. *Proceedings Symposium on Surface Mining Hydrology, Sedimentology and Reclamation, 1979 December 4-7, University of Kentucky*. Lexington, Kentucky: Office of Engineering Services, College of Engineering, University of Kentucky: 83-91.

Barker, M.L. 1984. Aboriginal land rights law in Australia: current issues and legislative solutions, 1984. In. Australian Mining and Petroleum Law Association. *Yearbook*. Melbourne: The Association: 483-524.

Barr, N.F. 1985. *Farmer perceptions of soil salting*. Melbourne: Department of Psychology, University of Melbourne.

Bates, G. 1983. *Environmental law in Australia*. Sydney: Butterworths.

Batie, S.S. 1982. Policies, institutions and incentives for soil conservation. In. Halcrow, H.G., Heady, E.O. and Cotner, M.L., eds. *Soil conservation policies, institutions and incentives*. Ankeny, Iowa: Soil Conservation Society of America.

Bator, F.M. 1958. The anatomy of market failure. *Quarterly Journal of Economics*. 72(3): 351-379.

Baumol, W.J. 1972. On taxation and the control of externalities. *American Economic Review*. 62(3): 307-322.

Baumol, W.J. and Oates, W.E. 1975. *The theory of environmental policy*. New Jersey: Prentice-Hall.

Baumol, W.J. and Oates, W.E. 1979. *Economics, environmental policy and the quality of life*. Englewood Cliffs, NJ: Prentice-Hall.

Bennett, D. and Thomas, J.F. 1982. *On rational grounds: systems analysis in catchment land use planning*. Amsterdam: Elsevier Scientific.

Bettenay, E., Blackmore, A.V., and Hingston, F.J. 1964. Aspects of the hydrological cycle and related salinity in the Belka valley, Western Australia. *Australian Journal of Soil Research*. 2: 187-210.

Bird, P.R., Lynch, J.J. and Obst, J.M. 1984. Effect of shelter on plant and animal production. *Proceedings of the Australian Society of Animal Production*. 15: 270-273.

Blaikie, P.M. 1984. The political economy of land degradation: Workshop on Land Degradation and Agrarian Change: Approaches to Understanding, 1984 February, Department of Human Geography, Australian National University.

Blong, R.J. 1985. Hillslope erosion: a review. In. Loughran, R.J., ed. *Drainage basin erosion and sedimentation*. Newcastle, NSW: University of Newcastle. v.2.: 23-32.

Blyth, M.J. and Kirby, M.G. 1985. The impact of government policy on land degradation in the rural sector. In. Jakeman, A.J., Day, D.G. and Dragun, A.K., (eds). *Policies for environmental quality control*. Canberra: CRES, ANU. (CRES monograph; no.15).

Bolton, G. 1981. *Spoils and spoilers*. Sydney: George Allen and Unwin.

Booth, C.A. and Barker, P.J. 1981. Shrub invasion in the sand plain country west of Wanaaring. *Journal of the Soil Conservation Service of New South Wales*. 37: 65-70.

Boyce, R.C. 1975. Sediment routing with sediment-delivery ratios. In. *Present and prospective technology for predicting sediment yields and sources*. US. Department of Agriculture: 61-73. (ARS-S-40).

Braden, J.B. 1982. Some emerging rights in agricultural land. *American Journal of Agricultural Economics*. 64(1): 19-27.

Bradsen, J.R. 1984. Soil conservation:

legislative measures. In. *Third National Environmental Law Symposium, 1984 August, University of Adelaide.*

Brennan, G. and Lomasky, L. 1983. Institutional aspects of 'Merit goods' analysis. Centre for Study of Public Choice (VPI). Mimeo. Forthcoming in Finanzarchiv.

Broecker, W.S. and Peng, T-H. 1982. *Tracers in the sea.* Palisades NY: Columbia University.

Bromfield, S.M., Cumming, R.W., David, D.J. and Williams, C.H. 1983. Change in soil pH, manganese and aluminium under subterranean clover pasture. *Australian Journal of Experimental Agriculture.* 23: 181-191.

Bromley, D.W. 1982a. The rights of society versus the rights of landowners and operators. In. Halcrow, H.G., Heady, E.O. and Cotner, M.L., eds. *Soil conservation policies: institutions and incentives.* Ankeny, Iowa: Soil Conservation Society of America.

Bromley, D.W. 1982b. Land and water problems: an institutional perspective. *American Journal of Agricultural Economics.* 64(5): 834-844.

Brown, J.A.H. 1972. Hydrologic effects of a bushfire in a catchment in south-eastern New South Wales. *Journal of Hydrology.* l5: 77-96.

Buchanan, J.M. 1975. *The limits of liberty.* Chicago: University of Chicago Press.

Bultena, G., Nowak, P.J., Hoiberg, E. and Albrecht, D. 1981. Farmers attitudes toward land use planning. *Journal of Soil and Water Conservation.* 36(4): 37-41.

Bunce, A. C. 1942. *The economics of soil conservation.* Lincoln: University of Nebraska Press.

Bunker, R.C. 1984. Soil conservation policy in Australia. In. *Third National Environmental Law Symposium, 1984 August, University of Adelaide.*

Burch, G.J. 1986. Land clearing and vegetation disturbance. In. Russell, J.S. and Isbell, R.F., eds. *Australian soils: The human impact.* Brisbane: University of Queensland Press for Australian Society of Soil Science. 24:

159-184.

Burch, G.J., Bath, R.K., Spate, A.P., Nicholls, A.O., and O'Loughlin, E.M. 1983. Soil water store, infiltration and runoff characteristics of forest and grassland catchments at Puckapunyal in central Victoria. In. *Hydrology and Water Resources Symposium, 1983 November, Hobart, Tasmania.* Institution of Engineers, Australia: 293-299.

Burch, G.J., Mason, I.B., Fischer, R.A., and Moore, I.D. 1986. Tillage effects on soils: physical and hydraulic responses to direct drilling at Lockhart, NSW. *Australian Journal of Soil Research.* 24: 377-391.

Burton, J.R. 1983. General summary. In. Australian Water Resources Council. *Proceedings: Workshop on Non-point Sources of Pollution in Australia, 1983 March 7-10, Monash University, Melbourne.* Canberra: AGPS: 267-275. (AWRC. Conference series; no.9).

Caldwell, L. 1970. *Environment.* New York: Doubleday.

Campbell, K. O. 1948. The development of soil conservation programmes in Australia. *Land Economics.* 24: 63-78.

Canada. Senate. Standing Committee on Agriculture, Fisheries and Foresty. 1984. *Soil at risk: Canada's eroding future.* Ottawa: The Committee. Chairman: H. O. Sparrow.

Cannell, R.Q., and Finney, J.R. 1973. Effect of direct drilling and reduced cultivation on soil conditions for root growth. *Outlook on Agriculture.* 7: 184-189.

Carder, D.J. and Humphry, M.G. 1983. The costs of land degradation. *Journal of Agriculture, Western Australia.* 24(2): 50-53.

Cary, J.W. 1982. Human dimensions to salinity management: psychological and economic aspects of individual salinity management decisions on farms: paper presented to 52nd ANZAAS Conference, Sydney.

Chamala, S. 1985. Need for extension of

sociological research in successfully transferring soil conservation technology. *Erosion Research Newsletter.* 11: 2-3.

Chamala, S. and Rickson, R.E. 1985. *Farmers' perception and knowledge of soil erosion and conservation practices in Australia: an overview.* St Lucia: Department of Agriculture, University of Queensland.

Chamala, S., Keith, K.J. and Quinn, P. 1982. *Adoption of commercial and soil conservation innovations in Queensland: information exposure, attitudes, decisions and actions.* St Lucia: Department of Agriculture, University of Queensland.

Chamala, S., Rickson, R.E. and Singh, D. 1985. *Annotated bibliography of socioeconomic studies on adoption of soil and water conservation methods in Australia.* St Lucia: Department of Agriculture, University of Queensland; Nathan: Institute of Applied Social Research, Griffith University.

Chartres, C.J. 1982. The role of geomorphology in land evaluation for tropical agriculture. *Zeitschrift fur Geomorphologie.* Supplement. 44: 21-32.

Chartres, C.J., Mabbutt, J.A., Johnston, D., Stanley, R.J. and Walker, P.J. 1982. Land system mapping as a basis for desertification assessment and mapping in NSW, Australia: report presented to the 3rd Expert Consultation on Desertification, FAO, Rome.

Chepil, W.S. 1961. The use of spheres to measure lift and drag on wind-eroded soil grains. *Proceedings of the Soil Science Society of America.* 25: 243-245.

Chepil, W.S. and Woodruff, N.P. 1963. The physics of wind erosion and its control. *Advances in Agronomy.* 15: 211-302.

Cherry, G. 1976. *Rural planning problems.* London: Leonard Hill.

Chisholm, A.H. 1973. Conservation and recreation in arid Australia: an economics perspective. In. Hyder, D.N., ed. *Arid shrublands: proceedings of the 3rd Workshop of the United States/Australia Rangelands Panel, 1973 March 26 - April 5, Tucson, Arizona.* Denver, Colorado: Society for Range Management.

Chisholm, A.H. 1978. Controlling non-point sources of pollution with special reference to soil erosion: paper presented at the 22nd Annual Conference of the Australian Agricultural Economics Society, February 7-9, Sydney.

Chisholm, A.H. 1984. Land degradation in Australia: an economics perspective: paper presented at a conference on Soil Degradation: the Future of Our Land? November, Australian Academy of Science, Canberra. Revised paper, December.

Chisholm, A.H. 1985. The choice of pollution control policies under uncertainty. In. Jakeman, A.J., Day, D.G. and Dragun, A.K., eds. *Policies for environmental quality control.* Canberra: CRES, ANU: 1-21. (CRES monograph; no.15).

Chisholm, A.H. 1986. Sustainable resource use: uncertainty, irreversibility and rational choice: paper prepared for Fourth World Congress of Social Economics, August, Toronto.

Chisholm, A.H. and Tyers, R. 1985. Agricultural protection and market insulation policies: applications of a dynamic multisectoral model. In. Piggott, J. and Whalley, J., eds. *New developments in applied general equilibrium analysis.* Cambridge University Pres: 189-220.

Chisholm, A.H., Walsh C. and Brennan, G. 1974. Pollution and resource allocation. *Australian Journal of Agricultural Economics.* 18(1): 1-21.

Cicchetti, C.J. and Freeman, A.M. 1971. Option demand and consumer surplus: further comment. *Quarterly Journal of Economics.* 85(3): 523-539.

Ciriacy-Wantrup, S.V. 1968. *Resource conservation: economics and policies.* 3rd ed. Berkeley, California: Division of Agricultural Sciences, University of California.

Clark, B., Gilad, A., Bisset, R. and Tomlinson, E., eds. 1984. *Perspectives on environmental impact assessment.* Dordrecht, Holland: D. Reidel.

Clark, C.M.H. 1955. *Select documents in Australian history, 1851-1900.* Sydney: Angus & Robertson.

Clark, E.H. 1985. The off-site costs of soil erosion. *Journal of Soil and Water Conservation.* 40(1): 19-23.

Clark, E.H., Haverkamp, J.A. and Chapman, W. 1985. *Eroding soils: the off-farm impacts.* Washington, DC: The Conservation Foundation.

Clark, W.C. 1985. Scales of climate impacts. *Climatic Change.* 7(1): 5-27.

Coase, R.H. 1960. The problem of social cost. *Journal of Law and Economics.* 3(3): 1-44.

Cole, P.J. 1985. *The River Murray irrigation and salinity investigation programme: results and future directions.* Adelaide: SA. Department of Agriculture. (Technical report; no. 69).

Commonwealth of Australia. 1911. Land tenure and settlement. In. *Official year book of the Commonwealth of Australia 1901-1910.* Melbourne: Commonwealth Bureau of Census and Statistics. no.4.

Conacher, A.J. and Murray, I.D. 1973. Implications and causes of salinity problems in the Western Australian wheatbelt: the York-Mawson area. *Australian Geographical Studies.* 11: 40-61.

Condon, R.W. 1978. Land tenure and desertification in Australia's arid lands with special reference to western New South Wales. *Search.* 9(7): 261-264.

Condon, R.W., Newman, J.C. and Cunningham, G.M. 1969a,b,c,d. Soil erosion and pasture degeneration in central Australia. *Journal of the Soil Conservation Service of NSW.* 25: 47-92, 161-182, 225-250, 295-321.

Cooke, J.W. 1985. Long term effects of farming practices on soil deterioration and revenue. In. Smith, I.S., ed.

Proceedings of Field Crops Training Workshop: Crop Modelling, Production Forecasting, Yield Estimation, 1984 May 2-3, Dookie Agricultural College. Melbourne: Department of Agriculture.

Costa, J.E. 1975. Effects of agriculture on erosion and sedimentation in the Piedmont Province, Maryland. *Geological Society of America Bulletin.* 86: 1281-1286.

Costin, A.B. 1980. Runoff and soil and nutrient losses from an improved pasture at Ginninderra, Southern Tablelands, New South Wales. *Australian Journal of Agricultural Research.* 31(4): 533-546.

Costin, A.B. 1983. Planning to save our soil in drought and flood. *Habitat.* 11(4): 28-29.

Craig, R.A. and Phillips, K.J. 1983. Agrarian ideology in Australia and the United States. *Rural Sociology.* 48(3): 409-421.

Crane, W.J.B. 1983. Perpetual productivity of pinus radiata in Australia re-examined. In O'Shaughnessy, P.J., Curry, C.J., Flinn, D.W. and Oates, N.M., eds. *Facing forestry's future: 10th Triennial Conference of the Institute of Foresters of Australia, 1983 August 29 - September 2, University of Melbourne.* Melbourne: Institute of Foresters of Australia: 43-51.

Crosson, P.R. 1981. *Conservation tillage and conventional tillage: a comparative assessment.* Ankeny, Iowa: Soil Conservation Society of America.

Crosson, P.R. 1984. Erosion: how big a threat? *Resources.* 75: 20-21.

Crosson, P.R. and Miranowski, J. 1982. Soil protection: why, by whom, and for whom? *Journal of Soil and Water Conservation.* 37(1): 27-29.

Crosson, P.R. and Stout, A.T. 1983. *Productivity effects of cropland erosion in the United States.* Washington, DC: Resources for the Future.

CSIRO. Division of Soils. 1983. *Soils: an Australian viewpoint.* Melbourne: CSIRO; London: Academic Press.

Cummins, V.G., Robinson, I.B., Pink, H.S. and Roberts, M.H. 1973. *A land use study of the Wyreema-Cambooya area of the eastern Darling Downs*. Department of Primary Industries, Queensland. (DLU technical bulletin; no.10).

Cunningham, O.R. and Jenkins, Q.A.L. 1982. Natural disasters and farmers: a neglected area of research by rural sociologists. *The Rural Sociologist*. 2(5): 325-330.

Daniel, P.R. 1969. The effects of sheet erosion on productivity. *Journal of the Soil Conservation Service of New South Wales*. 25(4): 322-329.

Davidson, B.R. 1965. *The northern myth*. Melbourne: Melbourne University Press.

Davidson, B.R. 1969. *Australia wet or dry*. Melbourne: Melbourne University Press.

Davidson, B.R. 1974. Irrigation economics. In. Frith, H.J. and Sawer, G., eds. *The Murray Waters*. Sydney: Angus and Robertson: 193-212.

Davidson, B.R. 1981. *European farming in Australia*. Amsterdam: Elsevier Scientific.

Davidson, B.R. and Graham-Taylor, S. 1982. *Lessons from the Ord*. St Leonards, NSW: Centre for Independent Studies. (CIS policy monograph; 2).

Davies, H.L. 1958. Milk yield of Australian merino ewes and lamb growth under pastoral conditions. *Proceedings of the Second Annual Conference of the Australian Society of Animal Production*: 15-21.

Davies, H.L. and Humphries, A.W. 1965. Stocking rate and wool production at Kojonup: 1. Wether sheep. *Journal of Agriculture, Western Australia*. 4th series. 6(7): 409-413.

Davis, B. 1980. The struggle for south-west Tasmania. In. Scott, R., ed. *Interest groups and public policy: case studies from the Australian States*. Melbourne: Macmillan: 152-169.

Davis, B.W. 1981. Characteristics and influence of the Australian conservation movement: an examination of selected conservation controversies. Hobart: University of Tasmania. Thesis. Ph.D.

Davis, K. 1976. *Land use*. New York: McGraw Hill.

De Boer, A.J. and Gaffney, J. 1976. A note on a mandatory land-use program: soil conservation on the Darling Downs. *Australian Journal of Agricultural Economics*. 20(1): 37-43.

Demsetz, H. 1967. Toward a theory of property rights. *American Economic Review*. 57(2): 347-359.

Demsetz, H. 1969. Information and efficiency: another viewpoint. *Journal of Law and Economics*. 12(1): 1-22.

Dillon, M.C. 1985. *Pastoral resource use in the Kimberley: a critical overview*. Canberra: CRES, ANU. (East Kimberley working paper; 1985/4).

Doe, B.R. 1983. The past is the key to the future. *Geochimica et Cosmochimica Acta*. 47: 1341-1354.

Donald, C.M. 1970. Innovation in Australian agriculture. In. Williams, D.B., ed. *Agriculture in the Australian economy*. Sydney: Sydney University Press: 57-86.

Donald, C.M. 1982. Innovation in Australian agriculture. In. Williams, D.B., ed. *Agriculture in the Australian economy*. 2nd ed. Sydney: Sydney University Press: 55-82.

Douglass, G.K., ed. 1984. *Agricultural sustainability in a changing world order*. Boulder, Colorado: Westview Press.

Downes, R.G. 1971. Land, land use and soil conservation. In. Costin, A.B. and Frith, H.J., eds. *Conservation*. Harmondsworth, England: Penguin Books: 43-70.

Downs, A. 1967. *Inside bureaucracy*. Boston: Little Brown.

Dregne, H. 1977. Studies of desertification in the hot arid regions. In. *UN Conference on Desertification, 29 August-9 September, Nairobi, Kenya*: 4-6. (A/Conf 74/31).

Dregne, H. 1983. *Desertification of arid*

lands. New York: Harwood Academic.

Dumsday, R.G. 1971. Evaluation of soil conservation policies by systems analysis. In. Dent, J.B. and Anderson, J.R., eds. *Systems analysis in agricultural management.* Sydney: John Wiley and Sons: 152-172.

Dumsday, R.G. 1973. *The economics of some soil conservation practices in the wheat belt of northern New South Wales and southern Queensland: a modelling approach.* Armidale: University of New England. (Farm management bulletin; no.19).

Dumsday, R.G. and Edwards, G.W. 1984. Economics of conservation in agriculture. In. Gilmour, G., Hamer, I. and Bourchier, J., eds. *Agriculture and conservation achieving a balance: proceedings of a conference, 1984 September 10-11, Clyde Cameron College, Wodonga, Victoria.* Benalla, Victoria: Australian Institute of Agricultural Science, (Northern Victorian Sub-branch): 58-64. (AIAS occasional paper; no.15).

Dumsday, R.G. and Seitz, W.D. 1985. A model for quantifying incentive payments for soil conservation in cropping regions subject to water erosion. In. El-Swaify, S.A., Moldenhauer, W.C. and Lo, A., eds. *Soil erosion and conservation.* Ankeny, Iowa: Soil Conservation Society of America. Ch.27: 296-306.

Dumsday, R.G., Edwards, G.W., Lumley, S.E., Oram, D.A. and Papst, W.A. 1985. *Economic aspects of dryland salinity control in the Murray River basin of south-eastern Australia.* Canberra: Department of Resources and Energy. (AWRC Research Project 80/137. Final report).

Dumsday, R.G., Oram, D.A. and Lumley, S.E. 1983. Economic aspects of the control of dryland salinity. *Proceedings of the Royal Society of Victoria.* 95(3): 139-145.

Dumsday, R.G., Walls, K.G., Arch, A.M.J. and Sutton, N.G. 1984. *Computation of indices for rainfall erosivity: RAINEI.* Bundoora, Victoria: School of Agriculture, La Trobe University. (Occasional paper; no.7).

Earle,T.R., Brownlea, A.A. and Rose, C.W. 1981. Beliefs of a community with respect to environmental management: a case study of soil conservation beliefs on the Darling Downs. *Journal of Environmental Management.* 12(2): 197-219.

Easdown, G. and King, M. 1985. The Murray under threat. *Melbourne Herald.* July 16-20.

Edwards, G.W. and Lumley, S. 1985. Dryland salting: conceptual characterisation and policy preamble: paper presented at the 22nd Annual Conference of the Australian Agricultural Economics Society, University of New England, Armidale.

Edwards, K. 1980. Run off and soil loss in the wheat belt of New South Wales. In. *Preprints of papers: Agricultural Engineering Conference, 1980 September 30 - October 2, Geelong.* Barton, ACT: Institution of Engineers, Australia: 94-98.

Edwards, K. and Charman, P.E.V. 1980. The future of soil loss prediction in Australia. *Journal of the Soil Conservation Service of New South Wales.* 36(4): 211-218.

El-Swaify, S.A., Dangler, E.W. and Armstrong, C.L. 1982. *Soil erosion by water in the tropics.* Honolulu, Hawaii: College of Tropical Agriculture and Human Resources, University of Hawaii. (Research Extension series 024).

El-Swaify, S.A., Moldenhauer, W.C. and Lo, A., eds. 1985. *Soil erosion and conservation.* Ankeny, Iowa: Soil Conservation Society of America.

Enright, N.F. 1983. Quarterly research report, January-March. Inverell Research Centre, Soil Conservation Service of NSW. Unpublished report.

Ervin, C.A. and Ervin, D.E. 1982. Factors affecting the use of soil conservation practices: hypotheses, evidence, and policy implications. *Land Economics.* 58(3): 277-292.

Fels, H.E. and Quinlivan, B.J. 1978. *Value for money in rural research*. Western Australia. Department of Agriculture. Miscellaneous publication.

Felveson, H., Sinden, F. and Socolow, R., eds. 1976. *Boundaries of analysis: an inquiry into the Tocks Island Dam controversy*. Cambridge, Mass: Ballinger.

Fisher, A.C. 1981. *Resource and environmental economics*. Cambridge: Cambridge University Press.

Fisher, A.C. and Krutilla, J.V. 1975. Resource conservation, environmental preservation and the rate of discount. *Quarterly Journal of Economics*. 89(3): 358-370.

Fitzpatrick, E.A. 1982. Recent changes in extent of mallee. In. Mabbutt, J.A., ed. *Threats to mallee in New South Wales*. Sydney: Department of Planning and Environment.

Fitzpatrick, E.N. 1974. Why new farmers need a super bounty. *Journal of Agriculture, Western Australia*. 15(4): 95-100.

Fliegel, F.C. and van Es, J.C. 1983. The diffusion-adoption process in agriculture: changes in technology and changing paradigms. In. Summers, G.F., ed. *Technology and social change in rural areas*. Boulder, Colorado: Westview Press: 13-28.

Food and Agriculture Organization of the United Nations (FAO)a. 1980. *Production yearbook*. Rome: FAO.

Food and Agriculture Organization of the United Nations (FAO)b. 1980. *Trade yearbook*. Rome: FAO.

Food and Agriculture Organization of the United Nations (FAO). 1984. *Provisional methodology for the assessment and mapping of desertification*. Rome: FAO/UNESCO.

Formby, J. 1977. *The environmental impact statement: problems and alternatives*. Canberra: CRES, ANU. (CRES general paper JRF/GP1).

Forster, D.L. and Abrahim, G. 1985. Sediment deposits in drainage ditches: a cropland externality. *Journal of Soil and Water Conservation*. 40(1): 141-144.

Foster, G.R. and Hakonson, T.E. 1984. Predicted erosion and sediment delivery of fallout plutonium. *Journal of Environmental Quality*. 13: 595-602.

Foster, G.R. and Lane, L.J. 1982. Estimating sediment yield from rangeland with CREAMS. In. *Proceedings Workshop on Estimating Erosion and Sediment Yield on Rangelands, 1981 March, Tucson, Arizona*. US. Department of Agriculture: 115-119. (ARM-W-26).

Foster, G.R. and Meyer, L.D. 1975. Mathematical simulation of upland erosion by fundamental erosion mechanics. In. *Present and prospective technology for predicting sediment yields and sources*. US. Department of Agriculture: 190-207. (ARS-S-40).

Foster, I.D.L., Dearing, J.A., Simpson, A., Carter, A.D., and Appleby, P.G. 1985. Lake catchment based studies of erosion and denudation in the Merevale Catchment, Warwickshire, UK. *Earth Surface Processes and Landforms*. 10: 45-68.

Freebairn, J.W. 1983. Drought assistance policy. *Australian Journal of Agricultural Economics*. 27(3): 185-199.

Freeman, A.M. 1979. *The benefits of environmental improvement: theory and practice*. Baltimore: Johns Hopkins University Press for Resources for the Future.

Gasteen, J.D.H., Page, R. and Davis, S., eds. 1985. *Agriculture and conservation in inland Queensland*. Brisbane: Wildlife Preservation Society of Queensland.

Gifford, R.M., Kalma, J.D., Aston, A.R. and Millington, R.J. 1975. Biophysical constraints in Australian food production: implications for population policy. *Search*. 6: 212-223.

Gilmour, D.A. 1977. Effect of rainforest logging and clearing on water yield and quality in a high rainfall zone of north-east Queensland. *Hydrology Symposium, 1977 June 28-30, Brisbane, Queensland*. Institution of Engineers,

Australia: 155-160. (National Conference publication; no.77/5).

Gilpin, A. 1980a. *Environment policy in Australia.* St Lucia: Queensland University Press.

Gilpin, A. 1980b. *The Australian environment: twelve controversial issues.* Melbourne: Sun Books.

Goddard, B., Humphrey, M. and Carter, D. 1982. *Wind erosion in the Jerramungup area 1980-81.* Soil Conservation and Services Branch, Division of Resource Management, Department of Agriculture, Western Australia. (Technical report; 3).

Graetz, R.D., Peck, R.P., Gentle, M.R. and O'Callaghan, J.F. 1986. The application of Landsat image data to rangeland assessment and monitoring: the development and demonstration of a land image-based resource information system (LIBRIS). *Journal of Arid Environments.* 10: 53-80.

Graf, W.H. 1971. *Hydraulics of sediment transport.* New York: McGraw Hill.

Graham, T.W.G., Webb, A.A., and Waring, S.A. 1981. Soil nitrogen status and pasture productivity after clearing of brigalow (Acacia harpophylla). *Australian Journal of Experimental Agriculture and Animal Husbandry.* 21: 109-118.

Gray, D.H. and Leiser, A.T. 1982. *Biotechnical slope protection and erosion control.* New York: Van Nostrand Reinhold.

Greenland, D.J. 1971. Changes in the nitrogen status and physical condition of soils under pastures, with special reference to the maintenance of the fertility of Australian soils used for growing wheat. *Soils and Fertilizers.* 34: 237-251.

Greig, P.J. 1983a. Recreation evaluation using a characteristics theory of consumer behaviour. *American Journal of Agricultural Economics.* 65(1): 90-97.

Greig, P.J. 1983b. Contributions of economics to forestry planning. *Australian Forestry.* 47(1): 16-27.

Greig, P.J. and Devonshire, P.G. 1981.

Tree removals and saline seepage in Victorian catchments: some hydrologic and economic results. *Australian Journal of Agricultural Economics.* 25(2): 134-148.

Hadley, R.F. and Schumm, S.A. 1961. *Hydrology of the upper Cheyenne River basin.* (US Geological Survey. Water supply paper 1531-8).

Hadley, R.F. and Shown, L.M. 1976. Relation of erosion to sediment yield. In. *Proceedings 3rd Federal Inter-Agency Sedimentation Conference, 1976 March 22-25, Denver, Colorado.* Washington, DC: Sedimentation Committee, Water Resources Council: 132-139.

Hagan, J. 1972. The politics of pollution. *Politics.* 7: 136-148.

Hakanson, L. and Jansson, M. 1983. *Principles of lake sedimentology.* Berlin: Springer-Verlag.

Halcrow, H.G., Heady, E.O. and Cotner, M.L., eds. 1982. *Soil conservation policies, institutions and incentives.* Ankeny, Iowa: Soil Conservation Society of America.

Ham, C. and Hill, M.J. 1984. *The policy process in the modern capitalist state.* Brighton, Sussex: Wheatsheaf Books.

Hamblin, A.P. 1980. Changes in aggregate stability and associated organic matter properties after direct drilling and ploughing on some Australian soils. *Australian Journal of Soil Research.* 18: 27-36.

Hamblin, A.P., Tennant, D. and Cochrane, H. 1982. Tillage and the growth of a wheat crop in a loamy sand. *Australian Journal of Agricultural Research.* 33: 887-897.

Hamilton, G.J. 1970. The effect of sheet erosion on wheat yield and quality. *Journal of the Soil Conservation Service of New South Wales.* 26(2): 118-123.

Harberger, A.G. 1971. Three basic postulates for applied welfare economics: an interpretive essay. *Journal of Economic Literature.* 9: 785-797.

Harrington, G.N., Friedel, M.H.,

Hodgkinson, K.C. and Noble, J.C. 1984. Vegetation ecology and management. In. Harrington, G.N., Wilson, A.D. and Young, M.D., eds. *Management of Australia's rangelands*. Melbourne: CSIRO: 41-62.

Harrington, G.N., Oxley, R.F. and Tongway, D.J. 1979. The effect of European settlement and domestic livestock on the biological system in poplar box (Eucalyptus populnea). *Australian Rangeland Journal*. 1: 271-279.

Harrington, G.N., Wilson, A.D. and Young, M.D. 1984. Management of rangeland ecosystems. In. Harrington, G.N., Wilson, A.D. and Young, M.D., eds. *Management of Australia's rangelands*. Melbourne: CSIRO: 3-14.

Harris, S.F., Crawford, J.G., Groen, F.H. and Honan, N.D. 1974. *The principles of rural policy in Australia: a discussion paper*. Canberra: AGPS. (Rural green paper).

Hawkins, K. 1984. *Environment and enforcement*. Oxford: Clarendon.

Haynes, J.E. and Sutton, M. 1985. *Taxation measures and soil conservation*. Canberra: AGPS. (BAE occasional paper; no.93).

Heady, E.O. 1975. The basic equity problem. In. Heady, E.O. and Whiting, L.R., eds. *Externalities in the transformation of agriculture*. Ames, Iowa: Iowa State University Press: 3-22.

Heathcote, R.L. 1965. *Back of Bourke: a study of land appraisal and settlement in semi-arid Australia*. Melbourne: Melbourne University Press.

Heffernan, W.D. and Green, G. 1981. Soil conservation attitudes and practices, paper presented at a meeting of the Rural Sociological Society, San Francisco.

Henderson, C.W. 1985. Compaction of sandy soils and its effects on crop yields. In. Proceedings 4th Australian Soil Conservation Conference, 1985, Queensland.

Henry, C. 1974. Option values in the economics of irreplaceable assets.

Review of Economic Studies. Supplement: 89-104. Symposium on the Economics of Exhaustible Resources.

Hewitt, K. 1983. The idea of calamity in a technocratic age. In. Hewitt, K., ed. *Interpretations of calamity*. London: Allen and Unwin: 3-22.

Hingston, F.J., Dimmock, G.M. and Turton, A.G. 1981. Nutrient distribution in a jarrah (*Eucalyptus marginata* Donn ex Sm.) ecosystem in south-west Western Australia. *Forest Ecology and Management*. 3: 183-207.

Hodge, I. 1982. Rights to cleared land and the control of dryland seepage salinity. *Australian Journal of Agricultural Economics*. 26: 185-201.

Hodgkinson, K.C., Harrington, G.N., Griffin, G.F., Noble, J.C. and Young, M.D. 1984. Management of vegetation with fire. In. Harrington, G.N., Wilson, A.D. and Young, M.D., eds. *Management of Australia's rangelands*. Melbourne: CSIRO: 141-156.

Hodgson, C.P. 1846. *Reminiscences of Australia with hints on the squatter's life*. London: W.N. Wright.

Holling, C.S., ed. 1978. *Adaptive environmental assessment and management*. Chichester: John Wiley and Sons.

Holmes, A.N. 1978. Studies of forms of nitrogen in streams near Adelaide, South Australia. Adelaide: University of Adelaide. Thesis. B. Sc.

Hood, C. 1976. *The limits of administration*. London: John Wiley and Sons.

Hoover, H. and Wiitala, M. 1980. *Operator and landlord participation in soil erosion control in the Maple Creek Watershed in northeast Nebraska*. Washington DC: US Department of Agriculture. Economics, Statistics, and Cooperative Service. (Staff report NRED 80-4).

Hore, H. and Sims, H.J. 1954. Loss of top soil: effect of yield and quality of wheat. *Journal of Agriculture, Victoria*. 56(6): 241-250.

Hudson, N.W. 1982. Non-technical

constraints on soil conservation. In. Tingsanchali, I. and Eggers, H., eds. *Proceedings of the Southeast Asian Symposium on Problems of Soil Erosion and Sedimentation, 1981, Bangkok, Thailand*. Bangkok, Thailand: Asian Institute of Technology.

Hughes, K.K. 1981. Assessment of dryland salinity in Queensland. In. *Proceedings Salinity Water Quality Symposium, 1980, Toowoomba, Australia*: 15-30.

Hunt, J.S. 1980. Structural stability of mallee soils under cultivation. *Journal of the Soil Conservation Service of New South Wales.* 36: 16-22.

International Institute of Agriculture. 1931. *International yearbook of agricultural statistics, 1930-31.* Rome.

Jackson, E.M. 1982. *Replenish the earth.* Albury: Catchment Education Trust.

Jarvis, R.J. 1985. *Direct drilling: 8 years results from 6 soil types.* Department of Agriculture, Western Australia. Internal publication.

Jarvis, R.J., Hamblin, A.P. and Delroy, N.D. 1985. *Continuous cereal cropping with alternative tillage systems in Western Australia.* Department of Agriculture, Western Australia. (Technical bulletin; no.71). In press.

Jasper, R. and Richards, R. 1983. Projected land release in Western Australia. In. *Proceedings National Arid Lands Conference, 1982 May 21-25, Broken Hill, NSW.* Hawthorn, Victoria: Australian Conservation Foundation: 99-101.

Jenkin, J.J. 1981. Terrain, groundwater and secondary salinity in Victoria, Australia. In. *Agricultural water management 4: proceedings Land, Stream Salinity Seminar, WA.* Amsterdam: Elsevier: 143-171.

Jenkins, A. 1978. *Policy analysis.* London: Martin Robertson.

Jones, C. and Matthes, D. 1983. Policy formation. In. Nagel, S., ed. *Encyclopedia of policy studies.* New York: Marcel Dekker: 117-142.

Jones, H.R. 1961. Runoff and soil loss studies at Wagga Research Station. *Journal of the Soil Conservation Service of New South Wales.* 17(3): 156-169.

Junor, R.S., Marston, D. and Donaldson, S.G. 1979. *A situation statement of soil erosion in the lower Namoi area.* Soil Conservation Service of NSW.

Kaleski, L.G. 1945. The erosion survey of NSW. *Journal of the Soil Conservation Service of New South Wales.* 1: 12-20.

Kellow, A.J. 1981. Policy science and political theory. *Politics.* 16: 33-45.

Kellow, A.J. 1985. Managing an ecological system: the politics of administration. *Australian Quarterly.* 57: 107-127.

Kelly, B. 1985. Time to turn cheap water off at the tap. *Australian.* July 29.

Khanbilvardi, R.M. and Rogowski, A.S. 1984. Mathematical model of erosion and deposition on a watershed. *Transactions of the American Society of Agricultural Engineers.*

King, C.J. 1957. An outline of closer settlement in New South Wales. *Review of Marketing and Agricultural Economics.* 25(3/4): 1-290.

Kirkby, M.J. 1980. The problem. In. Kirkby, M.J. and Morgan, R.P.C., eds. *Soil erosion.* Chichester: John Wiley & Sons: 1-16.

Klingebiel, A.A. and Montgomery, P.H. 1961. *Land capability classification.* Washington, DC: Soil Conservation Service, US Department of Agriculture. (USDA. Agriculture handbook no. 210).

Kneese, A.V. and Schultze, W.D. 1985. Ethics and environmental economics. In. Kneese, A.V. and Sweeney, J.L., eds. *Handbook of natural resource and energy economics.* Amsterdam: North Holland. v.1.

Kneese, A.V., Rolfe, S.E. and Harned, J.W., eds. 1971. *Managing the environment: international economic co-operation for pollution control.* New York: Praeger: 255-274.

Knisel, W.G., ed. 1980; *CREAMS: a field-scale model for chemicals, runoff, and erosion from agricultural management systems.* Washington, DC: US.

Department of Agriculture. (Conservation research report no.26).

Koide, M., Bertine, K.K., Chow, T.J. and Goldberg, E.D. 1985. The 240$_{Pb}$/239$_{Pu}$ ratio, a potential geochronometer. *Earth and Planetary Science Letters*. 72: 1-8.

Kovda, V.A. 1977. Soil loss: an overview. *Agroecosystems*. 3: 205-224.

Krutilla, J.V. 1967. Conservation reconsidered. *American Economic Review*. 57(4): 777-786.

Lang, M.G. 1980. Economic efficiency and policy comparisons. *American Journal of Agricultural Economics*. 62(4): 772-777.

Langford, K.J. and O'Shaughnessy, P.J. 1977. Some effects of forest change on water values. *Australian Forestry*. 40(3): 192-218.

Laut, P., Firth, D. and Paine, T.A. 1980. *Provisional environmental regions of Australia*. Melbourne: CSIRO. 2v.

Leeper, G.W. 1970. Soils. In. Leeper, G.W., ed. *The Australian environment*. 4th ed. Melbourne: Melbourne University Press.

Lemon, J. 1983. Reclaiming gully erosion. *Journal of Agriculture, Western Australia*. 24(2): 61.

Liddell Hart, B.H. 1970. *History of the First World War*. London: Pan Books.

Lightfoot, L.C. 1961. Soil erosion on the Gascoyne River catchment: report of the Gascoyne River Erosion Committee. Department of Agriculture, Western Australia. Mimeo.

Lind, R.C. and others, eds. 1982. *Discounting for time and risk in energy policy*. Baltimore: Johns Hopkins University Press for Resources for the Future.

Lindsay, C. 1976. A theory of government enterprise. *Journal of Political Economy*. 84(5): 1061-1077.

Lloyd, A.G. 1970. The Little Desert project. *Victoria's Resources*. 12(1): 6-9.

Loch, J. 1984. The role of rainfall simulation in erosion research. In. *Soil erosion research techniques: workshop held in Toowoomba, 1983 April 12-14*. Brisbane: Department of Primary Industries: 23-25. (Conference and Workshop series).

Loch, J. and Donnollan, T.E. 1983. Field rainfall studies on two clay soils of the Darling Downs, Queensland: 1. The effect of plot length and tillage orientation on erosion processes and runoff and erosion rates. *Australian Journal of Soil Research*. 21: 33-46.

Logan, J.M. 1960. Runoff and soil loss studies at Wellington Research Centre: II. Soil loss. *Journal of the Soil Conservation Service of New South Wales*. 16(4): 214-227.

Longmore, M.E. 1985. Caesium-137 and soil redistribution in Australian drainage basins. In. Loughran, R.J., ed. *Drainage basin erosion and sedimentation*. Newcastle, NSW: University of Newcastle. v.2.: 59–68.

Longworth, J. and Rudd, D. 1975. Plant pesticide economics with special reference to cotton insecticides. *Australian Journal of Agricultural Economics*. 19: 210-227.

Lourensz, R.S. and Abe, K.E. 1983. A dust storm over Melbourne. *Weather*. 38: 272-275.

Lovett, J.V. 1973. *The environmental, economic and social significance of drought*. Sydney: Angus & Robertson.

Lucy, R., ed. 1983. *The pieces of politics*. 3rd ed. Melbourne: Macmillan.

Lumley, S.E. 1983. The economic implications of dryland salting and associated stream salinity in north-central Victoria. Bundoora, Victoria: La Trobe University. Thesis. M. Agric. Sci.

Mabbutt, J.A. 1978. *Desertification in Australia*. Kingsford, NSW: Water Research Foundation. (Water Research Foundation. Report no. 54).

Mabbutt, J.A., ed. 1982. *Threats to mallee in New South Wales*. Sydney: NSW. Department of Environment and Planning.

McCallum, A. and Blyth, M. 1985. The measurement of salinity degradation in

Australia for government policy formulation: BAE paper presented at the 29th Annual Conference of the Australian Agricultural Economics Society, February 12-14, Armidale.

McConnell, K.E. 1983. An economic model of soil conservation. *American Journal of Agricultural Economics.* 65(1): 83-89.

McKean, R. 1972. Property rights within government, and devices to increase governmental efficiency. *Southern Economic Journal.* 39(4): 177-186.

McRae, S.G. and Burnham, C.P. 1981. *Land evaluation.* Oxford: Clarendon.

Mant, J., ed. 1985. Managing an ecological system: the Murray/Darling River. *Australian Quarterly.* 57: 105-164.

Marglin, S.A. 1963. The social rate of discount and the optimal rate of investment. *Quarterly Journal of Economics.* 77(1): 95-111.

Marsden, J.S., Martin, G.E., Parham, D.J., Ridsdill Smith, T.J. and Johnston, B.G. 1980. *Returns on Australian agricultural research.* Melbourne: CSIRO.

Marsh, B. a'B. 1979. *Magnitude of wind erosion and its economic effect.* Soil Conservation Service Branch, Department of Agriculture, Western Australia.

Marsh, B. a'B. 1983. Wind erosion in relation to tillage systems. In. *Proceedings of a National Workshop on Tillage Systems for Crop Production, Roseworthy College, South Australia.*

Marshall, C.J. 1977. Soil erosion hazards in private forest operations: Bombala. *Journal of the Soil Conservation Service of NSW.* 33(4): 244-253.

Mason, I.B., and Fischer, R.A. 1986. Tillage practices and the growth and yield of wheat in southern New South Wales: Lockhart, a 430mm rainfall site. *Australian Journal of Experimental Agriculture.*

Mathews, R.L., ed. 1985. *Federalism and the environment.* Canberra: Centre for Research on Federal Financial Relations, Australian National University.

Meade, R.H. 1982. Sources, sinks and storage of river sediment in the Atlantic drainage of the United States. *Journal of Geology.* 90: 235-252.

Meadows, D.H., Meadows, D.L. and Anders, J. 1972. *The limits to growth.* London: Earth Island.

Meinig, D.W. 1962. *On the margins of the good earth: the South Australian wheat frontier, 1869-1884.* Adelaide: Rigby.

Melbourne and Metropolitan Board of Works. 1980. *Water supply catchment hydrology research: summary of technical conclusions to 1979.* Melbourne: The Board.

Messer, J. and Mosley, G., eds. 1983. *What future for Australia's arid lands? Proceedings of the National Arid Lands Conference, 1982 May 21-25, Broken Hill, New South Wales.* Hawthorn, Victoria: Australian Conservation Foundation.

Meyer, L.D. 1984. Evaluation of the universal soil loss equation. *Journal of Soil and Water Conservation.* 39(2): 99-104.

Michael Read and Associates. 1984. *Financing salinity control in Victoria.* A discussion paper prepared for the Salinity Committee of the Victorian Parliament.

Miller, A. 1985. Technological thinking: its impact on environmental management. *Environmental Management.* 9(3): 179-190.

Milton, L.E. 1971. *A review of gully erosion and its control.* Soil Conservation Authority, Victoria.

Miranowski, J.A. and Hammes, B.D. 1984. Implicit prices of soil characteristics for farmland in Iowa. *American Journal of Agricultural Economics.* 66(5): 744-749.

Mishan, E.J. 1982. *Cost-benefit analysis.* 3rd ed. London: George Allen & Unwin.

Molnar, I. 1955. Effects of soil erosion on land values and production. *Journal of the Australian Institute of Agricultural Science.* 21(3): 163-166.

Molnar, I. 1964. Soil conservation: economic and social considerations. *Journal of the Australian Institute of*

Agricultural Science. 30(4): 247-257.

Morton, R., and Cunningham, R.B. 1985. Longitudinal profile of trends in salinity in the River Murray. *Australian Journal of Soil Research.* 23(1): 1-13.

Moss, A.J., Walker, P.H. and Hutka, J. 1979. Raindrop-stimulated transportation in shallow water flows: an experimental study. *Sedimentary Geology.* 20: 81-139.

Mott, J.J. and Tothill, J.C. 1984. Tropical and subtropical woodlands. In. Harrington, G.N., Wilson, A.D. and Young, M.D., eds. *Management of Australia's rangelands.* Melbourne: CSIRO

Mulcahy, M.J. 1978. Salinisation in the southwest of Western Australia. *Search.* 9: 269-272.

Mullins, J.A. 1981. Untitled report. Brisbane: Department of Primary Industries, Queensland. (Division of Land Utilisation report 1981-2).

Munro, C.H. 1974. *Australian water resources and their development.* Sydney: Angus and Robertson.

Musgrave, W. 1984a. Creative marketing and management. Keynote address to the Annual Conference of the NSW Soil Conservation Service, July 11.

Musgrave, W. 1984b. The economics of sustainable development: paper to Conservation and the Economy Conference, 1984 September, Sydney.

Musgrave, W. 1984c. Supplementary submission to the Senate Committee of Science and the Environment Inquiry into Land Use Policy.

Nash, R. 1972. American environmental history: a new teaching frontier. *Pacific Historical Review.* 41: 363-372.

Neher, P.A. 1976. Democratic exploitation of a replenishable resource. *Journal of Public Economics.* 5: 361-371.

Nelson, P. 1985. The role of lupins in integrating systems of crop and livestock production in northern agricultural areas of Western Australia. In. Proceedings of a Conference: Thinking Rotations, Kondinin Farm Improvement Group - Hyden.

Nethercote, J., ed. 1982. *Parliament and bureaucracy.* Sydney: Hale and Iremonger.

New South Wales. Parliament. Joint Select Committee to Enquire into the Western Division of NSW. 1983. *First report of the Joint Select Committee of the Legislative Council and Legislative Assembly to enquire into the Western Division of New South Wales.* Sydney: Government Printer.

New South Wales. Parliament. Joint Select Committee to Enquire into the Western Division of NSW. 1984a. *Second report of the Joint Select Committee of the Legislative Council and Legislative Assembly to enquire into the Western Division of New South Wales.* Sydney: Government Printer.

New South Wales. Parliament. Joint Select Committee to Enquire into the Western Division of NSW. 1984b. *Third report of the Joint Select Committee of the Legislative Council and Legislative Assembly to enquire into the Western Division of New South Wales.* Sydney: Government Printer.

New South Wales. Parliament. Joint Select Committee to Enquire into the Western Divison of NSW. 1984c. *Fourth report of the Joint Select Committee of the Legislative Council and Legislative Assembly to enquire into the Western Division of New South Wales.* Sydney: Government Printer.

New South Wales. Soil Conservation Service (n.d.). *Community costs of soil erosion.*

Newbery, D.M.G. and Stiglitz, J.E. 1981. *The theory of commodity price stabilisation: a study of the economics of risk.* Oxford: Clarendon Press.

Newman, J.C. and Condon, R.W. 1969. Land use and present condition. In. Slatyer, R.O. and Perry, R.A., eds. *Arid lands of Australia.* Canberra: ANU Press.

Newsome, A.E. 1975. An ecological comparison of two arid zone kangaroos of Australia, and their anomalous prosperity since the introduction of ruminant stock to their environment.

Quarterly Review of Biology. 50: 389-424.

Noble, J.C. and Tongway, D.J. 1986. Landscape degradation following pastoral settlement in Australian arid and semi arid rangelands. In. Russell, J.S. and Isbell, R.F., eds. *Impact of human activities on Australian soils.* Brisbane: Queensland University Press for Australian Society of Soil Science.

Northcote, K.H. and Skene, J.K.M. 1972. *Australian soils with saline and sodic properties.* CSIRO. (Soil publication no. 27).

Northcote, K.H., Hubble, G.D., Isbell, R.F., Thompson, C.H. and Bettenay, E. 1975. *A description of Australian soils.* Melbourne: CSIRO.

Northern Territory. Committee to Inquire into the Most Appropriate Form of Tenure for Pastoral Land. 1980. *Pastoral land tenure in the Northern Territory: a report to the Minister for Lands and Housing by a Committee appointed to inquire into the most appropriate form of tenure for pastoral land in the Northern Territory.* Darwin: Department of Lands.

Nowak, P.J. 1982. *Phase one final report of the selling of soil conservation: a test of the voluntary approach.* Ames, Iowa: Iowa State University, Department of Sociology and Anthropology.

Nowak, P.J. 1983. Adoption and diffusion of soil and water conservation practices. *The Rural Sociologist.* 3(2): 83-92.

Nulsen, R.A. and Baxter, I.N. 1982. The potential of agronomic manipulation for controlling salinity in Western Australia. *Journal of the Australian Institute of Agricultural Science.* 48: 222-226.

Oates, N.M. et al, eds. 1981. *Focus on farm trees: the decline of trees in the rural landscape: proceedings of a National Conference, 1980 November 23-26, University of Melbourne.* Box Hill, Victoria.

Old, K.M., Kile, G.A., and Ohmart, C.P. 1981. *Eucalypt dieback in forests and woodlands.* Melbourne: CSIRO

Oldfield, F. 1977. Lakes and their drainage basins as units of sediment-based ecological study. *Progress in Physical Geography.* 1: 460-504.

Olive, L.J. 1983. Impact of European-style land use: paper presented to a Workshop on Soil Management under Drought and Flood, Australian Society of Soil Science (ACT Branch) and the Australian Institute of Agricultural Science, March, Canberra.

Olive, L.J. and Walker, P.H. 1982. Processes in overland flow: erosion and production of suspended material. In. O'Loughlin, E.M. and Cullen, P., eds. *Prediction in water quality: proceedings of a Symposium on the Prediction in Water Quality sponsored by the Australian Academy of Science and the Institution of Engineers, Australia, 1982 November 30-December 2, Canberra.* Canberra: Australian Academy of Science.

O'Loughlin, E.M. 1986. Prediction of surface saturation zones in natural catchments by topographic analysis. *Water Resources Research.* 22(5): 794-804.

Olsen, G. 1975. Option value. *Australian Journal of Agricultural Economics.* 19: 197-209.

O'Riordan, T. 1976. *Environmentalism.* London: Pion Books.

O'Riordan, T. and Sewell, W.R.D., eds. 1981. *Project appraisal and policy review.* Chichester: John Wiley and Sons.

Osborne, R.C. and Rose, C.W. 1981. Retrospective and prospective: degradation of Australian land resources in response to world food needs. School of Australian Environmental Studies, Griffith University. Unpublished manuscript.

Packer, I.J. 1981. Soil Conservation Service of New South Wales. Unpublished report.

Packer, I.J., Hamilton, G.J. and White, I. 1984. Tillage practices to conserve soil and improve soil conditions. *Journal of* the Soil Conservation Service of NSW. 40: 78–87.

Page, T. 1977. *Conservation and economic efficiency: an approach to materials policy*. Baltimore: Johns Hopkins University Press.

Pampel, F.C. and van Es, J.C. 1977. Environmental quality and issues of adoption research. *Rural Sociology*. 42(2): 57-71.

Parish, R.M. 1976. The scope of benefit-cost analysis. *The Economic Record*. 52: 302-314.

Parry, M.L. 1985. Estimating the sensitivity of natural ecosystems and agriculture to climatic change. *Climatic Change*. 7: 1-3.

Passmore, J. 1980. *Man's responsibility for nature: ecological problems and western traditions*. 2nd ed. London: Duckworth.

Passmore, J. 1981. *The limits of government*. Sydney: Australian Broadcasting Commission. (Boyer lecture).

Paterson, J.P. 1985. Coordination in government: decomposition and bounded rationality as a framework for user friendly statute law. Melbourne: Department of Water Resources, Victoria. Mimeograph.

Pearce, D.W. 1976. *Environmental economics*. London: Longman.

Pearce, D.W. 1982. *Ethics, irreversibility, future generations and the social rate of discount*. Seminar presented at School of Economics, La Trobe University, September.

Pearce, D.W. and Nash, C.A. 1981. *The social appraisal of projects*. London: Macmillan.

Pearce, H., and Bradby, K. 1983. The threat to the environment of the W.A. land release programme. In. *Proceedings National Arid Lands Conference, 1982 May 21-25, Broken Hill, NSW*. Hawthorn, Victoria: Australian Conservation Foundation: 102-105.

Pearse, R.A. and Cowie, A.J. 1985. On-farm analysis of soil conservation options: paper presented at the 29th Annual Conference of the Australian Agricultural Economics Society, February 12-14, Armidale.

Peck, A.J. 1978. Salinization of non-irrigated soils and associated streams: a review. *Australian Journal of Soil Research*. 16: 157-168.

Peck, A.J., Thomas, J.F. and Williamson, D.R. 1983. *Salinity issues: effects of man on salinity in Australia*. Canberra: AGPS. (Water 2000. Consultant's report; no. 8).

Pennock, J.R. and Chapman, J.W., eds. 1982. *Ethics, economics and the law*. New York: New York University Press. (Nomos; 24).

Perry, R.A. 1968. Australia's arid rangelands. *Annals of Arid Zone*. 7(2): 243-249.

Peterson, F.M. and Fisher, A.C. 1977. The exploitation of extractive resources: a survey. *Economic Journal*. 87(4): 681-721.

Philp, D.M. 1982. Waste disposal to Australian inland rivers. In. O'Loughlin, E.M. and Cullen, P. eds. *Prediction in water quality: proceedings of a Symposium on the Prediction in Water Quality sponsored by the Australian Academy of Science and the Institution of Engineers, Australia, 1982 November 30 - December 2, Canberra*. Canberra: Australian Academy of Science: 417-435.

Pickup, G. In Press. The erosion cell: a geomorphic approach to landscape classification in range assessment. *Ecological Modelling*.

Pigou, A.C. 1932. *The economics of welfare*. 4th ed. London: Macmillan.

Pigou, A.C. 1947. *A study in public finance*. 3rd ed. London: Macmillan.

Pittock, A.B. 1975. Climatic change and the patterns of variation in Australian rainfall. *Search*. 6(11/12): 498-504.

Porter, A. 1985. *Environmental impact assessment*. St Lucia: Queensland University Press.

Posner, R.A. 1972. *Economic analysis of law*. Boston: Little Brown.

Posner, R.A. 1974. Theories of economic

regulation. *Bell Journal of Economics.* 5(2): 335-358.

Powell, J.M. 1970. *The public lands of Australia Felix.* Melbourne: Oxford University Press.

Powell, J.M. 1976. *Environmental management in Australia.* Oxford: Oxford University Press.

Powell, R.A. 1974. Technological change in Australian agriculture 1920-21 to 1969-70. Armidale: University of New England. Thesis. Ph.D.

Pressland, A.J. 1984. Productivity and management of northern Queensland's rangelands. *Australian Rangeland Journal.* 6: 26-45.

Quarterly Review of the Rural Economy. 1979. Canberra: AGPS. 1(2). May.

Quarterly Review of the Rural Economy. 1985. Canberra: AGPS. 7(3). August.

Queensland. Department of Primary Industries. 1983. *Queensland Planning Committee for Soil Conservation report.* Brisbane.

Queensland. Department of Primary Industries. 1984. *Soil erosion research techniques: workshop held in Toowoomba, 1983 April 12-14.* Brisbane: Department of Primary Industries. (Conference and Workshop series).

Queensland. Planning Committee for Soil Conservation. 1983. *Report 1983.* Brisbane: Department of Primary Industries.

Quiggin, J. 1983. *Discount rates for farm investments.* Canberra: Bureau of Agricultural Economics. (Working paper; no.1984/29).

Quiggin, J. 1984. *Common property, private property and regulation: the case of dryland salinity.* Canberra: CRES, ANU. (CRES working paper 1984/29).

Quiggin J. 1986a. Imperfections and failures in policy analysis: paper presented to the 30th Annual Conference of the Australian Agricultural Economics Society, February, Canberra.

Quiggin, J. 1986b. Common property,

private property and regulation: the case of dryland salinity. *Australian Journal of Agricultural Economics.* In press.

Raison, R.J. 1980. Possible forest site deterioration associated with slash-burning. *Search.* 11: 68-72.

Randall, A. 1981. Property entitlements and pricing policies for a maturing water economy. *Australian Journal of Agricultural Economics.* 25: 195-220.

Randall, A. 1983. The problem of market failure. *Natural Resources Journal.* 23(1): 131-148.

Rauschkolb, R.S. 1971. *Land degradation.* Rome: FAO. (Soil bulletin 13).

Rawls, J. 1971. *A theory of justice.* Cambridge, Mass.: Harvard University Press.

Reynolds, I.K. 1978. The relationship of land values to site characteristics: some implications for scenic quality management. *Journal of Environmental Management.* 6: 99-106.

Richards, K. 1982. *Rivers.* London: Methuen.

Richardson, J. and Jordan, A. 1979. *Governing under pressure.* London: Martin Robertson.

Rickson, R.E. and Stabler, P.J. 1985. Community responses to non-point pollution from agriculture. *Journal of Environmental Management.* 20(3): 281-294.

Ring, P.J. 1982. Soil erosion, rehabilitation and conservation: a comparative agricultural engineering/economic case study. In. *Proceedings of the Agricultural Engineering Conference, Armidale, New South Wales.*

Roberts, B.R. 1984. Land ethics: a necessary addition to Australian values: paper presented to a conference on Soil Degradation: the Future of Our Land? November, Australian Academy of Science. Australian National University Public Affairs Conference.

Roberts, B.R. 1985a. *How free is freehold? The social ecology of soil conservation.*

Toowoomba: Toowoomba Erosion Awareness Movement, School of Applied Science, Darling Downs Institute of Advanced Education.

Roberts, B.R. 1985b. Land ethics. *Erosion Research Newsletter.* 12: 2-6.

Roberts, S.H. 1924. *History of Australian land settlement.* Melbourne: Macmillan in association with Melbourne University Press.

Robertson, G. 1984. *Soil conservation districts, their role and aims.* Perth: Commissioner of Soil Conservation.

Robinson, G.W. 1934. Soils of Wales. *The Empire Journal of Experimental Agriculture.* 2: 258.

Robinson, I.B. 1982. Drought relief schemes for the 'Pastoral zone'. *Australian Rangeland Journal.* 4: 67-77.

Roddewig, R.J. 1978. *Green bans: the birth of Australian environmental politics.* Sydney: Hale and Iremonger.

Rose, C.W. 1984a. Recent advances in research on soil erosion processes. In. *Proceedings 2nd National Soils Conference, Brisbane.* Australian Society of Soil Science: 2-26.

Rose, C.W. 1984b. Progress in research on soil erosion processes and a basis for soil conservation practices. In. Creswell, E.T., Remenyi, J.V. and Nallana, L.G., eds. *Soil erosion management: ACIAR proceedings.* Canberra: 32. (Series no.6).

Rowan, J.N. 1971. *Salting on dryland farms in north-western Victoria.* Melbourne: Soil Conservation Authority, Victoria: 1-53.

Ruttan, V.W. 1982. *Agricultural research policy.* Minneapolis: University of Minnesota Press.

Sagoff, M. 1981. Do we need a land use ethic? *Environmental Ethics.* 3(4): 293-308.

Sallaway, M.M., Yule, D.F., and Ladwig, J.H. 1983. Effect of agricultural management on catchment runoff and soil loss in a semi-arid tropical environment. In. *Hydrology and Water Resources Symposium, 1983 November, Hobart, Tasmania.* Institution of Engineers, Australia: 252-256.

Samuel, S.N., Kingma, O.T. and Crellin, I.R. 1983. *Government intervention in rural research: some economic aspects of the Commonwealth's role.* Canberra: AGPS. (BAE occasional paper; no.76).

Sandbach, F. 1980. *Environment ideology and policy.* Oxford: Basil Blackwell.

Schaeffer, D.J., Kerster, H.W., Perry, J.A. and Cox, D.K. 1985. The environmental audit: I. Concepts. *Environmental Management.* 9(3): 191-198.

Schmalensee, R. 1972. Option demand and consumer's surplus: valuing price changes under uncertainty. *American Economic Review.* 62(5): 813-824.

Schnaiberg, A. 1980. *The environment: from surplus to scarcity.* New York: Oxford University Press.

Schultz, T.W. 1974. Is modern agriculture consistent with a stable environment? In. *The future of agriculture: technology, policies and adjustment: papers and reports Fifteenth International Conference of Agricultural Economists, 1973 August 19-30, Sao Paolo, Brazil.* Oxford: Agricultural Economics Institute: 235-242.

Schuster, C.J. 1979. *Rehabilitation of soils damaged by logging in south west Western Australia.* Forests Department of Western Australia. (Research paper; no.54).

Scott, R., ed. 1980. *Interest groups and public policy: case studies from the Australian states.* Melbourne: Macmillan.

Sen, A.K. 1967. Isolation, assurance and the social rate of discount. *Quarterly Journal of Economics.* 8(1): 112-124.

Sen, A.K. and Williams, B. 1982. *Utilitarianism and beyond.* New York: Cambridge University Press.

Shea, S.R., Peet, G.B. and Cheney, N.P. 1981. The role of fire in forest management. In. Gill, A.M., Groves, R.H. and Noble, I.R., eds. *Fire and the Australian biota.* Canberra: Australian

Academy of Science: 443-470.

Sieper, E. 1982. *Rationalising rustic regulation*. St Leonards, NSW: Centre for Independent Studies. (Research studies in government regulation; 2).

Sinden, J.A. 1974. A utility approach to the valuation of recreational and aesthetic resources. *American Journal of Agricultural Economics*. 56: 61-72.

Sinden, J.A. 1984. Estimation of the opportunity costs of natural resource management practices. Unpublished paper prepared for the Department of Arts, Heritage and Environment.

Sinden, J.A. and Worrell, A.C. 1979. *Unpriced values: decisions without market prices*. New York: Wiley Interscience.

Singh, A. 1972. How reliable is the factor of safety in foundation engineering. In. *Proceedings Southeast Asian Conference on Soil Engineering, 1972 November 6-10, Hong Kong*. Hong Kong: Organizing Committee III for the Southeast Asian Society of Soil Engineering: 390-424.

Smith, R.C.G., Mason, W.K., Meyer, W.S. and Barrs, H.D. 1983. Irrigation in Australia: development and prospects. In. Hillel, D., ed. *Advances in irrigation*. New York: Academic Press: 99-153.

Soil Conservation Society of America. 1979. *Soil conservation policies: an assessment*. Ankeny, Iowa: The Society. Based on material presented at the National Conference on Soil Conservation, 1979 November 15-16, Washington DC.

Solow, R.M. 1974. What do we owe the future? *Nebraska Journal of Economics and Business*. 13: 3-16. Text of the Gerald L. Phillippe Memorial Lecture delivered at the University of Nebraska-Lincoln on May 1, 1973.

Sornarajah, M., ed. 1983. *The South West Dam dispute: the legal and political issues*. Hobart: University of Tasmania. Sir Elliott Lewis memorial publication.

South Australia. Department of Lands. 1983. Freeholding of land used for primary production. Adelaide. Unpublished.

South Australia. Department of Lands. 1984a. *Arid lands review, 1984 February 24th, Quorn*. Adelaide: Department of Lands.

South Australia. Department of Lands. 1984b. *Arid Lands Review Conference, 1984 April 13th, Adelaide*. Adelaide: Department of Lands.

South Australia. Interdepartmental Working Group. 1981. *The administration, management and tenure of South Australia's pastoral lands: a report prepared by an Interdepartmental Working Group appointed to inquire into the Pastoral Act and the Dog Fence Act*. Adelaide: Department of Lands.

Speight, J.G. (n.d.) Definition of 'Land'. Manuscript. CSIRO. Division of Land and Water Resources. Unpublished.

Spigelman, J. 1972. *Secrecy: political censorship in Australia*. Sydney: Angus and Robertson

Stace, H.C.T., Hubble, G.D., Brewer, R., Northcote, K.H., Sleeman, J.R., Mulcahy, M.J. and Hallsworth, E.G. 1968. *A handbook of Australian soils*. Glenside, SA: Rellim Technical Publications.

Stanley, R.J. 1983. *Soils and vegetation of the Australian arid lands*. Australian Conservation Foundation.

Stocking, M.A. 1984. The universal soil loss equation for Australia? *Erosion Research Newsletter*. 10: 2-3.

Stoneman, T.C. 1973. Soil structure changes under wheatbelt farming systems. *Journal of Agriculture, Western Australia*. 14(3): 209-214.

Summers, D. 1976. *The great level: a history of drainage land reclamation in the Fens*. Newton Abbot, UK: David and Charles.

Taylor, C.R. and Frohberg, K.K. 1977. The welfare effects of erosion controls, banning pesticides, and limiting fertilizer application in the corn belt. *American Journal of Agricultural Economics*. 59(1): 25-36.

Thatcher, A. 1984. Strategy to reverse tree

decline in Victoria. In. *Focus on farm trees 2: National Conference, 1984 May, Armidale, University of New England.*

Thompson, D., ed. 1972. *Politics, policy and natural resources.* New York: The Free Press.

Thompson, D.F. 1982. Soil Conservation Service of NSW. Unpublished report.

Tisdell, C.A. In Press. Conserving and planting trees on farms: lessons from Australian cases. *Review of Marketing and Agricultural Economics.*

Trebeck, D.B. 1982. Primary producer organisations. In. Williams, D.B., ed. *Agriculture in the Australian economy.* 2nd ed. Sydney: Sydney University Press.

Tribe, L., Schelling, C. and Voss, J., eds. 1976. *When values conflict: essays on environmental analysis, discourse and decision.* Cambridge, Mass: Ballinger.

Trimble, S.W. 1975. A volumetric estimate of man-induced soil erosion on the southern Piedmont plateau. In. *Present and prospective technology for predicting sediment yields and sources.* US. Department of Agriculture: 142-152. (ARS-S-40).

Trimble, S.W. 1983. A sediment budget for Coon Creek basin in the driftless area, Wisconsin, 1853-1977. *American Journal of Science.* 283: 454-474.

Tuckwell, H.T. 1984. Transferability of water rights in South Australia. Water Resources Branch, Engineering and Water Supply Department, South Australia. Unpublished.

UNESCO/UNEP/UNDP. 1980. *Case studies on desertification.* Paris: Unesco. (Natural Resources Research; 18).

United Nations. Industrial Development Organization. 1972. *Guidelines for project evaluation.* New York: United Nations.

United States. Department of Agriculture. 1981. *America's soil and water: conditions and trends.* Washington, DC: USDA.

United States. Department of Agriculture.

Economic Research Service. 1982. *Economic indicators of the farm sector: production and efficiency statistics. 1980.* Washington, DC: USDA. Economic Research Service. (Statistical bulletin; no. 697).

van Beer, P.G.H. and Steentsma, W. 1981. *Benefits and costs of soil conservation.* Queensland. Department of Primary Industries. (Farm note: no.F113).

Van der Linden, P. 1983. Soil erosion in central Java (Indonesia). In. de Ploey, J., ed. *Rainfall simulation, runoff and soil erosion.* Cremlingen: Catena Verlag: 141-160. (Catena. Supplement; 4).

van Es, J.C. 1983. The adoption/diffusion tradition applied to resource conservation: inappropriate use of existing knowledge. *The Rural Sociologist.* 3: 76-82.

Vanclay, F. and Rickson, R.E. 1985. Socioeconomic correlates of farmer adoption of environmental practices: a discriminant analysis: paper read at the annual meetings of the Sociological Association of Australia and New Zealand, Brisbane, Queensland.

Vandersee, B.E. 1975. *Land inventory and technical guide, eastern Downs area.* Brisbane: Queensland Department of Primary Industries, Division of Land Utilization. (Technical bulletin no.7).

Vanoni, V.A., ed. 1975. *Sedimentation engineering.* New York : American Society of Civil Engineers.

Victoria. Parliament. Salinity Committee. 1984. *Salt of the earth: final report on the causes, effects and control of land and river salinity in Victoria.* Melbourne: Government Printer.

Victoria. State Conservation Strategy Task Force. 1983. *Conservation in Victoria: a discussion paper on a state conservation strategy.* East Melbourne, Victoria: Ministry for Conservation.

Vigon, B.W. 1985. The status of nonpoint source pollution: its nature, extent and control. Water Resources Bulletin. 21: 179-184.

Wade, J.C. and Heady, E.O. 1977. Controlling nonpoint sediment sources with cropland management: a national economic assessment. *American Journal of Agricultural Economics.* 59(1): 13-24.

Walker, J. 1981. Fuel dynamics in Australian vegetation. In. Gill, A.M., Groves, R.H. and Noble, I.R., eds. *Fire and the Australian biota.* Canberra: Australian Academy of Science: 102-127.

Walker, P.H., Kinnell, P.I.A. and Green, P. 1978. Transport of a noncohesive sandy mixture in rainfall and runoff experiments. *Soil Science Society of America. Journal.* 42: 793-801.

Walling, D.E. and Webb, B.W. 1982. Sediment availability and the prediction of storm-period sediment yields. In. Walling, D.E., ed. *Recent developments in the explanation and prediction of erosion and sediment yield.* Wallingford, Oxfordshire: International Association of Hydrological Sciences: 327-340. (IAHS publication; no.137).

Walters, C.J. and Hilborn, R. 1978. Ecological optimization and adaptive management. *Annual Review of Ecology and Systematics.* 9: 157-188.

Warhurst, J. 1982. *Jobs or dogma? The Industries Assistance Commission and Australian politics.* St Lucia, Queensland: University of Queensland Press.

Warr, P.G. and Wright, B.D. 1981. The isolation paradox and the discount rate for benefit-cost analysis. *Quarterly Journal of Economics.* 96(1): 129-145.

Wasson, R.J. and Clark, R.L. 1985. Environmental history for explanation and prediction. *Search.* 16(9/12): 258-263.

Wasson, R.J. and Galloway, R.W. Forthcoming. Sediment yield in the Barrier Range before and after European settlement. In preparation.

Wasson, R.J. and Nanninga, P.M. 1986. Estimating sand transport rates by wind on vegetated surfaces. *Earth Surface Processes and Landforms.*

11(3): 505-514.

Watson, R.A. 1969. Explanation and prediction in geology. *Journal of Geology.* 77: 488-494.

Watson, W. and Rose, R. 1980. Irrigation issues for the eighties: focussing on efficiency and equity in the management of agricultural water supplies: paper presented at the 24th Annual Conference of the Agricultural Economics Society, Adelaide.

Weisbrod, B.A. 1964. Collective-consumption services of individual-consumption goods. *Quarterly Journal of Economics.* 78(3): 471-477.

Western Australia. Department of Lands and Surveys. 1979. *The present and future pastoral industry of Western Australia.* Perth: The Department.

Western Australia. Department of Regional Development and the North West. 1985. *Kimberley pastoral industry inquiry: an industry and government report on the problems and future of the Kimberley pastoral industry.* Perth: The Department.

Wigley, T.M.L. 1985. Impact of extreme events. *Nature.* 316: 106-107.

Wilcox, D.G. and McKinnon, E.A. 1972. *A report on the condition of the Gascoyne catchment.* Perth: Department of Lands and Surveys, Western Australia.

Williams, C.H. 1980. Soil acidification under clover pasture. *Australian Journal of Experimental Agriculture and Animal Husbandry.* 20: 561-567.

Williams, J.R. 1975. Sediment yield prediction with universal equation using runoff energy factor. In. *Present and prospective technology for predicting sediment yields and sources.* US. Department of Agriculture: 244-252. (ARS-S-40).

Williams, M. 1978. Desertification and technological adjustment in the Murray mallee of South Australia. *Search.* 9: 265-268.

Williams, M. 1979. The perception of the hazard of soil degradation in South Australia: a review. In. Heathcote, R.L. and Thom, B.G., eds. *Natural hazards in Australia*. Canberra: Australian Academy of Science: 275-289.

Williams, O.B. 1979. Ecosystems of Australia. In. Goodhall, D.W. and Perry, R.A., eds. *Arid land ecosystems*. Cambridge: Cambridge University Press.

Williams, O.B., Suijdendorp, H. and Wilcox, D.G. 1980. The Gascoyne basin. In. Biswas, M.R. and Biswas, A.K., eds. *Desertification: associated case studies presented for the United Nations Conference on Desertifiction, 1977 August 29 - September 9, Nairobi, Kenya*. Oxford: Pergamon: 3-106.

Wills, I. 1985. Agricultural land use, environmental degradation and alternative property rules: paper presented at the 29th Annual Conference of the Australian Agricultural Economics Society, University of New England, Armidale.

Wilson, B.N., Barfield, B.J., Moore, I.D. and Warner, R.C. 1984. A hydrology and sedimentology watershed model: part II. Sedimentology component. *Transactions of the American Society of Agricultural Engineers*. 27: 1378-1384.

Wilson, B.N., Barfield, B.J., Ward, A.D. and Moore, I.D. 1984. A hydrology and sedimentology watershed model. Part I: Operational format and hydrologic component. *Transactions of the American Society of Agricultural Engineers*. 27: 1370-1377.

Wilson, S.J. and Cooke, R.U. 1980. Wind erosion. In. Kirkby, M.J. and Morgan, R.P.C., eds. *Soil erosion*. Chichester : John Wiley and Sons: 217-252.

Wischmeier, W.A. 1976. Use and misuse of the universal soil loss equation. *Journal of Soil and Water Conservation*. 31: 5-9.

Wischmeier, W.H. and Smith, D.D. 1978. *Predicting rainfall erosion losses*. Washington DC: US. Department of Agriculture. (Agricultural handbook; no.537).

Wolman, M.G. and Miller, J.P. 1960. Magnitude and frequency of forces in geomorphic processes. *Journal of Geology*. 68: 54-74.

Woods, L.E. 1983. *Land degradation in Australia*. Canberra: AGPS.

Woods, L.E. 1984. *Land degradation in Australia*. 2nd ed. Canberra: AGPS.

Wright, J. 1977. *The coral battleground*. Melbourne: Nelson.

Working Party on Dryland Salting in Australia. 1982. *Salting of non-irrigated land in Australia*. Melbourne: Soil Conservation Authority, Victoria.

Wunderlich, G. 1984. Fairness in landownership. *American Journal of Agricultural Economics*. 66(5): 802-806.

Wylie, F.R. and Johnston, P.J.M. 1984. Rural tree dieback. *Queensland Agricultural Journal*. 110: 3-6.

Wylie, F.R. and Landsberg, J. 1985. The impact of rural dieback on remnant farm woodlots. In. *CSIRO Division of Wildlife and Rangeland Research Symposium, Busselton, Western Australia*. In press.

Wynen, E. 1984. Variables affecting wheat yields in New South Wales 1945-46 to 1968-69. *Journal of the Australian Institute of Agricultural Science*. 50: 125-196.

Young, M.D. 1979a. Influencing land use in pastoral Australia. *Journal of Arid Environments*. 2: 279-288.

Young, M.D. 1979b. *Differences between states in arid land administration*. Melbourne: CSIRO. (CSIRO. Land resources management series; no.4).

Young, M.D. 1980. Resource management of Australian arid lands. In. Golang, G., ed. *Desert planning: international lessons*. London: Architectural Press.

Young, M.D. 1981. A legislative history of the South Australian Pastoral Act 1893-1980. In. South Australia. Interdepartmental Working Group. *The administration, management and tenure of South Australia's pastoral lands: a report prepared by an Interdepartmental Working Group appointed to inquire into the Pastoral Act and the Dog Fence Act*. Adelaide: Department of Lands.

Young, M.D. 1983. An analysis of the

main arid land tenure options in Australia: paper presented to Social Science Symposium, May, CSIRO, Deniliquin.

Young, M.D. 1984. Rangeland administration. In. Harrington, G.N., Wilson, A.D. and Young, M.D., eds. *Management of Australia's rangelands.* Melbourne: CSIRO.

Young, M.D. 1985a. Pastoral land tenure options in Australia. *Australian Rangeland Journal.* 7(1): 43-46.

Young, M.D. 1985b. The influence of farm size on vegetation condition in an arid area. *Journal of Environmental Management.* 7(1): 43-46.

Young, O. 1982. *Resource regimes, natural resources and social institutions.* Berkeley, California: University of California Press.

Index

United States of America
(USA), 28, 59, 60, 64,
85, 98, 104, 114, 134,
136, 161, 229, 233, 234,
243, 247(n), 269, 276,
307, 318, 323, 324, 331,
332, 355, 357, 361
see also North America
United States Department
of Agriculture (USDA),
25, 309
Universal Soil Loss
Equation (USLE), 59, 60,
62, 67, 74, 85, 90, 95,
96, 97, 125, 231, 232,
233, 246, 247(n), 322,
323, 333, 334
Upstill, Garrett, xi, 119,
120, 123, 323, 324, 368
urban and peri-urban
development, 108, 114–
115, 121, 223, 277, 280,
283, 361

vegetable growers, 357,
358
vegetation, native, 14, 17,
27, 30, 32, 53, 66, 67,
101, 108, 110, 112, 113,
115, 116, 177, 180, 238,
290, 291, 292, 307, 316,
322, 339
clearance or degradation
of, 9, 10, 13, 20, 23,
50, 73, 102, 111, 117,
129, 134, 155, 234,
241, 242, 280, 315
effect of fire on, 40, 44
effect on wind erosion,
58
halophytic, 15
see also flora and fauna;
forestry and forests;
land clearance; weeds
vehicles, 36, 97, 108, 110,
111, 309
'veil of ignorance'
approach, 253
vermin, 153, 161, 232,
269, 290
see also rabbits
Victoria, 13, 15, 16, 28,
34, 35, 38, 71, 84, 110,

111, 123, 135, 137, 141,
145, 152, 257, 266, 283,
295
see also names of
Victorian towns e.g.
Eppalock
*Victoria Soil Conservation
and Land Utilisation
Act,* 1958, 285

Wagga Wagga (NSW), 36
Wakefieldian land
settlement, 135
Walker, Dr. Brian, 368
Walker, Dr. Pat, 368
Wasson, Dr. Robert, xii,
49, 73, 74, 75, 305, 307,
308, 310, 312, 316, 368
waste disposal, 114, 116
waste, law of, 131–32,
133, 161
water erosion
see erosion, water
water quality, 7, 31, 40,
107, 122, 123, 188, 190,
215, 220, 297
Water Resource Program,
281
Water Resources
Commission, NSW, 113,
291
water storages, 38, 215,
220, 296
water table, 15, 32, 38, 50,
203, 215
waterlogging, 15, 34, 36,
50, 70, 73, 187, 307,
309, 316
Watkins, Warwick, xii,
172, 275, 295, 298, 368
Watson, Bill, 368
weather, 18
Webb, Adrian, xii, 70, 368
weeds, 140, 153, 154, 161,
221, 223, 232, 269
Western Australia, 13, 14,
15, 17, 27, 34, 42, 71,
73, 98, 135, 140, 142,
146, 151, 152, 155, 175,
176, 179, 185, 269, 272,
273, 285, 307, 309, 313,
318, 320, 359

Western Australia State
Labor Party, 273
Western Australian Water
Authority, 98
Western Division of NSW,
147, 148, 149, 165, 175,
275, 359
Western Lands Act
(NSW), 175, 182
wetlands, 113
Whalan, Prof. Douglas J,
vi, 368
wheat, 13, 14, 17, 18, 28,
36, 61, 62, 69, 80, 81,
82, 84, 85, 88, 247, 319,
334, 357, 358, 359
Whitlam government, 265
wildlife
see flora and fauna
Wimmera (Vic), 64, 357
wind erosion
see erosion, wind
Wind Erosion Equation,
96
windspeed, 60, 61, 62
Wisconsin (USA), 233
woodchip industry, 116,
159, 223, 243, 276, 278
Woods, Dr. Lance, xii,
108, 119, 121, 147, 323,
327, 360, 368
Working Party on
Dryland Salinity in
Australia (1982), 15, 71,
84, 94, 363
World Conservation
Strategy, 1980, 278, 282,
283

Yapp, Timothy, xii, 119,
120, 123, 323, 324, 368
Young, Michael, xii, 147,
169, 248, 249, 254, 330,
362, 368

zinc, 116